The Iguanid Lizards of Cuba

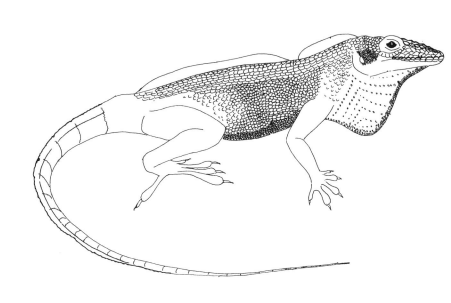

THE
IGUANID LIZARDS
OF CUBA

EDITED BY

LOURDES RODRÍGUEZ SCHETTINO

Contributing editors: Alberto Coy Otero,
Georgina Espinosa López, Ada R. Chamizo Lara,
Mercedes Martínez Reyes, Luis V. Moreno García

University Press of Florida
Gainesville · Tallahassee · Tampa · Boca Raton
Pensacola · Orlando · Miami · Jacksonville

LIBRARY OF CONGRESS CATALOGING-IN-PUBLICATION DATA
The iguanid lizards of Cuba / edited by Lourdes Rodríguez Schettino;
contributing editors, Alberto Coy Otero . . . [et al.].
p. cm.
Includes bibliographical references and index.
ISBN 0-8130-1647-9 (alk. paper)
1. Iguanas—Cuba. I. Rodríguez Schettino, Lourdes.
QL666.L25I383 1999
597.95—dc21 99-10011

The University Press of Florida is the scholarly publishing agency for
the State University System of Florida, comprising Florida A&M
University, Florida Atlantic University, Florida International University, Florida State University, University of Central Florida, University
of Florida, University of North Florida, University of South Florida,
and University of West Florida.

University Press of Florida
15 Northwest 15th Street
Gainesville, FL 32611-2079
http://www.upf.com

To my parents, husband, and sons

CONTENTS

MAPS

FIGURES

PLATES

1. *Leiocephalus carinatus.*
2. *Leiocephalus cubensis.*
3. *Leiocephalus raviceps.*
4. *Leiocephalus macropus.*
5. *Leiocephalus stictigaster.*
6. *Leiocephalus onaneyi.*
7. *Cyclura nubila nubila* (adult male).
8. *Cyclura nubila nubila* (juvenile).
9. *Chamaeleolis chamaeleonides.*
10. *Chamaeleolis porcus.*
11. *Chamaeleolis barbatus.*
12. *Anolis equestris.*
13. *Anolis equestris* (juvenile).
14. *Anolis luteogularis.*
15. *Anolis noblei.*
16. *Anolis smallwoodi.*
17. *Anolis baracoae.*
18. *Anolis pigmaequestris.*
19. *Anolis porcatus.*
20. *Anolis porcatus* (female).
21. *Anolis isolepis.*
22. *Anolis allisoni.*
23. *Anolis angusticeps.*
24. *Anolis paternus.*
25. *Anolis guazuma.*
26. *Anolis loysiana.*
27. *Anolis argillaceus.*
28. *Anolis centralis.*

TABLES

PREFACE

The iguanids constitute 51.7% of the Cuban terrestrial reptiles; they include four genera and sixty-two species, of which fifty-seven (91.9%) are endemic. They are present in almost all terrestrial ecosystems and form complex communities, with one or two abundant species and many other less abundant ones. They are mostly secondary consumers, devouring a great quantity of arthropods and serving as food for other vertebrates; thus, they form an important part of the trophic webs, not only in natural ecosystems but also in those modified by humans.

Because of their biological importance, they have been the object of various taxonomic, ecological, and zoogeographical studies, the results of which have been presented in about 350 papers, books, and oral dissertations. The compilation of these and other works related to the biology of the iguanids is valuable for anyone who studies the Cuban herpetofauna. If this book—besides providing the largest possible amount of information—is able to stimulate the study of unknown subjects, the authors will have achieved their purpose.

The structure of the book allows one to read each chapter independently. Chapter 1, the introduction, deals with the taxonomic and historical aspects of the group. In chapter 2, "Morphology," the principal characteristics of external and internal morphology, morphological variation, and eco-morphology are described, and identification of the Cuban species is set out by means of dichotomous keys. Chapter 3, "Ecology and Behavior," offers a general view of the distribution of the iguanids in space and time, their trophic and climatic relations, reproduction, and the composition of communities.

Chapter 4, "Genetics," presents an analysis of the known information about the group's biochemical genetics and karyology. An outline of ecto-

and endoparasites that concern the Cuban species is found in chapter 5, "Parasites." The historical and ecological biogeography of iguanids is dealt with in chapter 6, "Biogeography."

In preparing these six chapters, the authors used a bibliography related to the Cuban segment of the group as well as to other species, mainly from the Caribbean region. However, chapter 7, "Systematic Accounts of the Species," which provides information about the taxonomy, geographic distribution, description, morphological variation, and natural history of each of the Cuban iguanids, refers only to them.

The information that could be gathered about Cuban iguanids is not of the same scope for all species because of the unequal attention they have received. In general, we have included much more information about systematics and geographic distribution than about reproduction, feeding, population, or community structure; hence, these latter elements should be objects for future study, especially in places that have been or will be modified by humans. Such information will be extremely valuable in proposing necessary actions for the management of these species, which are so important in Cuban ecosystems.

The authors are grateful for the unselfish support of many people who contributed time and effort during the preparation of this book.

The late Mario S. Buide and Miguel L. Jaume provided a great portion of the bibliography and offered their valuable knowledge about the group in the first stage of this work. Orlando H. Garrido has been dedicated to the study of the Cuban herpetology, and his papers were an important source of information.

Lázaro González Pino drew the color illustrations and most of the black-and-white figures and maps; Heriberto Rodríguez Guerra and Teresa Regalado Calero, the other black-and-white graphs and maps. Jorge Hernández Ávila measured the specimens used for gender comparisons of almost every species. Erena Valdés Monteagudo compiled the data about altitudinal distribution.

Julio Novo Rodríguez collected the animals that served as live models for the color illustrations. Riberto Arencibia Preces and Arturo Hernández Marrero helped obtain material for ecological studies of some species.

Fernando M. González Bermúdez and Pedro Pérez Álvarez, directors of the Institute of Zoology and the Institute of Ecology and Systematics, respectively, encouraged the accomplishment of this work, as did Alcides Sampedro Marín of the University of Havana.

Miguel A. Vales García, Yobana Figueredo, Inés García Valdés, and Felipe Pérez Álvarez provided logistical support.

Each chapter was critically reviewed by many persons whose opinions and advice we acknowledge. Julio Novo Rodríguez reviewed the chapter on morphology; Armando Pérez Martínez, the one on ecology; Vicente Berovides Álvarez and Alberto R. Estrada, the chapter on genetics; María Teresa del Valle, the chapter on parasites; and Alcides Sampedro Marín, the entire book. The following researchers of the Institute of Ecology and Systematics examined certain chapters: Hiram González Alonso, Natalia Manójina, Ada Camacho Pérez, Luis F. de Armas Chaviano, Jorge L. Fontenla Rizo, Alberto García González, Bárbara Sánchez Oria, Vilma Rivalta González, and Liana Bidart Cisneros. We are also very grateful to Jonathan B. Losos for his logistical support, and to Kevin de Queiroz and Paul E. Hertz for their useful comments about the manuscript. Finally, our thanks to Janet Perodín Hernández, who translated the book into English.

ABBREVIATIONS

AMNH	American Museum of Natural History, New York
ANSP	Academy of Natural Sciences, Philadelphia
BMNH	British Museum (Natural History), London
CM	Carnegie Museum, Pittsburgh
CZACC	Zoological Collections, Institute of Ecology and Systematics, Havana
ChM	Charleston Museum, Charleston
HZM	Universität Hamburg, Zoologische Museum, Hamburg
MCZ	Museum of Comparative Zoology, Harvard University, Cambridge
MNHN	Le Muséum National d'Histoire Naturelle, Paris
MNHNCU	National Museum of Natural History of Cuba, Havana
USNM	National Museum of Natural History, Washington, D.C.
ZMB	Museum für Naturkunde, Humboldt-Universität, Berlin

1

Introduction

LOURDES RODRÍGUEZ SCHETTINO

The iguanids constitute a group of great interest in herpetology. Because of their wide geographic distribution throughout the greater part of the Western Hemisphere and the Madagascar and Fiji Islands, and their huge ecological diversity, their study has taken on great importance. Throughout the world, studies on iguanids have increased markedly since the 1960s, not only in number but also in importance, and include theoretical, experimental, and synthetical points of view. These studies have served as the basis for the development of general theories in ecology, physiology, and behavior.

In the Neotropical region, the genus *Anolis* alone has a great morphological, ecological, and ethological diversity: there are more than 250 species living in forests, coastal zones, savannas, and cultivated and urban areas, including jumping, running, and crawling species. All these factors, together with the high endemicity of species in the genus, even more notable in the Caribbean Islands, make this group appropriate for studies on taxonomy, population change, zoogeography, evolution, and conservation of biodiversity. In Cuba, the iguanids are found throughout the national territory in wild, cultivated, and urban areas. There are fifty-seven (91.9%) endemic species, of which twelve are endemic at the national level, twenty-five at the regional level, and eighteen at the local level.

As secondary, mainly insectivorous, consumers, iguanids help maintain the ecological equilibrium and are, together with diurnal birds, the principal agents of biological control for insects potentially noxious to agricultural crops, as well as to humans and animals. They compose an important part of the natural trophic webs, their abundance increases the biomass in diverse ecosystems, and they are used by other vertebrates as food.

In spite of the theoretical and practical importance that this group has, many of the studies about Cuban iguanids deal only with taxonomy and geographic distribution; there are few studies about different populations or community-level ecology, the species' relationships to each other and to their environment, or other interesting matters that take into account the group's potential value in indicating changes in biodiversity and ecosystem conservation.

Taxonomy

The family Iguanidae (iguanas, giant anoles, chameleons, lizards, and other anoles) is included in the class Reptilia, subclass Lepidosauria, order Squamata, suborder Sauria, and infraorder Iguania. It is the most diverse family of the order Squamata; there are about 60 genera and almost 600 recognized species at present (Etheridge and de Queiroz 1988). Many attempts have been made to subdivide it taxonomically to reflect the great morphological differences that exist among the numerous genera and species. However, such divisions were not always formally categorized nor were they always accepted by the systematists of the group.

In this direction, Cope (1900) proposed the recognition of three subfamilies: Iguaninae, Anolinae, and Basiliscinae; however, these subfamilies were almost never used. Savage (1958) added a group, the sceloporines, which included various genera from North America, but without giving it a taxonomic category. Etheridge (1960, 1964, 1967) recognized several generic groups which he proposed to name as sceloporines, tropidurines, anolines, basiliscines, and iguanines, although he did not consider them formal subfamilies.

Later, Varona (1985) proposed to formally establish the generic groups of Etheridge (1967), treating them as subfamilies: Sceloporinae, Tropidurinae, Anolinae, Basiliscinae, and Iguaninae.

By means of phylogenetic analysis of forty-nine characters, mainly osteological ones, for fifty-seven genera of the family, Etheridge and de Queiroz (1988) concluded that each species can be included in one of eight well-defined monophyletic groups: morunasaurs, anoloids, oplurines, tropidurines, iguanines, basiliscines, crotaphytines, and sceloporines. They also suggested that the family Iguanidae may be paraphyletic. However, they did not propose to assign these groups to one of the Linnean taxonomic categories and were uncertain about the relationships between the groups because of the weakness of evidence.

Frost and Etheridge (1989) reevaluated the available data about phylogenetic relationships among the components of the infraorder Iguania; they phylogenetically analyzed sixty-seven characters of thirty-five genera or species groups. As a result, they confirmed the eight previously mentioned generic groups and proposed to recognize them as *sedis mutabilis* families, arguing that Iguanidae *sensu lato* is paraphyletic since there was no evidence of its monophyly.

These families are Corytophanidae Fitzinger, 1843 (basiliscines), Crotaphytidae Smith and Brodie, 1982 (crotaphytines), Hoplocercidae Frost and Etheridge, 1989 (morunasaurs), Iguanidae Oppel, 1811 (iguanines), Opluridae Moody, 1983 (oplurines), Phrynosomatidae Fitzinger, 1843 (sceloporines), Polychridae Fitzinger, 1843 (anolines), and Tropiduridae Bell, 1843 (tropidurines), the last one composed of three subfamilies (Tropidurinae Bell, 1843; Leiocephalinae Frost and Etheridge, 1989; and Leiolaeminae Frost and Etheridge, 1989).

However, the adoption of the taxonomic category of family for these generic groups is controversial because it destabilizes the classification scheme for iguanians. For instance, Lazell (1992) severely criticized the splitting of Iguanidae *sensu lato* into the families of Frost and Etheridge (1989) because he found no new information, clarification, or improvement from such fragmentation, which does not benefit him as a field biologist. In replaying Lazell's arguments, Frost and Etheridge (1993) explained that they considered their recognized families to be useful because they are consistent with recovered evolutionary history, although they also recognized that there are some unresolved problems with the taxonomy that they proposed. Furthermore, Schwenk (1994) considered that the identification of monophyletic groups, referring to Frost and Etheridge's (1989) paper, must be applauded despite some unresolved relationships between the groups. On the other hand, Macey *et al.* (1997) suggested that the Iguanidae should be restored to its previous content because their results on genetics and phylogeny of Iguania indicate its monophyly; they also suggested that the families of Frost and Etheridge (1989) be considered subfamilies.

Conversely, as a result of the analysis of indirect estimates of the divergence in serum albumin, in terms of both immunological distance units and direct DNA sequence information, Hass *et al.* (1993) have reported that the genus *Chamaeleolis* is a recent derivative within the genus *Anolis* and have proposed that these genera must be synonyms. Since I believe that the morphological and osteological data for both genera are enough to separate them, my colleagues and I prefer to keep their taxonomic status unchanged until more evidence is available.

Taking into account the several divergent opinions on this still unresolved problem of the taxonomy of Iguania, the authors of the present book prefer, for the time being, to continue including Cuban iguanian species in a single family, Iguanidae. That is why, having no intention of evaluating the phylogenetic relationships of the genera and species in question, we do not recognize the families proposed by Frost and Etheridge (1989) or the placement of *Chamaeleolis* within the genus *Anolis*. Thus, iguanids are represented in Cuba by four genera and sixty-two species.

Geographic Range

The family Iguanidae *sensu lato* is widely distributed all over the Americas, from 48° N to the southern part of South America, including the Lesser and Greater Antilles, Bahama Islands, and Galápagos Islands. Furthermore, the genus *Brachylophus* is found on the Fiji Islands and the genera *Oplurus* and *Chalarodon* on Madagascar and the Comores Archipelago. In Cuba, the family is present throughout the national territory.

Description

The Cuban members of the family Iguanidae *sensu lato* are described as follows: Snout-vent length means from 28.5 to 405.0 mm; extensible dewlap in species of the genera *Anolis* and *Chamaleolis*; relatively large head; presence of pineal eye; conic and tricuspid tooth; generally, pleurodont dentition; a little protrusible tongue; eyes with eyelids; panoramic and stereoscopic vision; circular pupil; distinguishable cervical region; slender body; regenerable, long, and slender tail, with either circular or vertically oval section, sometimes with caudal crest; caudal autotomy in most species; metachromatic changes in most species; usually either small and granular scales or imbricate scales, often keeled; subdigital lamellae in species of the genera *Anolis* and *Chamaeleolis*; superior temporal arch present; parasternal ribs present, not all connected to the bony ribs, continuous midventrally; procoelous vertebrae; eight cervical vertebrae.

Classification

Classification is the most frequent theme in literature about Cuban iguanids. Nevertheless, the taxonomic treatment in several of these papers has been typological, and the original descriptions of many species and subspecies have been based on one or a few specimens, sometimes comprising only one sex or subadult individuals. Consequently, published descriptions characterizing some taxa do not succeed in expressing the differences between them in color, size, or both. For instance, the several subspecies of *Leiocephalus macropus*, *L. stictigaster*, *Anolis equestris*, *A. jubar*, and other species exhibit minor differences in color, a characteristic that has great individual and population variation. On the other hand, subspecific differences in size have been generally considered according to the maximal measures found in a certain locality, which has led to the conclusion that a "giant" characterizes a population whose mean length can be smaller than another population with a smaller giant. An extreme case is that of *A. jubar balaenarum* (males to 62 mm), a subspecies located within the geographic range of *A. jubar cuneus*, whose males do not reach the same maximal size (males to 58 mm).

Despite such deficiencies, having summarized here the family Iguanidae in Cuba, we have dealt with all currently recognized species and subspecies. Because of this work's complexity and the necessities of obtaining additional material and using current techniques applied to animal taxonomy, an exhaustive review of the taxonomy of the three polytypic genera is not the objective.

The taxonomic ordering followed by Gundlach (1867, 1880), Barbour (1914), Barbour and Ramsden (1916b, 1919), and Alayo Dalmau (1955) in their lists and catalogs is based on morphological similarities. Other authors followed an alphabetic ordering (Barbour 1930a, b, 1935, 1937; Schwartz and Thomas 1975; Schwartz et al. 1978; Garrido and Jaume 1984; Schwartz and Henderson 1985, 1988, 1991; Powell et al. 1996).

Etheridge (1960) grouped the species of the family Iguanidae according to their caudal vertebrae. He placed the genus *Leiocephalus* in group I, *Cyclura* in group II, and *Anolis* and *Chamaeleolis* in group III. Within the genus *Anolis* he separated the species into two large groups: the alpha section (absence of transverse processes in the autotomic caudal vertebrae) and the beta section (presence of transverse processes; fig. 1). At the same time he divided each section into series, including in each series species with similar osteological characteristics.

Ruibal (1964) followed this ordering, but only in the introduction of his paper; he did not adopt it in the list. Williams (1976) embraced the scheme of Etheridge (1960) but made some modifications within the genus *Anolis*, proposing many informal categories based on morphological, osteological, karyological, and biochemical characteristics of the species. The taxonomic arrangement he established, which attempts to reflect relationships between species, has been followed by the majority of herpetologists.

In his taxonomic arrangement of iguanids, Varona (1985) grouped them into five subfamilies, three of which included Cuban genera: Iguaninae (*Cyclura*), Tropidurinae (*Leiocephalus*), and Anolinae (*Anolis* and *Chamaeleolis*). Within the last subfamily he considered two tribes, Anolini for the species of the alpha section of *Anolis* and *Chamaeleolis*, and Noropini for the species of the beta section of *Anolis*, to which he assigned the genus *Norops*. Moreover, he proposed to divide the genus *Anolis* into eight subgenera (*Anolis*,

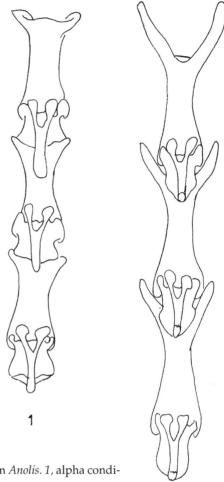

Fig. 1. Caudal vertebrae conditions in *Anolis*. *1*, alpha condition; *2*, beta condition.

Pseudoequestris, Draconura, Gekkoanolis, Acantholis, Brevicaudata, Xiphosurus, and *Deiroptyx*) and the genus *Norops* into two subgenera (*Norops* and *Trachypilus*).

Subsequently, Guyer and Savage (1986) proposed to formalize sections, assigning the species of the alpha section to the genera *Anolis, Dactyloa, Semiurus,* and *Ctenonotus* and those of the beta section to *Norops*. This proposal was followed by Schwartz and Henderson (1988) but not by Schwartz and Henderson (1991).

Cannatella and de Queiroz (1989) concluded that the data used by Guyer and Savage (1986) preclude recognition of the genera proposed by those authors because of the lack of evidence of monophyly in some of them.

Williams (1989a) presented a very exhaustive analysis of each of the genera proposed by Guyer and Savage (1986) and concluded that, although the classification of the genus *Anolis sensu lato* needs to be reexamined, the informal taxonomic changes are preferable as a basis for a more rigorous future classification.

Moreover, according to the analysis of twelve slow-evolving proteins in forty-nine Antillean species of the genus *Anolis*, Burnell and Hedges (1990) proposed that the taxa assigned to the categories of section and subsection be abandoned, demonstrating that they are not monophyletic and that the taxa of the category of series for the species groups of each island are better since they agree with each intraisland radiation.

However, the results obtained by Quintana Ferrer (1987) regarding morphology, and by Espinosa López and Posada García *et al.* (1990); Espinosa López, Sosa Espinosa, and Berovides (1990); and Burnell and Hedges (1990) regarding electrophoresis of several protein systems of many Cuban species of the genus *Anolis* did not support the taxonomic classification of Williams (1976) above the species group level. This has been analyzed by Gorman, Buth, and Wyles (1980); Wyles and Gorman (1980); Shochat and Dessauer (1981); and Burnell and Hedges (1990) in relation to other Caribbean species of the genus.

Because there is not yet enough knowledge about phylogenetic relationships between genera within the family Iguanidae *sensu lato* and between species within the genera, in the present work we follow the ordering of genera presented by Etheridge (1960), and the arrangement of Williams (1976) for the species of the genus *Anolis*. However, we follow the chronological order of their original descriptions, since such a system is commonly used in many of the papers about lizards, despite the controversy about the use of the categories of series, subseries, section, and subsection. In the following list, each endemic taxon is marked with an asterisk (*):

Family Iguanidae Oppel, 1811
    Genus   *Leiocephalus* Gray, 1827
    Leiocephalus carinatus Gray, 1827
            L. c. carinatus Gray, 1827*
            L. c. aquarius Schwartz and Ogren, 1956*
            L. c. zayasi Schwartz, 1959*
            L. c. mogotensis Schwartz, 1959*
            L. c. labrossytus Schwartz, 1959*
            L. c. cayensis Schwartz, 1959*
            L. c. microcyon Schwartz, 1959*
    Leiocephalus cubensis (Gray), 1840*
            L. c. cubensis (Gray), 1840*

     L. c. paraphrus Schwartz, 1959*
     L. c. gigas Schwartz, 1959*
     L. c. pambasileus Schwartz, 1959*
     L. c. minor Garrido, 1970*
  Leiocephalus macropus Cope, 1862*
     L. m. macropus Cope, 1862*
     L. m. inmaculatus Hardy, 1958*
     L. m. hoplites Zug, 1959*
     L. m. hyacinthurus Zug, 1959*
     L. m. koopmani Zug, 1959*
     L. m. aegialus Zug, 1959*
     L. m. phylax Schwartz and Garrido, 1967*
     L. m. asbolomus Schwartz and Garrido, 1967*
     L. m. lenticulatus Garrido, 1973*
     L. m. felinoi Garrido, 1979*
     L. m. torrei Garrido, 1979*
  Leiocephalus raviceps Cope, 1862*
     L. r. raviceps Cope, 1862*
     L. r. uzzelli Schwartz, 1960*
     L. r. klinikowskii Schwartz, 1960*
     L. r. jaumei Schwartz and Garrido, 1968*
     L. r. delavarai Garrido, 1973*
  Leiocephalus stictigaster Schwartz, 1959*
     L. s. stictigaster Schwartz, 1959*
     L. s. sierrae Schwartz, 1959*
     L. s. exotheotus Schwartz, 1959*
     L. s. astictus Schwartz, 1959*
     L. s. lucianus Schwartz, 1960*
     L. s. parasphex Schwartz, 1964*
     L. s. ophiplacodes Schwartz, 1964*
     L. s. naranjoi Schwartz and Garrido, 1968*
     L. s. lipomator Schwartz and Garrido, 1968*
     L. s. celeustes Schwartz and Garrido, 1968*
     L. s. gibarensis Schwartz and Garrido, 1968*
     L. s. septentrionalis Garrido, 1975*
  Leiocephalus onaneyi Garrido, 1973*
  Genus *Cyclura* Harlan, 1824
  Cyclura nubila (Gray), 1831
     C. n. nubila (Gray), 1831*
  Genus *Chamaeleolis* Cocteau, 1838*
  Chamaeleolis chamaeleonides Duméril and Bibron, 1837*
  Chamaeleolis porcus Cope, 1864*
  Chamaeleolis barbatus Garrido, 1982*
  Chamaeleolis guamuhaya Garrido, Pérez-Beato, and Moreno, 1991*
  Genus *Anolis* Daudin, 1802
  Section alpha Etheridge, 1960

Subsection *carolinensis*
Series *carolinensis*
Species group *equestris*
Anolis equestris Merrem, 1820*
    A. e. equestris Merrem, 1820*
    A. e. thomasi Schwartz, 1958*
    A. e. buidei Schwartz and Garrido, 1972*
    A. e. persparsus Schwartz and Garrido, 1972*
    A. e. juraguensis Schwartz and Garrido, 1972*
    A. e. verreonensis Schwartz and Garrido, 1972*
    A. e. potior Garrido, 1975*
    A. e. cincoleguas Garrido, 1981*
Anolis luteogularis Noble and Hassler, 1935*
    A. l. luteogularis Noble and Hassler, 1935*
    A. l. hassleri Barbour and Shreve, 1935*
    A. l. nivevultus G. Peters, 1970*
    A. l. delacruzi Schwartz and Garrido, 1972*
    A. l. sectilis Schwartz and Garrido, 1972*
    A. l. coctilis Schwartz and Garrido, 1972*
    A. l. calceus Schwartz and Garrido, 1972*
    A. l. jaumei Schwartz and Garrido, 1972*
    A. l. sanfelipensis Garrido, 1975*
Anolis noblei Barbour and Shreve, 1935*
    A. n. noblei Barbour and Shreve, 1935*
    A. n. galeifer Schwartz, 1964*
Anolis smallwoodi Schwartz, 1964*
    A. s. smallwoodi Schwartz, 1964*
    A. s. palardis Schwartz, 1964*
    A. s. saxuliceps Schwartz, 1964*
Anolis baracoae Schwartz, 1964*
Anolis pigmaequestris Garrido, 1975*
Species group *carolinensis*
Anolis porcatus Gray, 1840*
Anolis isolepis (Cope), 1861*
    A. i. isolepis (Cope), 1861*
    A. i. altitudinalis Garrido, 1985*
Anolis allisoni Barbour, 1928
Anolis angusticeps Hallowell, 1856
    A. a. angusticeps Hallowell, 1856*
Anolis paternus Hardy, 1966*
    A. p. paternus Hardy, 1966*
    A. p. pinarensis Garrido, 1975*
Anolis guazuma Garrido, 1983*
Anolis alayoni Estrada and Hedges, 1995*
Anolis garridoi Díaz, Estrada, and Moreno, 1996*
Species group *argillaceus*

Anolis loysiana Duméril and Bibron, 1837*
Anolis argillaceus Cope, 1862*
Anolis centralis G. Peters, 1970*
   A. c. centralis G. Peters, 1970*
   A. c. litoralis Garrido, 1975*
Anolis pumilus Garrido, 1988*
Series *lucius*
Species group *lucius*
Anolis lucius Duméril and Bibron, 1837*
Anolis argenteolus Cope, 1861*
Species group *vermiculatus*
Anolis vermiculatus Duméril and Bibron, 1837*
Anolis bartschi (Cochran), 1928*
Series *alutaceus*
Species group *alutaceus*
Anolis alutaceus Cope, 1861*
Anolis clivicola Barbour and Shreve, 1935*
Anolis anfiloquioi Garrido, 1980*
Anolis inexpectata Garrido and Estrada, 1989*
Anolis macilentus Garrido and Hedges, 1992*
Anolis vescus Garrido and Hedges, 1992*
Anolis alfaroi Garrido and Hedges, 1992*
Anolis cyanopleurus Cope, 1861*
   A. c. cyanopleurus Cope, 1861*
   A. c. orientalis Garrido, 1975*
Anolis cupeyalensis Garrido, 1975*
Anolis mimus Garrido, 1975*
Anolis fugitivus Garrido, 1975*
Anolis juangundlachi Garrido, 1975*
Anolis spectrum W. Peters, 1863*
Anolis vanidicus Garrido and Schwartz, 1972*
   A. v. vanidicus Garrido and Schwartz, 1972*
   A. v. rejectus Garrido and Schwartz, 1972*
Section beta Etheridge, 1960
   Series *sagrei*
Species group *sagrei*
Anolis sagrei Duméril and Bibron, 1837
   A. s. sagrei Duméril and Bibron, 1837
   A. s. greyi Barbour, 1914*
Anolis bremeri Barbour, 1914*
   A. b bremeri Barbour, 1914*
   A. b. insulaepinorum Garrido, 1972*
Anolis homolechis (Cope), 1864*
   A. h. homolechis (Cope), 1864*
   A. h. turquinensis Garrido, 1973*
A. quadriocellifer Barbour and Ramsden, 1919*

A. jubar Schwartz, 1968*
    A. j. jubar Schwartz, 1968*
    A. j. cuneus Schwartz, 1968*
    A. j. balaenarum Schwartz, 1968*
    A. j. oriens Schwartz, 1968*
    A. j. yaguajayensis Garrido, 1973*
    A. j. gibarensis Garrido, 1973*
    A. j. maisiensis Garrido, 1973*
    A. j. albertschwartzi Garrido, 1973*
    A. j. santamariae Garrido, 1973*
    A. j. cocoensis Estrada and Garrido, 1990*
Anolis guafe Estrada and Garrido, 1991*
Anolis confusus Estrada and Garrido, 1991*
Anolis mestrei Barbour and Ramsden, 1916*
Anolis allogus Barbour and Ramsden, 1919*
Anolis ahli Barbour, 1925*
Anolis delafuentei Garrido, 1982*
Anolis rubribarbus Barbour and Ramsden, 1919*
Anolis imias Ruibal and Williams, 1961*
Anolis birama Garrido, 1991*
Anolis ophiolepis Cope, 1861*

History

FOSSIL RECORD

The family Iguanidae has appeared in the fossil record of the Gondwana Plate since the Middle to Late Jurassic, which suggests that the group is at least 160 my BP (Estes 1983). Iguanids were also found in South America from the Eocene (Porter 1972). The most ancient remains found to date are closely related to recent forms. Fragments of *Anolis* skin in amber have been found in Mexican deposits from the Oligocene or Miocene, and in the Dominican Republic a specimen of *Anolis dominicanus* in amber belonging to the beginnings of the Early Miocene has been found (Rieppel 1980), indicating the presence in the Antilles of anolines closely related to extant forms earlier than the Pleistocene, from which fossils of that genus have been reported (Etheridge 1965, 1966*a*; Morgan 1977; Pregill 1981*b*, 1982).

    Fossils are not known of the genus *Chamaeleolis*, a Cuban endemic and considered a primitive anoline that evolved *in situ* from ancestors of tropical America at the end of the Mesozoic age (Williams 1969; Guyer and Savage 1986; Cannatella and de Queiroz 1989; Williams 1989*b*). However, based on immunological distance units and on DNA sequencing, Hass *et al.* (1993) have considered that *Chamaeleolis* is a relatively recent addition to the anoline fauna of the Antilles.

Although fossils of the genus *Leiocephalus* in the Antilles are known only from the Late Pleistocene in the Bahamas, Barbuda, Hispaniola, and Jamaica (Etheridge 1966*a, b*); Cayman Islands (Morgan 1977); Puerto Rico (Pregill 1981*b*); and Anegada and Guadeloupe (Pregill 1992), the presence of the genus in the Early Miocene from Florida (Estes 1963) suggested to Etheridge (1966*b*) that its specializations were fixed, at least, at the mid-Cenozoic, and that this should have happened in the Antilles, from where some species invaded North America later. However, Pregill (1992) pointed out that the characteristics of such fossils and of others described after 1963 are from *Leiocephalus* but not unique to this genus; that is why he believed their classification must still stay inconclusive.

Fossils of the genus *Cyclura* have been reported only from the Pleistocene of the Cayman Islands (Morgan 1977) and Puerto Rico (Pregill 1981*b*), and from the Late Holocene of Grand Cayman (Morgan *et al.* 1993). All of them belong to living species (*C. nubila, C. pinguis,* and *C. nubila,* respectively).

Few studies of the Cuban iguanids have reported fossil species. Koopman and Ruibal (1955) found some dentaries and maxillae of *Anolis lucius, A. equestris,* and *Leiocephalus* sp. in caves of Sierra de Cubitas, Camagüey province. Acevedo *et al.* (1975) found remains of a large iguanian in Pleistocenic deposits of Cueva del Túnel, La Salud, La Habana province; Varona and Arredondo (1979) reported two places with Pleistocenic remains of iguana: Cueva de José Brea, Pan de Azúcar, Pinar del Río province and Arroyo del Palo, Mayarí, Holguín province. In addition, burned bones of iguanas have been found in some remains of Cuban aborigines (Aguayo 1951; Acevedo *et al.* 1975), which indicates that this lizard was used as food.

Summary of the Studies of the Cuban Species

Whereas information about Cuban fossil iguanids is extremely scarce, extant species have been widely studied since the last century. The Frenchmen André Marie Constant Duméril and Gabriel Bibron published a series of volumes on general herpetology between 1834 and 1854, and in the 1837 volume they described *Chamaeleolis* and four species of *Anolis.* Between 1838 and 1843, Jean Theodore Cocteau and G. Bibron published in different parts the fourth volume of the Historia Física, Política y Natural de la Isla de Cuba, which deals with reptiles. In these two great works, new and known species of that time were carefully characterized. They have an incalculable value from the taxonomic, historical, and aesthetic points of view, since they are exquisitely illustrated.

The Englishman John Edward Gray published many papers on amphibians and reptiles and described four new species of Cuban iguanids between 1827 and 1840. Edward Drinker Cope, an American, described ten more species between 1861 and 1864. Works on Cuban herpetology were published in 1867 and 1880 by Johannes Christopher Gundlach, a German resident in Cuba. He compiled up-to-the-moment information, not only taxonomic but also from his own observations of natural history and behavior. His papers contain some of the first news about the habits of Cuban species.

The American Thomas Barbour and Charles T. Ramsden, a Cuban, published many papers on amphibians and reptiles. Among them was a work on the herpetology of Cuba from 1919, in which they summarized the species known at that time and described four new ones.

During the second half of the twentieth century the taxonomy of the group was studied by Albert Schwartz, an American, who described five species and many subspecies between 1958 and 1972 and published lists of amphibians and reptiles from the Antilles in 1975, 1978, 1988, and 1991. Also analyzed in his papers were some zoogeographic and evolutive aspects of the group.

In 1964 Rodolfo Ruibal published an annotated list about Cuban anolines. Orlando H. Garrido, the Cuban author who has most thoroughly studied the classification and geographic distribution of Cuban reptiles from 1967 to date, has described twenty species and many subspecies.

Since 1965 the American George C. Gorman has published papers on karyology and phylogenetic relationships; Georgina Espinosa López and Octavio Pérez-Beato de la Vega, Cuban researchers, have published their results about protein electrophoresis and relationships between species.

Jerry D. Hardy Jr. in 1957, and Bruce B. Collette and Rodolfo Ruibal, both in 1961, offered the first ecological data on Cuban iguanids; however, it was not until the end of the 1970s that the ecology of the group began to be treated in greater depth and with more regularity by several Cuban specialists and their students.

Studies of ectoparasites of Cuban species have been conducted mainly by Idelfonso Pérez Vigueras, who reported on them in the 1930s, and, more recently, by Jorge de la Cruz Lorenzo. Endoparasites have been studied by Alberto Coy Otero and the Czech Vlastimil Baruš since 1969.

On the other hand, many legends, beliefs, and traditions have ascribed a noxious character to the Cuban iguanids. Misunderstandings based on such folklore have resulted in unfair rejection of these lizards; in some cases they have been pursued until death.

For example, in the last century Cocteau and Bibron (1837–1843) reported that iguanas could not be found in Cuban markets anymore, although their meat was considered very delightful. Gundlach (1880) said that iguanas were being forgotten as food items. Barbour and Ramsden

(1919) explained that the Cuban iguana was no longer eaten because it was considered poisonous, a belief based on the fact that when iguanas were hanged alive they salivated a great deal, reminiscent of yellow fever patients. However, iguana meat is edible and its ingestion is not dangerous in any way. Ramos Guadalupe (1997) commented that according to old stories, iguanas were used to cause intentional harm to humans, although he did not mention how, to whom, or when the damages were made.

Cuban farmers believed that the bite of giant anoles causes fever. The belief seems to have its origins in the lack of hygiene and medical care prevalent among farmers, under which conditions even a small bite wound from one of these lizards would become infected. Nowadays it is senseless to sustain this belief; moreover, no Cuban iguanid inoculates poisonous substances while biting.

Traditionally, children have played with small lizards, getting them to fight; some kids use the lizards to hunt "hairy spiders" (*Eurypelma spinicrum*) by tying them to sticks and placing them in spider caves.

To date it has not been demonstrated that Cuban iguanids transmit either disease or parasitism, nor has any damage to wild vegetation or cultures that could be caused by phytophagous or omnivorous species been verified. On the contrary, the species that live in association with humans eat many noxious insects such as domestic flies, mosquitoes, and cockroaches.

Conservation

From a conservationist perspective, it must be considered that although the majority of iguanid species are abundant and not geographically restricted, there are subspecies whose geographic range is very limited, making them more vulnerable to ecological change from natural or human sources. Nevertheless, many populations of the 120 species of terrestrial reptiles living in Cuba have diminished during the last centuries, although there are no papers in which such decreases were calculated.

The first attempt to establish threat criteria for Cuban vertebrates was made by Buide González *et al.* (1974). These authors proposed eight different degrees of threat, determined by the magnitude of numeric and spatial reduction of the populations, based on several sources. According to their research, there were three Cuban iguanids with some degree of threat in group II (very restricted, scarce), seven in group III (very restricted, somewhat scarce) and one in group VI (restricted, somewhat scarce).

Perera *et al.* (1994) updated this information using the categories of the I.U.C.N. (International Union for the Conservation of Nature and Natural

Resources) Red List (after Mace *et al.* 1992). They proposed a list of fifty species of reptiles threatened to some degree by extinction. Of these fifty species, twenty-eight were iguanids; all were placed in the Vulnerable (V) category, that is, facing a high probability of extinction in the wild in the medium-term future (Mace *et al.* 1992).

Berovides Álvarez (1995) considered seven species of reptiles threatened by extinction, *Cyclura nubila* and *Anolis vermiculatus* among them. He proposed three categories of threat for grouping these species: I, endemic species with just one area of distribution; II, species with a few areas of distribution; and III, species with several areas of distribution. However, this classification system is not being followed currently.

Taking into account the most recent data on endemicity, geographic distribution, and relative abundance of Cuban reptiles, Rodríguez Schettino and Chamizo Lara (1995) proposed to modify the list of Perera *et al.* (1994). They followed the criteria approved by the I.U.C.N. in December 1994, based on version 2.2 by Mace and Stuart (1994); that is, the one in effect. According to these researchers, four iguanids have a Critical (CR) status because they number fewer than 250 mature individuals each, all in only one population: *Leiocephalus onaneyi, Anolis pigmaequestris, A. fugitivus,* and *A. delafuentei.* Four are Endangered (EN) because they have fewer than 2,500 individuals each and their populations are fragmented (*Cyclura nubila*) or all the individuals live in one population (*Chamaeleolis guamuhaya, Anolis juangundlachi,* and *A. birama*). Twenty-two are considered Vulnerable (VU), mainly as a result of modification of their habitats. *Anolis loysiana* and *A. pumilus* are at Lower Risk (LR) because they are almost threatened.

Through the first Conservation Assessment and Management Plan Workshop for Cuban species (CAMP), Berovides Álvarez *et al.* (1996) changed the status of the Cuban iguana (*Cyclura nubila*) from Endangered to Vulnerable. During the second CAMP, Ruiz Urquiola *et al.* (1997*a, b*) considered *Chamaeleolis barbatus* and *C. guamuhaya* to be Endangered (EN); Fernández Méndez (1997*a*) regarded *Anolis pigmaequestris* as Critical (CR); and Rodríguez Schettino and Chamizo Lara (1997) recognized *A. vermiculatus* as being at Lower Risk (LR). In addition, Fernández Méndez *et al.* (1997) proposed Critical (CR) status for *A. equestris potior,* and Fernández Méndez (1997*b*) proposed the same for *A. equestris* ssp. from Cayo Las Brujas: these subspecies have very limited geographic ranges and are exposed to great modifications of their habitats. In the third CAMP, *A. juangundlachi* was classified as Critical (CR) by Rodríguez Schettino *et al.* (1998), as was *Leiocephalus raviceps klinikowskii* by Chamizo Lara and Rodríguez Schettino (1998), while *A. bartschi* was assigned Lower Risk (LR) status by Rodríguez Schettino and Chamizo Lara (1998).

Rodríguez Schettino and Chamizo Lara (in press), considering published papers on conservation, geographic distribution, endemicity, and the relative abundance of the Cuban species, are proposing a new list with sixty-one endangered taxa, of which thirty-five are iguanids: four are regarded as CR, four EN, twenty-two VU, and five LR. With the exceptions of the second and third CAMPs, the authors worked at the species level. At the subspecies level, much of the Cuban iguanid taxa can be placed into one of the categories of threat by extinction because of habitat reduction, alteration, or destruction, even with no intention of taking an extremely conservationist position. Accordingly, Cuban iguanid taxa must be classified as follows:

Extinct in the wild (EW) (presumed):
Leiocephalus onaneyi.
Critical (CR), facing an extremely high risk of extinction in the wild in the immediate future:
L. raviceps klinikowskii, Anolis equestris potior, A. pigmaequestris, A. fugitivus, A. juangundlachi, and *A. delafuentei.*
Endangered (EN), facing a very high probability of extinction in the wild in the near future:
Leiocephalus macropus hyacinthurus, L. stictigaster lipomator, Chamaeleolis barbatus, C. guamuhaya, Anolis equestris juraguensis, A. luteogularis sanfelipensis, A. l. coctilis, A. vanidicus rejectus, A. jubar balaenarum, A. j. santamariae, and *A. birama.*
Vulnerable (VU), facing a high probability of extinction in the wild in the medium-term future:
Leiocephalus cubensis pambasileus, L. c. minor, L. macropus lenticulatus, L. m. hoplites, L. m. aegialus, L. m. felinoi, L. m. torrei, L. r. delavarai, L. r. jaumei, L. stictigaster lucianus, L. s. parasphex, L. s. gibarensis, L. s. septentrionalis, L. s. astictus, L. s. naranjoi, L. s. ophiplacodes, Cyclura nubila nubila, Anolis equestris buidei, A. e. verreonensis, A. e. cincoleguas, A. luteogularis jaumei, A. l. calceus, A. l. delacruzi, A. l. nivevultus, A. baracoae, A. isolepis isolepis, A. i. altitudinalis, A. guazuma, A. alayoni, A. garridoi, A. clivicola, A. cupeyalensis, A. cyanopleurus orientalis, A. anfiloquioi, A. macilentus, A. vescus, A. alfaroi, A. mimus, A. inexpectata, A. vanidicus vanidicus, A. sagrei greyi, A. homolechis turquinensis, A. quadriocellifer, A. guafe, A. confusus, A. jubar albertschwartzi, A. j. maisiensis, A. j. cocoensis, A. ahli, A. rubribarbus, A. imias, Chamaeleolis chamaeleonides, and *C. porcus.*
Lower risk (LR), taxa that have been evaluated and do not qualify for any of the previous categories:
A. loysiana, A. pumilus, A. vermiculatus, A. bartschi, and *A. v. vanidicus.*

It is necessary to define more accurately the conditions of these different populations in order to propose adequate rules for their appropriate utilization as valuable natural resources and to help prevent their extinction.

2

Morphology

LOURDES RODRÍGUEZ SCHETTINO

Systematics is the principal subject treated in literature about Cuban iguanids, and morphology (particularly external morphology) occupies a very important place in it. Body and dewlap color patterns, measurements and ratios between pairs of them, and the shapes, sizes, and number of scales on different parts of the body have been the characteristics most frequently used to describe species. Generally, original descriptions of species and subspecies also include intraspecific variations in morphological characteristics and comparisons of these variations with those found in related taxa.

Internal morphology has not been treated in the same way. Only in some isolated publications (Schwartz 1959a; Etheridge 1960; González Martínez 1981; García de Francisco et al. 1984b) can be found descriptions of internal organs and systems. In the majority of these cases, the morphological descriptions are related to function or to possible adaptive value.

Accordingly, my purpose in writing this chapter is to provide general information about the main characteristics of morphology, not only external but also internal, of the iguanids that inhabit Cuba. At the same time, I wish to provide the reader with dichotomous keys to facilitate the identification of such species.

External Morphology

All Cuban iguanids follow a similar morphological pattern that conforms to a considerable degree with the principal characteristics of the group. Shared features are a slender, lengthened body with a distinguishable head and neck, a long tail, and four relatively long limbs.

However, the four genera are easily separated according to their peculiar morphological features. The species of the genus Leiocephalus have keeled dorsal scales and robust hindlimbs with thin, long toes lacking both digital pads and subdigital lamellae. *Cyclura* is of great size, with a dorsal crest of spines and fingers and toes with neither expansions nor subdigital lamellae.

The species of the genera *Anolis* and *Chamaeleolis* have a useful system of adaptation for living on trees, the subdigital lamellae (fig. 2, parts 3 and 4), and an efficient distinctive organ, the dewlap (fig. 3, part 11); in addition, *Chamaeleolis* has in the median longitudinal gular line two rows of differentiated scales (fig. 4, part 2) that are triangular, filamentous, or barbell-like.

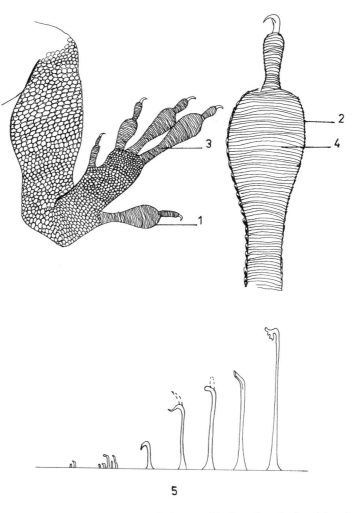

Fig. 2. Hindlimb of *Anolis* and *Chamaeleolis*. *1 and 2,* digital pads; *3 and 4,* subdigital lamellae; *5,* spines, setae, and spikes.

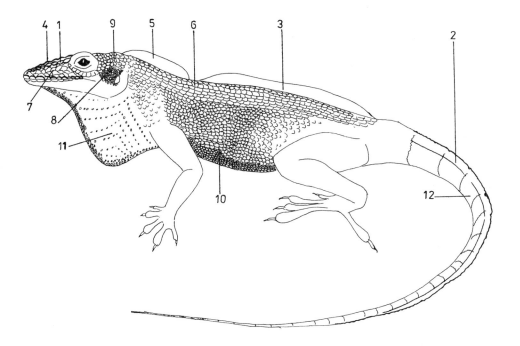

Fig. 3. Lateral view of an iguanid. *1*, canthal ridge; *2*, caudal crest; *3*, dorsal crest; *4*, frontal ridge; *5*, nuchal crest; *6*, dorsal scales; *7*, loreal scales; *8*, preauricular scales; *9*, temporal scales; *10*, ventral scales; *11*, dewlap; *12*, caudal verticils.

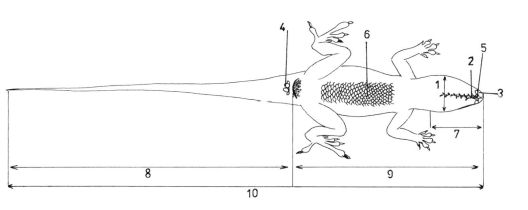

Fig. 4. Ventral view of an iguanid. *1*, head width; *2*, scales of the longitudinal median gular line; *3*, mental scales; *4*, postcloacal scales; *5*, postmental scales; *6*, ventral scales; *7*, head length; *8*, tail length; *9*, snout-vent length; *10*, total length.

The scalation shows a regular pattern of big scales or scutes on the head, and dorsal scales generally smaller than in the ventral region. The scales have different shapes—circular, oblong, rectangular, or lanceolate—depending on the species and the body region; furthermore, they can be smooth or keeled, flat or convex, tubercular or spine-like (fig. 5).

Predominant colors among Cuban iguanids are different shades of brown, gray, and green; some species also show shades of blue. The pattern of color distribution in each species is a generally constant characteristic that can be taken into account in taxonomic studies, although individual variations must always be considered.

Metachromatism is observed in the majority of the species; that is, the same individual is able to change the shade and sometimes the color of its body in response to external stimuli such as light, temperature, and substrate color. Similar changes in response to physiological stimuli are effected through hormonal control of chromatophores (Hadley 1931; Weber 1983).

Species with cryptic coloration are also very common; their color is similar to the substrate, which allows them to stay unnoticed. To obtain the opposite effect, males and some females of the genera *Anolis* and *Chamae-*

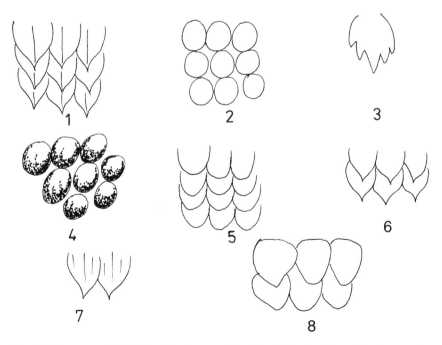

Fig. 5. Scale types. *1*, keeled scales; *2*, smooth scales; *3*, denticulate scales; *4*, granular scales; *5*, imbricate scales; *6*, lanceolate scales; *7*, mucronate scales; *8*, subcircular scales.

leolis have distinctly colored dewlaps; this coloration permits an individual to recognize its mate as well as other individuals from the same or different species.

The dewlap is a skin fold with contrasting colors, differing specifically or subspecifically; it is unfolded by the erection of calcified cartilage, the second ceratobranchial arches of the hyoid apparatus (fig. 6, part 2). Scales covering the dewlap are separated when it is unfolded and are scattered like a fan (fig. 7, part 5). Dewlap size is seemingly in correspondence with

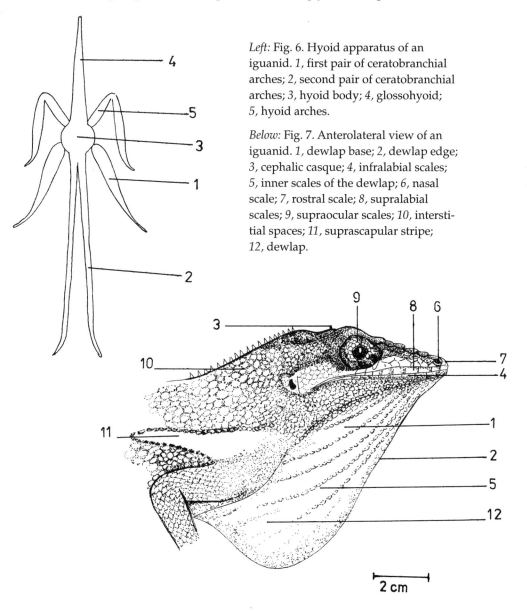

Left: Fig. 6. Hyoid apparatus of an iguanid. *1,* first pair of ceratobranchial arches; *2,* second pair of ceratobranchial arches; *3,* hyoid body; *4,* glossohyoid; *5,* hyoid arches.

Below: Fig. 7. Anterolateral view of an iguanid. *1,* dewlap base; *2,* dewlap edge; *3,* cephalic casque; *4,* infralabial scales; *5,* inner scales of the dewlap; *6,* nasal scale; *7,* rostral scale; *8,* supralabial scales; *9,* supraocular scales; *10,* interstitial spaces; *11,* suprascapular stripe; *12,* dewlap.

2 cm

the lengths of the ceratobranchial arches and not with the size of the animal, since some small species have very large dewlaps and vice versa.

The species of the genera *Anolis* and *Chamaeleolis* have expanded digital pads, and the ventral surfaces of all phalanxes with subdigital lamellae have in their free borders many microscopic spinules, spines, spikes, prongs, and setae (Ruibal and Ernst 1965; Peterson and Williams 1981) (fig. 2, part 5). Collette (1961) found that the more terrestrial the species, the fewer the lamellae it has, and the thinner its fingers and toes; on the contrary, the arboricolous species have a great number of lamellae and their fingers and toes are very wide. Furthermore, the bigger the individuals of a species, the greater the number of lamellae they have.

To characterize the size of a species, it is customary to take several measurements from different parts of the body. Snout-vent length (fig. 4, part 9) is obtained ventrally, from the tip of the snout to the posterior border of the vent opening. Head length (fig. 4, part 7) is measured laterally or ventrally, from the tip of the snout to the posterior border of the ear opening, or to the posterior part of the jaw. Tail length (fig. 4, part 8) is obtained ventrally, from the vent opening to the end of the tail. The snout-vent length plus the tail length make up the total length of the animal (fig. 4, part 10). Head width is measured transversely, at the broadest point (fig. 4, part 1). Fore- and hind-limb lengths and distances between different parts of the head, such as nostril-eye distance, ridge length, and interorbital distance, are also very commonly used. In addition, various ratios can be calculated for different pairs of these measurements.

Internal Morphology

The internal morphological characteristics are less variable than the external ones; consequently, what has been published about iguanians in general can be understood as a pattern of reference for Cuban iguanids. That is why the features offered under this subhead follow those used by such authors as Romer (1956) and Etheridge (1960) (osteology); Baird (1970) and Underwood (1970) (sensorial organs); Zug (1971) (circulatory system); and Porter (1972) (other systems).

The skull is flattened; the frontonasal region, elongate and triangular shaped; the superior temporal arch is present; there is one frontal, one premaxilla, and one parietal; the remaining skull bones are paired; the pineal foramen is generally found in the frontoparietal suture. In the jaw, the dentary, supra-angular, coronoid, and articular are present; the angular is absent only in the genus *Anolis*; it is very reduced in *Chamaeleolis* and present in *Cyclura* and *Leiocephalus*. The splenial is generally absent or very reduced in *Anolis* and *Chamaeleolis*, but present in *Cyclura* and *Leiocephalus*. The teeth are conic and tricuspid; the dentition is pleurodont.

The hyoid apparatus is well developed, formed by calcified cartilages; it consists of a hyoid body or basihyal (fig. 6, part 3), a lingual process or glossohyal (fig. 6, part 4), a pair of hyals or hypohyals that articulate with a pair of ceratohyals (fig. 6, part 5), a first pair of ceratobranchial arches (fig. 6, part 1), and a second pair of ceratobranchial arches (fig. 6, part 2).

The vertebrae are procelous; there are eight cervical vertebrae, twenty-one to twenty-four presacral vertebrae, and two sacral vertebrae. Caudal vertebrae are present in one of four conditions: (1) the transverse processes disappear after a few vertebrae (alpha condition in *Anolis* and *Chamaeleolis*; fig. 1, part 1); (2) the transverse processes are present in all vertebrae, but their direction abruptly changes forward (beta condition in *Anolis*; fig. 1, part 2); (3) almost all vertebrae have transverse processes and there are a few vertebrae at the end of the tail with two transverse processes at each side (*Cyclura*); or (4) all of the vertebrae have transverse processes, but after the first ones, they abruptly diminish in size and their direction changes to lateral (*Leiocephalus*).

The sternum is rhomboidal; there are five cervical ribs, two or four sternal ribs, and two xiphisternal ribs. Not all parasternal ribs are joined to the bony ribs; some are joined at the mid-ventral line. The clavicle is flat; the interclavicle can be T-shaped or arrow-shaped. All species have pentadactyl forelimbs.

There are always three pelvic bones; the preacetabular process of the ileum is well developed; there are always pentadactyl hindlimbs.

The muscular system is adapted to terrestrial life; the muscles and the spinal column support the body weight. The hypaxial musculature allows necessary movement of the ribs for breathing; the musculature of the scapular and pelvic girdles, and that of the limbs, permits locomotion (Porter 1972).

The tongue is broad, thick, papillous, and not protrusible. There are mucous and salivary glands in the mouth in palatal, lingual, sublingual, and labial positions (Schwenk 1988). The components of the digestive system follow the general pattern for vertebrates: esophagus, stomach, small intestine, and large intestine, as well as a short caecum that opens in the vent.

The nasal ducts are large and open in posterior choanas in the oral cavity. The lungs are large and located in the dorsal anterior region of the body cavity.

The heart has two auricles and one ventricle; neither the conus arteriosus nor the sinus venosus is present. There is a carotid arch on either side, which is divided into external and internal carotids; the subclavians start from the right systemic arch and, generally, below the heart. The superior and inferior mesenteric arteries are always present and arise from a common trunk that begins in the basal aorta (Zug 1971).

The kidneys are lobulated metanephros; there are two ureters that open in the vent. The urinary bladder is present; the glomeruli are of small size to limit urine volume and thus prevent excessive water loss. The kidneys' excretions consist mainly of uric acid (80% to 98%), plus urea and ammonia in different proportions (Porter 1972).

The testes are oval and their volume varies according to fluctuations in reproductive activity, which is under endocrine control. The epididymis is well developed and the vas deferens open in the vent, where there are two tubular, sacular hemipenes that evert by muscular action and filling of blood; they have no erectile tissue (Porter 1972). During copulation only one hemipenis is everted, depending on the side on which the male is in relation to the female. The ovaries are racemous. The oviducts are long, with broad anterior openings; they can have thick-walled final ends or a uterus that opens in the cloaca.

In the posterior ventral region two fat bodies are found. These are accumulations of grease that increase in size in inverse correlation with increases in reproductive activity, from which Licht and Gorman (1970) inferred that they act as energy storage units for reproduction in both sexes.

The brain hemispheres are small and lengthened; they are continued forward as fully differentiated olfactory tracts and bulbs. The mesencephalon is well developed and has large optic lobes; the cerebellum is slightly developed. There are twelve pairs of cranial nerves; the dorsal roots of the spinal nerves have only sensory fibers, and the ventral roots motor fibers, not only visceral but somatic ones (Porter 1972).

The ear provides the functions of hearing and equilibrium; the tympanic membrane and the columella (middle ear), visible from the outside, are in the bottom of a small depression of skin in the temporal region (external ear). The inner ear is organized as in other vertebrates; it is constituted of a bony labyrinth, membranous labyrinth, cochlea (or lagena), sacculus, utriculus, and semicircular canals (Baird 1970; Porter 1972).

Vision is the most important sense; the lens is susceptible to accommodation by means of ciliary muscle contractions; the pupil is round (Underwood 1970). Joining the sclerotic with the cornea is a ring of sigmoid profile that maintains the concavity of this joint (the sulcus) against intraocular pressure. It is composed of bony plates, the scleral oscicles, that are overlapped; their shape and number vary according to species (Underwood 1970). In the studied Cuban species (*Anolis equestris* and *Chamaeleolis cha-*

maeleonides) there are fourteen scleral oscicles (de Queiroz 1982), and there should also be fourteen in the other species.

There are three types of visual cells: major single cones, minor single cones, and double cones; there is a central fovea and a temporal fovea. The lower eyelid is more movable and bigger than the upper eyelid (Underwood 1970).

The pineal eye is present in the diencephalon and can be externally observed in the interparietal scale (fig. 8, part 4); it has the structure of a degen-

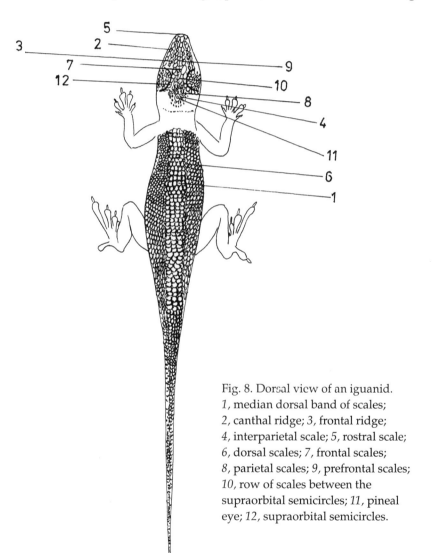

Fig. 8. Dorsal view of an iguanid.
1, median dorsal band of scales;
2, canthal ridge; *3*, frontal ridge;
4, interparietal scale; *5*, rostral scale;
6, dorsal scales; *7*, frontal scales;
8, parietal scales; *9*, prefrontal scales;
10, row of scales between the supraorbital semicircles; *11*, pineal eye; *12*, supraorbital semicircles.

erated eye with a lens, retina, and nerve that joins it to the brain, but it cannot accommodate (Porter 1972). Stebbins and Eakin (1958) proved that the pineal eye in North American iguanids has a photoreceptive function and regulates the amount of time an animal must be exposed to sunlight.

The nasal cavities are divided into two parts, the anterior (outer), or vestibule, and the posterior (inner), or olfactory chamber, whose walls have one or two small folds covered by sensitive epithelium (Porter 1972). The vomero-nasal organ, or Jacobson's organ, and the olfactory system are well developed. The tongue is functionally related to Jacobson's organ, to which environmental scent particles are carried through ducts that connect this organ with the mouth in the anterior palate (Simon 1983). The taste sense is of relatively little importance; the taste papillae are mainly in the pharynx and there are few in the tongue (Porter 1972; Simon 1983).

Knowledge about cutaneous receptors in reptiles is limited, but undoubtedly there are temperature, touch, and pain receptors (Porter 1972). In some species of *Anolis* and in *Chamaeleolis chamaeleonides*, sensorial organs constituted of a circular, smooth lens have been found in an elliptical depression of the scales (Etheridge 1960).

Morphological Variation

The differences in external morphology that have been observed in diverse allopatric or parapatric populations of the same species have been considered geographic variations, which has led to descriptions of races or subspecies. Coloration and size in individuals are the main characteristics that identify geographic subspecies.

Accordingly, of the sixty-two Cuban iguanid species, twenty are polytypic, with two or more described subspecies. *Cyclura n. nubila* and *Anolis a. angusticeps* are the only subspecies living in Cuba; the others live in the Cayman and Bahama Islands, respectively. *Anolis sagrei* and *Leiocephalus carinatus* have both Cuban and non-Cuban geographic races.

On the other hand, twenty-seven of the monotypic species are found in restricted localities or regions; their morphological variations are merely individual, within the limits of their specific definitions. The remaining fifteen monotypic species are widely distributed but their morphological characteristics are relatively constant, with no defined pattern of individual variation that could identify them as geographic subspecies in any of the involved populations.

Furthermore, clines are known in scalation of *Anolis angusticeps* (Schwartz and Thomas 1968) and that of western populations of *A. porcatus*

(Pérez-Beato and Berovides Álvarez 1979, 1981, 1984), although their ends have not been diagnosed as subspecies or geographic races.

The most important intrapopulation variations are those related to sex and age. In most of the species males differ from females in having more conspicuous color patterns, a dewlap in the genus *Anolis*, greater size of the head, body, and tail, and enlarged postcloacal scales.

A few species show sexual differences in scalation. Male iguanas (*Cyclura n. nubila*) have protuberant head scales and long, abundant spines in the nuchal, dorsal, and caudal crests, while the spines of females are small and scarce. Some species have head scales that are more or less rough or keeled depending on the sex. However, in species such as *Leiocephalus carinatus*, the *equestris* group, and species of the genus *Chamaeleolis*, the sexes differ only slightly in size.

Juveniles are smaller than adults and, with the naked eye, males cannot be differentiated from females because their coloration is the same, and similar to that of adult females in species with sexual dimorphism. When no evident sexual differences in coloration are evident, juveniles are similar to both male and female adults but have light-colored transverse bands on their bodies and tails. Youngsters of species of the genus *Chamaeleolis* have small, poorly defined cephalic casques. Newly born lizards have a small umbilical opening in the middle of the ventral region that remains for some days after eclosion.

According to Rodríguez Schettino and Martínez Reyes (1989a), in each studied population up to six age groups can be found on examination of the reproductive condition of its members. During the reproductive period adult males have voluminous testes and a well-developed epididymis, and numerous spermatozoa can be observed in histological preparations of testes and ducts. Adult females during the reproductive period have oviductal eggs or big ovarian follicles containing food yolk. In both cases, using large samples, it is possible to find the minimal size at which sexual maturity or the adult condition is reached, which enables recognition of adults during months of little or no reproductive activity.

During any season of the year, subadult males have poorly developed testes and epididymides and do not reach adult size. Subadult females do not have oviductal eggs and their ovarian follicles are small, without food yolk, and generally white. However, these two groups have typical adult external features. The gonads of both male and female juveniles are so small that it is very difficult to identify them macroscopically. Juvenile maximal sizes are reached when the gonads are easily identified as either male or female. The boundaries between juvenile and subadult groups cannot always be clearly detected; in such cases it is preferable to refer to such groups as immature stages or juveniles.

Ecomorphology

The first attempt to correlate the morphology of Cuban iguanids with their ecology was made by Collette (1961), who found that, among the six sympatric species of the genus *Anolis* observed at Bosque de La Habana, the most arboricolous ones had shorter hindlimbs and tails, greater size, and a larger number of subdigital lamellae than did the most terrestrial species. He considered that such morphological adaptations help several related species to coexist in the same geographic area by allowing them to use the structural resources of the habitat in different ways, thus diminishing the competition among them.

Along similar lines, other authors have found some patterns of ecological distribution among species of the genus *Anolis* that inhabit the Greater Antilles that agree with the morphological adaptations of such species. Rand (1964) was the first to recognize ecological types in Puerto Rico, and Rand and Williams (1969) defined their names according to the structural habitat in which at least the adult males of the species spend most of the day at a locality in the Dominican Republic.

In this system, crown anoles are large and usually live in the forest canopy; twig anoles are small and perch on twigs near the canopy; trunk-crown anoles are generally found on the highest parts of the trunk; trunk anoles are mainly confined to the trunk, traveling neither up to the canopy nor down to the ground; trunk-ground anoles are found on the lowest part of the trunk and go down to the ground; and grass-bush anoles perch on the stems of grasses and bushes. Williams (1972) gave the name ecomorphs to these ecological types or species with the same habitats or niches, similar in morphology and behavior but not necessarily phyletically related. He emphasized the morphological aspect of the similarities between types. Finally, Williams (1983; table 15.1) defined the main characteristics of the six ecological types or ecomorphs in terms of size, color, scalation, body proportions, modal perch, and foraging and defensive behavior of the species.

Ecomorphs are represented on each island of the Greater Antilles, and the species of each ecomorph are morphologically more similar among different islands than among other ecomorphs within the same island (Losos *et al.* 1998). The situation is different when DNA sequences are examined: the species are first related to each other within the same island and second by ecomorphs.

Williams (1976) placed the *equestris* group in the crown ecomorph, the remaining species of the *carolinensis* series in trunk-crown and twig ecomorphs, the *alutaceus* series in the grass-bush ecomorph, the *sagrei* series in the trunk-ground ecomorph, and the *lucius* series in a category called "rock and other." He took into account the morphological similarities of the species and their spatial distributions in their characteristic habitats. However, he included very morphologically different species of the *carolinensis* series in only two ecomorphs. In addition, in his paper of 1983 he did not consider the Cuban species but included two very distinct groups, the grass anoles (*alutaceus* group) and an unnamed ecomorph of the *argillaceus* group, in the grass-bush ecomorph.

I think it is more useful to separate the species of the *carolinensis* series by their appropriate ecomorphs, placing the *carolinensis* subgroup (*porcatus, isolepis,* and *allisoni*) in the trunk-crown ecomorph; the *angusticeps* subgroup (*angusticeps, paternus, guazuma, alayoni,* and *garridoi*) in the twig ecomorph; the *argillaceus* group (*argillaceus, loysiana, centralis,* and *pumilus*) in another ecomorph called the bush ecomorph; and the *lucius* series, except for *A. vermiculatus,* a trunk anole, in the rock ecomorph.

On the other hand, Moermond (1979) related the use of perches and movement type in seven Haitian species of the genus *Anolis* to four morphometric indexes that include, in different ways, snout-vent length, forelimb length, and tail length. Thus he divided them into "crawlers," which are equivalent to trunk-crown and twig ecomorphs; "runners," which correspond to trunk ecomorphs; and "jumpers," which are consistent with trunk-ground and grass-bush ecomorphs.

Pérez-Beato (1982a) simplified one of Moermond's (1979) coefficients in an attempt to relate the morphometry of anolines to their locomotive behavior; this relationship was refuted by Estrada and Silva Rodríguez (1984). The latter two authors analyzed twenty-three Cuban species of the genus *Anolis* and found that the intervals of the coefficients proposed by Moermond (1979) overlapped; therefore they tested a new one, by means of which the studied species were clustered in groups that matched Moermond's (1979) categories of locomotion.

According to this new General Index of Locomotion (= humerus length/ tibia length), *Anolis anfiloquioi, A. alutaceus, A. vanidicus, A. spectrum, A. cyanopleurus, A. clivicola,* and *A. fugitivus* are jumper species; *A. bartschi, A. rubribarbus, A. mestrei, A. allogus, A. lucius, A. sagrei, A. ophiolepis, A. bremeri,* and *A. homolechis* are runners; and *A. allisoni, A. paternus, A. angusticeps, A. porcatus, A. centralis, A. argillaceus,* and *A. loysiana* are crawlers.

Granda Martínez (1987) used a coefficient of expected arboreality, which is the sum of the three Moermond's coefficients, and, moreover, a coefficient of observed arboreality (CAo):

CAo = F (1—1/h) + Xe, where F is the frequency of individuals observed on trees or bushes; h is the mean height above ground of those individuals; and Xe is 0.5 if modally the species escapes downward, or 1 if modally it escapes upward. Applying both coefficients to some Cuban species, he concluded that they have adaptive ecomorphological characteristics for surviving in the environment where they live.

Identification

The use of dichotomous keys for identification of taxonomically related species is very common because they make arriving at a given taxa convenient, quick, and easy. Some keys have been elaborated for Cuban reptiles, into which from time to time new species have been incorporated. One of the first keys, by Barbour and Ramsden (1919), included twenty-three species; Alayo Dalmau (1955) considered thirty-one species; Ruibal (1964) used only the anolines (twenty-three species); and Schwartz and Henderson (1985) offered keys for fifty-five species. The characteristics used in the last paper are mainly coloration and size, which are easily distinguished; in the other papers, the authors considered scalation (Barbour and Ramsden 1919; Ruibal 1964) and habitat data (Alayo Dalmau 1955), characteristics that are more difficult to precisely identify.

Developing a key for Cuban iguanids is not an easy task. The patterns of body and dewlap coloration change at death and with the use of preserving liquids. Consequently, the characteristics that most obviously differentiate the species are not fixed in time in preserved specimens. Body size and proportions between different parts of the body should be used only for specimens of the same size class. Nevertheless, I provide here keys for the four genera living in Cuba and for sixty species of the polytypic genera (*Leiocephalus, Anolis,* and *Chamaeleolis*). *Anolis delafuentei* is not included because the only known specimen, a male, is not precisely described by Garrido (1982b); this precludes its proper comparison with the most closely related species, *A. ahli.* In developing the keys I used external features such as coloration, size, and scalation in live, adult animals, males in species with sexual dimorphism. Therefore, these keys should be used primarily for living individuals, or for dead animals that have been preserved for a very short time.

Key for Cuban Genera of the Family Iguanidae

1. Snout-vent length less than 200 mm 2

   Snout-vent length more than 200 mm .... *Cyclura*

2. With subdigital lamellae 3
   Without subdigital lamellae 4
3. With differentiated scales at the dewlap edge .... *Chamaeleolis*
   Without differentiated scales at the dewlap edge, or without
   dewlap .... *Anolis*
4. With smooth dorsal scales .... *Cyclura* (juveniles)
   With keeled dorsal scales .... *Leiocephalus*

Key for Cuban Species of the Genus *Leiocephalus*

1. With more or less defined longitudinal color zones 2
   Without longitudinal color zones .... *L. carinatus*
2. Dorsal region uniformly brown (zones 1, 2, and 3) 3
   Dorsal region with pale and dark brown lines 4
3. Postorbital stripe dark brown with black suprascapular patch,
   generally with a yellow bar in its center .... *L. macropus*
   Postorbital stripe dark brown, with black scattered
   spots .... *L. cubensis*
4. Gular region with brown, gray, or black dashes and
   spots .... *L. stictigaster*
5. Longitudinal color zones continuing to the first part of the
   tail .... *L. onaneyi*
   Longitudinal color zones less defined to the sacra region .... *L. raviceps*

Key for Species of the Genus *Chamaeleolis*

1. Dewlap edge scales small, conical .... *C. chamaeleonides*
   Dewlap edge scales filamentous or barbell-like 2
2. Dorsal crest from the superior border of the cephalic casque 3
   Dorsal crest from the nuchal region, not from the cephalic
   casque .... *C. porcus*
3. Dewlap edge scales short and thin, without reaching the level of the
   posterior edge of the orbit .... *C. guamuhaya*
   Dewlap edge scales elongate and gross, to the level of the posterior
   edge of the orbit .... *C. barbatus*

Key for Cuban Species of the Genus *Anolis*

1. Head without cephalic casque 2
   Head with cephalic casque; dewlap in both sexes 10
2. Head much longer than wide 3
   Head almost as long as wide 4

3. Canthal and frontal ridges slightly developed or absent 5
   Canthal and frontal ridges well developed 15
4. Tail much longer than the snout-vent length; with or without
   caudal crest; brown color with yellow, reddish, or greenish
   stripes, dashes, and spots 32
   Tail a little longer than the snout-vent length; without caudal crest;
   brown and gray coloration, with brown spots and dots 47
5. Tail much longer than the snout-vent length 6
   Tail a little longer or shorter than the snout-vent length, thick at the
   beginning 29
6. Habitus not slender 7
   Very elongate and slender habitus; very small size 16
7. Dewlap only in males 8
   Males with a transverse fold of skin in the gular region 9
8. Three blue, semitransparent scales in the lower eyelid; dewlap
   white, with gray and yellow semicircular stripes .... *A. lucius*
   Two blue, semitransparent scales in the lower eyelid; dewlap white
   or gray with narrow violet stripes .... *A. argenteolus*
9. Digits with the digital pads not expanded .... *A. vermiculatus*
   Digits with the digital pads very expanded .... *A. bartschi*
10. Suprascapular stripe weakly defined, with orange flecks on a
    black field .... *A. baracoae*
    Suprascapular stripe well defined 11
11. Tail with transverse bands .... *A. luteogularis*
    Tail without transverse bands in adults 12
12. Low number of vertical dorsal scales included in the distance
    between the tip of the snout and the anterior border of the
    orbit (9–19) 13
    High number of vertical dorsal scales included in the distance
    between the tip of the snout and the anterior border of the
    orbit (18–26) 14
13. Brown head and cephalic casque .... *A. equestris*
    Gray head and cephalic casque .... *A. pigmaequestris*
14. With yellow-spotted labials .... *A. smallwoodi*
    With green-spotted labials .... *A. noblei*
15. Canthal ridge higher than frontal ridge; triangular skin depression
    behind the ear opening; dark pink dewlap .... *A. allisoni*

Canthal ridge lower than frontal ridge; round or oval ear opening, without skin depression; strawberry dewlap .... *A. porcatus*

16. Labial stripe whose color differs from that of the body 17
    Without labial stripe; pale brown body; orange brown dewlap; brown iris; ten rows of scales in the mid-dorsal zone ...... *A. spectrum*

17. Labial stripe to the ear opening 18
    Labial stripe farther behind the ear opening 19

18. With postocular black blotch 20
    Without postocular blotch 21

19. Labial stripe to the forelimbs 23
    Labial stripe to the hindlimbs 24

20. Brown body; pale yellow dewlap; blue iris .... *A. alutaceus*
    Green or pale brown body; yellowish green dewlap; green iris .... *A. cyanopleurus*

21. Smooth dorsal scales; greenish brown body; ochraceous yellow dewlap, with greenish inner scales; green iris .... *A. clivicola*
    Keeled dorsal scales 22

22. With six scale rows in the median dorsal zone; brown body; pale yellow dewlap; green iris .... *A. mimus*
    With seven to nine scale rows in the median dorsal zone; olivaceous brown body; greenish ochre dewlap; green iris .... *A. vanidicus*

23. With postocular blotch 25
    Without postocular blotch 26

24. Round ear opening 27
    Oval ear opening; with eight scale rows in the median dorsal zone; brown body; pale yellow dewlap; blue iris .... *A. juangundlachi*

25. Smooth dorsal scales; olivaceous brown body; greenish yellow dewlap; blue iris .... *A. inexpectata*
    Keeled dorsal scales; pale brown body; grayish brown dewlap; deep blue iris .... *A. macilentus*

26. With suprascapular blotch 28
    Without suprascapular blotch; pale brown body; pale gray, very small dewlap; blue iris .... *A. alfaroi*

27. With a pale brown mid-ventral longitudinal line; pale brown body; pale yellow dewlap; blue iris .... *A. cupeyalensis*
    Without mid-ventral line; pale brown body; pale yellow dewlap; greenish blue iris .... *A. fugitivus*

28. Grayish brown body; ochraceous dewlap; brown iris .... *A. anfiloquioi*

    Brown body; pale brown dewlap; gray iris .... *A. vescus*

29. Brown or grayish body 30

    Green-and-black body; pale yellow dewlap .... *A. isolepis*

30. Tail longer than snout-vent length 31

    Tail shorter than snout-vent length 32

31. Yellow or reddish ventral region 33

    Grayish white ventral region; yellowish pink

    dewlap .... *A. angusticeps*

32. Snout elongated, overhanging the mental region; orange-red

    dewlap, with an ochraceous basal spot .... *A. guazuma*

    Snout short, does not overhang the mental region; dark reddish

    orange dewlap, centrally yellow .... *A. garridoi*

33. Carmine and yellow dewlap, with black lines in the gular

    region .... *A. paternus*

    Yellow dewlap .... *A. alayoni*

34. Keeled ventral scales 35

    Smooth ventral scales 37

35. Lanceolate, imbricate, keeled scales; red dewlap, with big and

    keeled inner scales .... *A. ophiolepis*

    Small scales, with a mid-dorsal longitudinal zone of bigger,

    keeled scales; pale or dark brown body, with yellow dots,

    spots, and dashes 36

36. Brick red or ochraceous dewlap, with a yellow edge and pale

    yellow or dark brown inner scales .... *A. sagrei*

    Ochraceous or terra cotta–colored dewlap, very big, with a yellow

    edge and black inner scales .... *A. bremeri*

37. Smooth or slightly keeled supracarpal scales 38

    Keeled or multicarinate supracarpal scales..................... 42

38. Smooth head scales; dark brown and grayish green crossbands in

    the body; saffron orange dewlap .... *A. imias*

    Keeled head scales 39

39. Gray or olive green body 40

    Dark brown or black body, with gray, yellow or reddish

    longitudinal dashes and stripes 41

40. Greenish gray body; dewlap red at the base with two or three

    greenish yellow thin inner bars and white edge .... *A. mestrei*

    Gray body with dark spots 45

41. Dewlap white, gray, or white with gray semicircular
    bars .... *A. homolechis*
    Dewlap dark or pale yellow, or dark or pale orange 46
42. Vertical oval ear opening.. 43
    Diagonal oval ear opening; brown body, with a black
    suprascapular blotch; orange yellow dewlap with three red
    transverse bars .... *A. quadriocellifer*
43. Brown body with reddish and yellow dots and spots................... 44
    Grayish brown body with black transverse bars; yellow dewlap,
    with three or four semicircular red bars and white
    edge .... *A. rubribarbus*
44. Yellow dewlap with three or four semicircular reddish
    bars .... *A. allogus*
    Yellow dewlap with a round basal reddish blotch.................. *A. ahli*
45. Yellow dewlap .... *A. birama*
    Yellowish white dewlap .... *A. guafe*
46. Orange or yellow dewlap .... *A. jubar*
    Very small pale yellow dewlap, or yellow with white posterior
    end .... *A. confusus*
47. With many spine-like scales; pinkish yellow dewlap ...... *A. loysiana*
    Without spine-like scales 48
48. Round ear opening 49
    Oval ear opening, with an external fold of skin; pale yellow or
    brick red dewlap .... *A. centralis*
49. Snout-vent length to 46.2 mm; pale orange dewlap with reddish
    dots .... *A. argillaceus*
    Snout-vent length to 39.2 mm; pale orange dewlap ...... *A. pumilus*

3

Ecology and Behavior

LOURDES RODRÍGUEZ SCHETTINO, MERCEDES MARTÍNEZ REYES,
AND LUIS V. MORENO GARCÍA

The first ecological papers about lizards from temperate zones were written by Cowles (1939), Fitch (1940), Bogert and Cowles (1947), Oliver (1948), and Bogert (1949). Some subsequent studies on tropical species were conducted by Collette (1961), Ruibal (1961), and Rand (1962, 1964) after which research on the autoecology of tropical iguanids increased, mainly in the Antillean islands (Rand 1967c; Sexton 1967; Jenssen 1970; Sexton *et al.* 1972; Fleming and Hooker 1975; Floyd and Jenssen 1983; Andrews 1991).

Because of the great diversity of species and the abundance of individuals in some communities of tropical iguanids, these animals have been used as models to characterize inter- and intraspecific relationships that allow them to coexist in sympatry without severe competitive interference (Rand 1967b; Rand and Humphrey 1968; Schoener and Gorman 1968; Rand and Williams 1969; Fitch 1975; Lister 1976; Pacala and Roughgarden 1982; Roughgarden *et al.* 1983; Duellman 1987).

The first observations on the use of some environmental resources and the behavior of Cuban iguanids appeared in the last century, with the classic works of Cocteau and Bibron (1837–1843) and Gundlach (1867, 1880). It is also possible to find some data on their habitat use in several later papers of a taxonomic nature, such as species or subspecies descriptions or lists in which taxa living in a given locality are mentioned. Examples of these papers and books can be found in Barbour and Ramsden (1919), Barbour (1930b), Schwartz and Ogren (1956), Ruibal (1964), Garrido and Schwartz (1968), Schwartz and Garrido (1968b), Garrido (1973b, c, 1980a, b), Garrido and Jaume (1984), and Schwartz and Henderson (1991).

However, autoecological studies about Cuban species truly began during the second half of the 1970s (Valderrama Puente *et al.* 1976; Llanes

Echevarría 1978; Ortiz Díaz 1978; González Bermúdez and Rodríguez Schettino 1982; Estrada and Novo Rodríguez 1986*a*; Rodríguez Schettino and Valderrama Puente 1986; Rodríguez Schettino and Martínez Reyes 1985, 1989*a, b*; Martínez Reyes *et al.* 1990). The first attempts were made to characterize some communities from the perspective of the partitioning of environmental resources among their components (Sampedro Marín *et al.* 1979, 1982; Rodríguez Schettino 1985; Rodríguez Schettino and Martínez Reyes 1985; García Rodríguez 1989; Quesada Jacob *et al.* 1991).

Thus in this chapter I reveal some aspects of the use of environmental resources, intra- and interspecific ecological relationships, and reproduction, based on published data for tropical iguanid species, not only Cuban but also from other countries.

Spatial Distribution

Iguanids are able to inhabit every Cuban terrestrial ecosystem by means of their morphological, physiological, and behavioral adaptations. Few species are restricted to only one type of vegetation; in cases where such restriction exists it is almost always due to limits imposed by geologic, geographic, or climatic barriers, and, occasionally, to limited knowledge of the distribution of those taxa. The cases of *Anolis vermiculatus*, which lives only in the gallery forests of Pinar del Río province, and of *A. bartschi*, which can be found solely at the *mogotes* of the same province, are remarkable. Likewise, *A. juangundlachi* is located only in the grasslands of the environs of Carlos Rojas, Matanzas province, and *A. imias* in the rocky outcrops of the southern, semidesertic coast of Guantánamo province. On the contrary, many species are found in several types of vegetation. *A. equestris, A. luteogularis, A. porcatus, A. angusticeps, A. alutaceus, A. homolechis,* and especially *A. sagrei* (table 3.1), are noteworthy.

It is possible to find only one or two species in habitats with the most homogeneous vegetation structure, while in those with the greatest structural diversity up to eleven species can coexist in sympatry, depending on their geographic and altitudinal distributions. Mangroves and cloud forests are the vegetation zones that shelter the fewest species (table 3.1) because they also have the least favorable climatic and structural conditions for reptile survivorship. Species that have been found in such places—for instance, *A. porcatus, A. angusticeps, A. sagrei,* and *A. homolechis*—are the most ecologically plastic and live in the majority of the terrestrial ecosystems in Cuba.

On the contrary, most of the Cuban species live in forests, mainly in semideciduous and evergreen forests (table 3.1), which contain 61.3% of

Table 3.1. Distribution of Cuban iguanids living in the vegetation types (after Capote *et al.*, 1989). *1*, cloud forest; *2*, montane rainforest; *3*, submontane rainforest; *4*, submontane mesophyllous evergreen forest; *5*, mesophyllous evergreen and semideciduous forest; *6*, pine forest; *7*, "*mogote*" complex of vegetation; *8*, swamp forest; *9*, mangrove forest; *10*, coastal and subcoastal microphyllous forest; *11*, semidesertic thorny shrubwood; *12*, serpentine xeromorphous shrubwood; *13*, cultures and pastures; *14*, urban zones.

Species	1	2	3	4	5	6	7	8	9	10	11	12	13	14
L. carinatus							X			X	X			X
L. cubensis										X			X	X
L. raviceps										X	X			
L. macropus					X					X	X			
L. stictigaster					X	X				X				
L. onaneyi											X			
C. nubila					X				X	X	X			
C. chamaeleonides			X	X	X					X			X	
C. porcus			X	X	X									
C. barbatus			X	X	X									
C. guamuhya				X										
A. equestris			X	X	X	X			X	X			X	X
A. luteogularis			X	X	X	X	X	X	X	X			X	X
A. noblei			X	X	X									
A. smallwoodi			X	X	X				X				X	
A. baracoae			X	X	X									
A. pigmaequestris										X				
A. porcatus				X	X					X	X	X	X	X
A. isolepis	X	X	X											
A. allisoni					X					X				
A. angusticeps	X	X	X	X	X	X				X	X	X		X
A. paternus													X	X
A. guazuma				X										
A. alayoni				X		X						X		
A. garridoi				X										
A. loysiana					X									
A. argillaceus					X	X				X		X		
A. centralis										X	X	X		
A. pumilus										X		X		X
A. lucius					X		X			X				
A. argenteolus					X					X	X		X	X
A. vermiculatus					X									
A. bartschi							X							
A. alutaceus			X	X	X		X			X		X		
A. clivicola	X	X												
A. anfiloquioi					X									
A. inexpectata			X	X		X								

Species	1	2	3	4	5	6	7	8	9	10	11	12	13	14
A. macilentus				X										
A. vescus				X										
A. alfaroi					X									
A. cyanopleurus				X										
A. cupeyalensis				X										
A. mimus				X	X									
A. fugitivus			X											
A. juangundlachi					X									
A. spectrum					X									
A. vanidicus			X	X	X									
A. sagrei			X	X	X	X	X	X	X	X	X	X	X	X
A. bremeri					X									
A. homolechis	X	X	X	X	X	X	X	X		X	X	X	X	X
A. quadriocellifer					X									
A. jubar					X					X	X			
A. guafe					X						X			
A. confusus					X									
A. mestrei					X		X							
A. allogus			X	X	X									
A. ahli			X	X										
A. delafuentei				X										
A. rubribarbus			X	X	X	X								
A. imias												X		
A. birama						X								
A. ophiolepis					X								X	X
Total	4	4	19	26	38	12	8	3	5	25	13	8	12	13

iguanids. The coastal shrubwoods also shelter many species because of their relatively stable climates and the abundance of adequate sites for performing lizard social activities, feeding, and reproduction.

Besides this horizontal distribution, there is a vertical spatial distribution, not only intraspecifically but interspecifically as well. Generally, males are situated higher above ground, on trunks, branches, or rocks, than are females and juveniles. For example, *A. vermiculatus* at Soroa occurs on tree trunks, branches, and rocks, males at a mean height above ground of 2.85 m and females and juveniles at a mean height of 1.02 m (Rodríguez Schettino et al. 1987). When two or more species coexist in sympatry or syntopy, the bigger ones occupy the highest parts of the trees (Williams 1983; Rodríguez Schettino 1985; Rodríguez Schettino and Martínez Reyes 1985, 1989a). Estrada and Novo Rodríguez (1986a) found that *A. sagrei* and *A. porcatus* at Cayo Inés de Soto, Pinar del Río province, differed in their height above ground: the first one was positioned lower, most frequently between 0 and

1.0 m, and the second one higher, most frequently between 1.0 and 1.5 m.

When asleep or resting, iguanids are sheltered among branches of crowns of trees or bushes, under stones, under the bark of trees, or among the grasses and leaves of bushes and trees. Species of the genus *Chamaeleolis* and those of the *equestris* group of the genus *Anolis* have been seen at night sleeping on the highest branches of trees, embracing the branches with their four legs and sometimes with their tails too. *A. porcatus* and *A. allisoni* use the high broad leaves of bushes and trees, sprawling over the surface with their legs and tails extended. *A. angusticeps* and its relatives sleep draped across the high, thin branches of shrubs and trees, embracing them with their legs and tails. Members of the *argillaceus* group customarily rest under the bark of trees or under leaf litter. Species living on rocks (for example, *A. lucius, A. bartschi,* and *A. mestrei*) enter holes and sleep inside them. The *alutaceus* and the *sagrei* groups mainly take refuge under leaf litter. Other species are buried under earth or sand or under fallen leaves; these species make their own burrows or use existing holes in the soil or stones. For instance, species of the genus *Leiocephalus* bury themselves under the ground in the afternoon and emerge when the sun is heating the soil; however, *L. cubensis* has also been seen in towns, emerging in the morning from holes beneath sidewalks.

All of these species can use diverse substrates; generally, however, a given species is most frequently found on one type or part of the substrate, which characterizes it. The species that resemble each other morphologically use the space in a similar way; Rand and Williams (1969) and Williams (1972, 1983) proposed the term *ecomorph* to describe such species. The variety of substrates used by Cuban iguanids, and those which give them the classification ecomorph, can be observed in table 3.2. *A. porcatus, A. vermiculatus, A. sagrei,* and *A. homolechis* widely use these structural resources, while *Leiocephalus stictigaster, L. onaneyi, A. guazuma, A. loysiana, A. rubribarbus,* and *A. birama* are found on only one given type of substrate.

Trunks of trees and bushes are the most frequently used perches; however, some species, such as those of the genus *Leiocephalus*, typically dwell on the ground, and those in the *alutaceus* group of the genus *Anolis* live among grasses. Nevertheless, other substrates, such as branches and rocks, are also frequently used. Species of the genus *Leiocephalus* and *Cyclura nubila* are almost exclusively ground and rock dwellers, and only on some occasions have *L. carinatus, L. macropus,* and *C. nubila* been observed climbing tree trunks or bushes.

Space, in both the vertical and the horizontal dimension, is defended in a certain range that depends on the ecomorphological and behavioral characteristics of each member of the lizard population or community. Gener-

Table 3.2. Distribution of Cuban iguanids according to the substrates used. *1*, crown; *2*, twig; *3*, trunk-crown; *4*, trunk; *5*, trunk-ground; *6*, grass-bush; *7*, grass; *8*, rock; *9*, ground; *(XX)*, the more used substrate.

Species	1	2	3	4	5	6	7	8	9
L. carinatus				X				XX	X
L. cubensis								X	XX
L. macropus				X				X	XX
L. raviceps								X	XX
L. stictigaster									XX
L. onaneyi									XX
C. nubila	X	X						XX	X
C. chamaeleonides		XX		X	X				
C. porcus		X		XX	X				
C. barbatus		X		XX	X				
C. guamuhaya				XX					
A. equestris	XX	X	X	X					
A. luteogularis	XX	X	X	X					
A. noblei	XX	X	X	X					
A. smallwoodi	XX	X	X	X					
A. baracoae	XX	X	X	X					
A. pigmaequestris	X	X	X	XX					
A. porcatus	X	X	X	XX	X	X			
A. isolepis		XX		X	X				
A. allisoni	X	X	X	XX					
A. angusticeps		XX		X					
A. paternus		X		XX					
A. guazuma		XX							
A. alayoni		XX		X					
A. garridoi		XX							
A. loysiana				XX					
A. argillaceus		X		X		XX			
A. centralis		X		X		XX			
A. pumilus		X		X		XX			
A. lucius				X				XX	
A. argenteolus		X	X	X	XX			X	
A. vermiculatus		X		XX		X		X	
A. bartschi				X				XX	
A. alutaceus		X				XX	X	X	
A. clivicola		X		X		XX			
A. anfiloquioi		X				XX	X		
A. inexpectata						XX	X		
A. macilentus		X				XX	X		
A. vescus						XX	X		
A. alfaroi						XX	X		
A. cyanopleurus						X	XX		

(continued on next page)

Table 3.2. (*continued*)

Species	1	2	3	4	5	6	7	8	9
A. cupeyalensis						X	XX		
A. mimus						X	XX		
A. fugitivus						X	XX		
A. juangundlachi						X	XX		
A. spectrum						X	XX		
A. vanidicus						XX	X		X
A. sagrei					X	XX	X	X	X
A. bremeri				X	XX	X			
A. homolechis				X	XX	X		X	X
A. quadriocellifer					XX	X		X	
A. jubar					XX	X			X
A. guafe				X				XX	
A. confusus				X	XX				
A. mestrei					X			XX	
A. allogus					XX			X	
A. ahli					XX			X	
A. delafuentei					XX				
A. rubribarbus					XX				
A. imias					X			XX	
A. birama					XX				
A. ophiolepis						X	XX		
Total	9	28	9	31	19	25	20	12	12
Percentage	14.5	45.2	14.5	50.0	30.6	40.3	32.2	19.3	19.3

ally, the bigger the species, the larger the territory it defends (Turner 1977); the same happens with the biggest individuals within a population, which almost always are males. Perching higher than females and juveniles, they have more opportunity to locate and obtain food, as well as to perceive other lizards that could enter their settled territories.

Territories of males need to be sufficiently large to guarantee food, shelter, basking sites, and, most importantly, the presence of several females; females require space only for sheltering, basking, feeding, and laying eggs. Nevertheless, the bigger the territory, the more difficult it is to defend it (Rand 1967d); that is why its range has to be fixed in equilibrium between benefits and costs of maintenance.

Males defend their territories against males of the same species and, rarely, against conspecific females and juveniles or individuals from other species. Rand (1967d) proved that successful territoriality brings the selective advantages of increasing the safety of environmental resources, mating opportunities, and the survival of descendants. Adult males do not need to exclude from their territories other members of the same species that are smaller in size because there is a correlation between lizard size and prey size (Schoener 1971); in this way, intraspecific competition is reduced. However, other adult males may interfere not only to obtain food, but also other resources and available females. The risk of competition with other sympatric species exists only when resource usage is similar, due to similarities in morphology and behavior, but such competition is not likely among females.

The assertion and aggressive displays of Cuban lizards have been described for *A. sagrei* (Ruiz García 1977; Regalado and Garrido 1993), *A. homolechis* (Ruiz García 1975), *A. bremeri* (Regalado and Garrido 1993), *A. vermiculatus* (Silva Lee 1985), and other species that have been observed occasionally in the field, such as *A. lucius, A. bartschi, A. porcatus, A. allisoni, A. angusticeps,* and *A. equestris.* All of these examples fit well with the patterns described above. The more frequently observed aggressive displays in Cuban iguanids reported by Ruibal (1967), Ruiz García (1977), and Garrido (1980a, 1988) consist of head bobbing, extension of the dewlaps, erection of the nuchal and dorsal crests, tail wagging, lateral compression of the body, and direct attack to try to intimidate and bite the rival. These displays are stereotyped in each species and the code is based on the frequency, sequence, and intensity of movements (Carpenter 1967, 1978; Ruibal 1967; Jenssen and Hover 1976; Jenssen 1978). Direct combat almost never ensues because generally the intruder does not insist; if it does, the fight is of short duration, usually ending with the victory of the resident and the escape of the intruder.

On the other hand, almost all Cuban iguanids are abundant throughout their geographic ranges (especially *L. carinatus, L. stictigaster, A. porcatus, A. lucius, A. argenteolus, A. sagrei,* and *A. homolechis*). However, some do not tolerate the close presence of individuals that compose populations and have wide territories (for instance, the species of the genus *Chamaeleolis* and of the *equestris* group, and *A. vermiculatus*). Nevertheless, some species, such as *A. angusticeps* and the *argillaceus* group, seem to be more scarce than they actually are because of their cryptic coloration, which makes detection difficult. Others, whose territories are very small, form populations with high densities.

Generally, an inverse correlation between population density and size of territory is observed (Turner 1977). *A. lucius,* the most gregarious species in Cuba, is found in big groups on limestone outcrops and in semideciduous forests on limestone rocks. *A. pigmaequestris, A. delafuentei,* and *Leiocephalus onaneyi* may be the most scarce because they were taxonomically described from very few specimens; since then they have not been observed, despite having been sought many times. Recently, Fernández Méndez (1997a) reported some sightings of *A. pigmaequestris* at Cayo Santa María, but no specimens were captured to confirm the classification.

Some males of species of the genus *Anolis* emit very acute sounds while fighting or being captured. Although Milton and Jenssen (1979) stated that sound emission in *Anolis grahami* is a preadaptation to social communication derived from a defensive response to predators, it is possible that this behavior also assists individuals in marking and defending their territories; for instance, by warning of the presence of an individual in a given place. *A. lucius, A. argenteolus, A. vermiculatus,* and *A. bartschi* show this behavior, females as well as males, in free life and while being handled. *A. isolepis* and *A. jubar* squeak when they are captured.

Besides being territorial, most species are also afraid of and flee humans and predators; on escaping, they climb the trees where they are, run to hide among grasses or stones, or rapidly bury themselves. When persecuted, some species face the enemy, open their mouths, extend their dewlaps, and try to bite; others stay motionless to avoid being noticed and only run away at the proximity of danger. Species of the genera *Leiocephalus* and *Cyclura* typically run and stop very rapidly for short distances before burying themselves or hiding in holes and caves. Species of the *equestris* group of the genus *Anolis* and those of *Chamaeleolis* crawl around the trunks where they are, climb upward, and finally jump to the tops of adjacent trees. The *angusticeps* and *argillaceus* groups stay motionless, masked by their cryptic coloration; members of the *alutaceus* series escape by jumping into the grass; the *carolinensis* group and the *lucius* series, while fleeing, climb up trunks or rocks; *A. vermiculatus* also jumps into the water and submerges or hides in bank holes of streams and rivers. Species of the *sagrei* series hide among stones or grasses and generally run down from their perches.

On the other hand, Cuban iguanids are altitudinally distributed—from sea level, on rocky and sandy coasts, to the greatest heights of the country, where they occupy every type of vegetation floor described for the mountain ranges of Cuba (Rodríguez Schettino 1985; Moreno and Valdés 1991). Of the sixty-two Cuban species, ten (16.1%) are found only in lowlands (group I); thirty (48.4%) are distributed throughout lowland and mountain areas (group II); and twenty-two (35.4%) live only in uplands (group III; table 3.3). That is, fifty-two species (83.8%) are able to inhabit different elevations, which indicates that these species are greatly adapted to diverse environmental conditions. *A. angusticeps* and *A. homolechis* are the two most altitudinally widespread species, living from the coast to the Pico Cuba (1,872 m asl) in Sierra Maestra, Santiago de Cuba province.

The most important factor affecting the diversity and spatial distribution

Table 3.3. Groups of Cuban iguanids according to altitudinal distribution. *Group I*, species live only in lowlands; *group II*, species live in lowlands and uplands; *group III*, species live only in uplands.

Group I	Group II		Group III
L. raviceps	L. carinatus	A. confusus	L. onaneyi
C. nubila	L. cubensis	A. jubar	C. barbatus
A. pigmaequestris	L. macropus	A. mestrei	C. guamuhaya
A. paternus	L. stictigaster	A. allogus	A. noblei
A. juangundlachi	C. chamaeleonides	A. ophiolepis	A. isolepis
A. spectrum	C. porcus		A. guazuma
A. quadriocellifer	A. equestris		A. alayoni
A. bremeri	A. luteogularis		A. garridoi
A. imias	A. smallwoodi		A. clivicola
A. birama	A. baracoae		A. anfiloquioi
	A. porcatus		A. inexpectata
	A. allisoni		A. macilentus
	A. angusticeps		A. vescus
	A. loysiana		A. alfaroi
	A. argillaceus		A. cyanopleurus
	A. centralis		A. cupeyalensis
	A. pumilus		A. mimus
	A. lucius		A. fugitivus
	A. argenteolus		A. vanidicus
	A. vermiculatus		A. ahli
	A. bartschi		A. delafuentei
	A. alutaceus		A. rubribarbus
	A. sagrei		
	A. homolechis		
	A. guafe		

of lizards is human activity in natural areas, undertaken for social expansion and improvement. Such activity can be either uncontrolled or managed so as to preserve the environment to the greatest extent possible. Unfortunately, during previous centuries and the first half of the current one, development of Cuban society was based on the destruction of vast forested territories for sowing crops, raising cattle, or woodcutting. In this way, the primary forests were fragmented and limited to mountains and other relatively inaccessible places such as peninsulas and keys, which were not as satisfactory for development purposes. Space and construction materials were also required for the establishment of cities and towns.

Since lizards are mainly forest dwellers, we can certainly presume that human development activity caused many of them to become extinct or restricted them to the remaining undeveloped places. In fact, we cannot verify this assumption because of the paucity of papers about Cuban lizards published in the last century or earlier. If we compare the number of species reported by the naturalists of that time (Duméril and Bibron 1837; Cocteau and Bibron 1837–1843; Gundlach 1867, 1880) with the sixty-two species known at present, we are surely led to believe that there are more living species now, irrespective of habitat transformation. But this is an artifact due to differences in the level of knowledge between the past and present.

The actual factors to take into account, more than the number of species, are historical spatial distribution and species abundance. For example, the iguana *Cyclura n. nubila* formerly lived along rocky and sandy coasts and in many of the keys (Gundlach 1880; Barbour 1914), but nowadays it is restricted to some keys and very few coastal areas of the island of Cuba, such as Península de Guanahacabibes (Rodríguez Schettino and Martínez Reyes 1985). This is a species that is affected by human activity, principally by the use of coastal areas for living sites and beaches.

The presence of sixteen species with local distributions (see chapter 6 for patterns of distribution) indicates that certain factors are responsible. However, almost all of these species live in the most preserved areas of the country, so urbanization cannot be the cause of their restricted distributions. On the other hand, many species live in a wide variety of habitats, including those modified by humans, such as cultivated areas, secondary vegetation, and urban zones. Such species are ecological generalists and are genetically adapted to survive under different environmental conditions; therefore, they have benefited from social development, finding more kinds of substrate and food in habitats that are more complex than some types of forests.

Nevertheless, since modification, fragmentation, or destruction of natural habitats by human expansion has negative effects on the diversity and abundance of living species, some actions are being adopted to stop or diminish Cuban forest transformation and its consequences on animal life. The main focus of these actions is reforestation, which has allowed surfaces covered with forest to increase from 14% in 1959 to 19% in 1985 (Arcia Rodríguez 1989). In addition, the Protected Areas System guarantees the controlled use of natural resources in several areas with different levels of protection. Since the 1980s approval has been obtained for four Reserves of the Biosphere, which cover the main forest zones of the country and have individual management plans.

In July 1997 environmental legislation was approved that includes all previous regulations and requires a potential developer to obtain an environmental license to begin, expand, or modify any construction that may affect the environment. Such a license may be obtained only after the results of an Environmental Impact Assessment for the proposed project have been evaluated. This law will mitigate as much as possible the negative consequences of human activity on natural habitats.

Climatic Relationships

Iguanids are extraordinarily dependent on climatic conditions for survival. Temperature has the greatest influence on the life mode of different species, since physiological processes depend on body temperature and, as ectotherms, these species use the sun as their principal means of obtaining the necessary thermic threshold. However, this threshold can also be achieved via the substrate or air warmed by the sun. Metabolic production of heat is insignificant because this heat is easily lost through the corporal surface (Brattstrom 1965).

However, the majority of species can maintain a relatively constant body temperature above that of the air through behavioral means such as basking and retreating to shade, postural changes on the substrate, and physiological adjustments for cooling by panting and vasomotor responses. These procedures, known collectively as thermoregulation, are considered by many researchers (including Cowles and Bogert 1944, Brattstrom 1965, and Ruibal and Philibosian 1970) to have great adaptive value.

Nevertheless, since the publication of papers by Huey (1974) and Huey and Slatkin (1976), thermoregulation has been considered in terms of costs and benefits. For instance, changes of location or postural shifts on the substrate to look for sun or shade result in energy loss; such movements can also increase exposure to predators. Furthermore, they may be disadvantageous for obtaining food or performing social relations. The cost of thermoregulation depends to a great extent on habitat: it is more difficult to increase body temperature in close forests than in more open areas. Accordingly, species that live in expensive habitats for thermoregulation generally have a passive response to environmental temperature changes; that is, they are thermoconformers. Thermoregulatory behavior also depends on altitude, climate, season, time of day, abundance of food, competition, and degree of predation (Huey and Slatkin 1976).

Few Cuban studies have analyzed the behavior of iguanids in relation to climatic conditions, but the obtained data indicate that some species (*Anolis porcatus, A. allisoni, A. sagrei,* and *A. homolechis*) are active thermoregulators, while others (*A. vermiculatus, A. bartschi,* and *A. allogus*) are thermoconformers (Ruibal 1961; Silva Rodríguez 1981; González Bermúdez and Rodríguez Schettino 1982; Estrada and Novo Rodríguez 1987; Valderrama Puente and Rodríguez Schettino 1988; Rodríguez Schettino and Martínez

Reyes 1989*a*). Some species, such as *A. lucius*, may even behave in one way or another depending on habitat or season of the year (Menéndez *et al.* 1986; Rodríguez Schettino and Valderrama Puente 1986).

Iguanids maintain a relatively high tolerance to changes in humidity. However, Bentley and Schmidt-Nielsen (1966) experimentally demonstrated that cutaneous evaporation is a very significant part of total water loss in reptiles; Gilles-Baillien (1981) affirmed that the permeability of reptile tegument is greater than has been thought, but that there are many ways through which osmoregulation of body liquids is achieved under different habitat and climatic conditions. Generally, species of the genus *Anolis* that live in dry habitats have lower body water loss rates than species that live in wet places; this was experimentally proved by Hillman and Gorman (1977). Leclair (1978) reached the conclusion that environmental water content acts as a selective pressure to physiologically exclude species that do not have the adaptive means to decrease water loss. Some Cuban iguanids, such as species of the genera *Leiocephalus* and *Cyclura*, can live in extremely dry places, for instance, coastal zones and open areas; other species that inhabit only moist forests, such as *A. allogus, A. ahli, A. rubribarbus, A. isolepis* and *Chamaeleolis guamuhaya*, can live only in wet or very wet places.

Since the rate of dehydration is higher during basking, thermoregulation interferes with ways of diminishing water loss (Hertz 1980). Water loss also depends on body size: smaller species dehydrate more easily than larger ones.

Activity

All Cuban species are diurnal. Early in the morning they begin to leave their nocturnal shelters and sleeping places. However, time of emergence for the first individuals depends on the season of the year and the photoperiod; that is why dawn seems to stimulate diurnal activity. The number of active individuals gradually increases as air temperature rises, but above a certain temperature activity decreases. During the dry season this happens in the afternoon, favoring an activity pattern with only one peak—between approximately 11:00 A.M. and 2:00 P.M. However, during the wet season activity generally decreases at noon and increases again at the end of the afternoon; this activity pattern is bimodal, with a higher peak in the morning, near 10:00 A.M., than in the afternoon, at about 4:00 P.M. (Huey and Slatkin 1976; Rodríguez Schettino and Martínez Reyes 1989*a*).

Nevertheless, not all species, or even both sexes of the same species, behave in the same way. Arboricolous species living in sympatry, or both sexes of one such species, do not differ in their use of time; they can occupy one tree or another without interfering with each other (Berovides Álvarez and Sampedro Marín 1980; Rodríguez Schettino and Martínez Reyes 1989*a*). However, species that live on limestone cliffs have a continuous substrate on which they would frequently meet if they did not use it asynchronically (Schoener 1970*b;* Rodríguez Schettino and Martínez Reyes 1989*a*).

The number of active individuals also varies seasonally. In general, during the dry season fewer animals are observed; most of these are juveniles and subadults, whose surface/volume quotients allow for increases in body temperature with little cost when environmental temperatures are not high (Cowles 1945). The number of active individuals increases when the wet season begins, then slowly decreases at the end. In addition, activity patterns vary with locality. In western Cuba, where the dry season is a little colder than in the eastern region of the country, *A. quadriocellifer, A. bartschi, A. lucius,* and *A. vermiculatus* have been observed emerging from their nocturnal shelters after 9:00 A.M., while *A. argenteolus* from La Mula, which lives in a warmer, more stable place, awakens at 7:00 A.M.

Trophic Relationships

As secondary consumers, iguanids occupy a very important place in the trophic webs of Cuban terrestrial ecosystems; many species have numerous populations that directly affect the ecological equilibrium and contribute considerable biomass to the habitats they occupy. They eat a great variety of prey types, mainly insects and other arthropods. Vegetable material is also part of the diet of some species; it is the main source of food for the iguana, which is a phytophagous generalist and an opportunistic omnivore (Berovides Álvarez 1980; Rodríguez Schettino and Martínez Reyes 1985). Large species of the genus *Anolis* also consume small vertebrates.

Species of the genera *Anolis* and *Leiocephalus* eat one or two prey types in great amounts and many other types in small quantities; thus consumption seems to be related to the types of prey available as well as to ease of capture (Sampedro Marín *et al.* 1979; Berovides Álvarez 1980; Floyd and Jenssen 1983; Estrada 1984; Armas 1987; García Rodríguez 1989; Rodríguez Schettino and Martínez Reyes 1989*a;* Martínez Reyes *et al.* 1990).

The great diversity of species, genera, and even orders of prey that has been found in the digestive tracts of dissected lizards, coupled with the lizards' deficient taste capacity (Simon 1983), indicates that prey types are selected by size and not by taxonomic category. This has been demonstrated by Schoener (1968) and Schoener and Gorman (1968) for species of the genus *Anolis* from other Caribbean islands and by Sampedro Marín *et al.* (1979) for Cuban species of the genus *Leiocephalus*.

On the other hand, prey size and number are correlated with predator size. Larger individuals of a given species, as well as species of larger sizes, eat larger prey in smaller amounts than small individuals and species, which eat smaller prey in greater quantities (Schoener 1971; Sexton *et al.* 1972; Rodríguez Schettino and Novo Rodríguez 1985). Seasonal factors also influence type, size, and number of prey that can be obtained: in the dry season, diversity as well as size and number of food items consumed is diminished, while during the wet season consumption is generally higher, due to increased environmental productivity and availability of food (Sexton *et al.* 1972; Fleming and Hooker 1975; Rodríguez Schettino 1981*a;* García Rodríguez 1989; Rodríguez Schettino and Martínez Reyes 1989*a*).

Two extreme strategies for obtaining prey have been observed: sit-and-wait and foraging. Species using the first strategy spend energy and time only to catch and eat prey, while the foraging species also have to search to locate prey (Schoener 1971). Both strategies have adaptive advantages: sit-and-wait species spend less energy and are more protected against predators, but foraging species achieve a higher net gain of food (Huey and Pianka 1981). It is generally agreed that iguanids are sit-and-wait lizards; however, *Cyclura nubila* and species of the genus *Leiocephalus* are more commonly foragers than those of the genera *Anolis* and *Chamaeleolis*, which are typically sit-and-wait strategists. Nevertheless, each species' strategy for obtaining food varies according to its ecological plasticity and depending on prey availability and capturability. That is, sit-and-wait lizards can search for food when it is scarce near their perches, and foragers can find places within their territories where food is abundant, then spend more time in such places.

On the other hand, as part of the trophic webs of their ecosystems, species of small size are often captured by such birds as the Sparrow Hawk (*Falco sparverius*), Cuban Blackbird (*Dives atroviolaceus*), Cuban Lizard Cuckoo (*Saurothera merlini*), Loggerhead Flycatcher (*Tyrannus caudifasciatus*), Red-tailed Hawk (*Buteo jamaicensis*), House Sparrow (*Passer domesticus*), Greater Antillean Grackle (*Quiscalus niger*), and others (Silva Lee 1985; Kirkconnell and Posada 1987; Berovides Álvarez 1989; Wotzkow 1989); juveniles of larger species also succumb to these predators. Wild and domestic cats are also common predators of lizards (Iverson 1978). Almiquí (*Solenodon cubanus*) eat iguanids, not only in free life but also in captivity (Abreu and Cruz 1988), and they complement the diet of Conga Hutia (*Capromys pilorides;* Manójina *et al.* 1989). Almost all snakes, mainly species of the genus *Tropidophis* and the Cuban racer (*Alsophis cantherigerus*), feed on different species of lizards (Henderson and Crother 1989).

Saurophagy has not been frequently reported for Cuban iguanids, but it occurs in some species, in captivity as well as in the wild, principally toward species of small size. *Anolis vermiculatus, A. porcatus, A. allogus, A. imias,* and giant anoles occasionally incorporate into their diets small individuals of other species. Cannibalism is also known for *Leiocephalus carinatus, L. cubensis, L. stictigaster, Anolis porcatus, A. lucius, A. quadriocellifer,* and *A. imias,* which indicates that ingestion of lizards of the same or other species may be more common than is generally thought (Rodríguez Schettino and Martínez Reyes 1986*b;* Armas 1987; Martínez Reyes and Rodríguez Schettino 1987; Socarrás *et al.* 1988; Rodríguez Schettino and Martínez Reyes 1989*a*).

Reproduction

The reproductive characteristics of Cuban iguanids are little known, and complete reproductive cycles have been studied for only some species (Rodríguez González 1982; Silva Rodríguez and Estrada 1984; Valderrama Puente and Rodríguez Schettino 1988; Rodríguez Schettino and Martínez Reyes 1989*a*). However, numerous observations have been made regarding egg-laying dates and sites, and egg measurements taken (Garrido 1980*a*), and isolated data exist for other aspects of reproduction. This allows us to gather related information and verify its agreement, not only biologically but also ecologically, with data from other tropical species, especially Caribbean ones.

Females of species of the genus *Anolis* ovulate alternately; they lay only one egg each time, but in successive layings in each reproductive cycle (Licht and Gorman 1970; Sexton *et al.* 1971; Rodríguez Schettino and Martínez Reyes 1989*a*). Andrews and Rand (1974) considered that laying one egg each time provides two advantages: not only can each egg be bigger, but a higher frequency of layings is possible, even if a smaller number of eggs is produced. However, according to Sexton (1980), these females can retain eggs in their oviducts; the ability to retain two oviductal eggs allows not only greater production but also a temporal delay in reproduction until environmental conditions are propitious for oviposition.

Eggs are generally laid in wet places: under fallen leaves, in the interstices of bromeliad leaves, under earth or sand, or in the fissures and crevices of rocks and cave walls. Females of some species lay their eggs in solitude; those of other species lay at communal egg-laying sites for various females and different reproductive seasons (Dunn 1926; Hardy 1957; Ruibal

1964; Rand 1967*a;* Silva Rodríguez, Berovides Álvarez, and Estrada 1982; Novo Rodríguez 1985; Estrada 1987*b,* 1988).

Very little is known about reproduction in the genera *Leiocephalus* and *Chamaeleolis.* Martínez Reyes *et al.* (1990) and Martínez Reyes (1994*b*) have considered that females of the genus *Leiocephalus* probably lay more than one egg at a time, since the frequency of gravid females with two or more oviductal eggs of quite similar lengths ready for laying is very high. Iguana juveniles have been seen in the wild as early as April (Perera 1985*a;* Rodríguez Schettino and Martínez Reyes 1985), so it could be inferred that courtship and mating occur from the beginning of the year. Furthermore, clutches have been obtained in captivity from June to August (Shaw 1954; Rodríguez Schettino 1990). Consequently, it seems that reproductive activity is extended at least to August. This is the only Cuban species known to build nests for incubating about seventeen eggs, which are deposited in layers (Shaw 1954).

The behavioral aspects of reproduction (courtship, female receptivity, mating) have been described for some species (Hardy 1957; Carpenter 1967, 1978; Jenssen 1978; Garrido 1980*a,* 1988) and observed for many others in free life or in captivity. The most common courtship displays in males are quick bobbing with different frequencies and intensities, elevation of the forelimbs or of all four legs at the same time, tail wagging, and dewlap extension. Females respond with head bobbing and short runs. Each species shows a stereotypical courtship pattern whose code consists of different frequencies and sequences of movements. The time spent in courtship also differs according to species (Garrido 1980*a*).

Mating is similar among different species. After being accepted during courtship, the male positions himself laterally on the female, which allows intromission of one of his hemipenes into the female's vent. Usually the male bites the neck or shoulder region of the female during the first moments of copulation, and in some cases the female performs brief, tenuous head bobbings. The duration of matings observed in captivity varies with the species (Garrido 1980*a*). After mating, pair members in some species abruptly separate and rapidly move far away from each other; separation and moving away occur slowly in other species.

The productive periods of tropical iguanids are of different durations and can even last throughout the year. Physiologically, reproduction is always possible: males have spermatogenesis and hormonal secretions during the entire year (Licht and Gorman 1970). However, the majority of species exhibit some decrease in reproductive activity during the dry season. The length of egg production time in a given species depends on its sensitivity to the availability of environmental resources such as food and egg-laying sites, and, most importantly, to climatic factors such as temperature, rain, photoperiods, and air and soil humidity. Seasonal changes in these factors modify hormonal production, physiology, and behavior in tropical species; therefore only some display their potential to reproduce during the whole year.

It has been proved, both in nature and experimentally, that increased air temperature and day length, beginning in February or March, stimulate testicular activity (Licht and Gorman 1970; Licht 1971; Gorman and Licht 1974; Crews 1979) and that the female cycle corresponds to increases in precipitation, humidity, and air temperature (Licht and Gorman 1970; Brown and Sexton 1973; Silva Rodríguez and Estrada 1984; Valderrama Puente and Rodríguez Schettino 1988).

According to Huey and Pianka (1981) and Vitt (1990), the mode of obtaining food may be one of the factors mainly affecting seasonality of reproduction, since the majority of the foraging species reproduce continuously and those that sit and wait for their prey are seasonal to varying degrees. However, the genus *Anolis* is an exception: its general mode of obtaining food is the sit-and-wait strategy, while reproductively it displays a great capacity to respond during different climatic conditions (Vitt 1990).

Although testes size only gives an idea of reproductive intensity (Licht and Gorman 1970), because spermatogenesis also occurs in small testes, but with much less frequency, this characteristic has been used as an index of reproductive activity. According to different studies of reproductive cycles in species of the genus *Anolis*, testes of adult males start to increase in size before the beginning of the wet season in relation to increased air temperature, which activates testicular regeneration. In addition, Licht (1971) experimentally demonstrated that testicular regression occurs before the wet season ends and is closely related to diminishing day length, which ends the cycle. Activities related to reproduction (courtship, mating, defense of territory) are controlled by hormonal mechanisms that are regulated by environmental conditions, mainly photoperiods and temperature.

Females start to produce eggs before the wet season begins, in close relation with the increase in air temperature, and production decreases or ceases before the percentage of active males in populations diminishes; thus it seems that females impose some limitations on reproduction. Yet not all species that live together in sympatry have similar reproductive patterns, even when they are subject to the same seasonal climatic variations and to equal changes in food availability. In this regard, Gorman and Licht (1975) found genetic differences in such species, which means that they vary in the plasticity of response to environmental stimuli.

Crews (1979) demonstrated that sexual relations also influence reproductive cycles, which begin with environmental stimulation of testicular activity. It is male courtship behavior that stimulates ovarian growth; later, the acceptance behavior of the female influences testicular development. Furthermore, female sexual receptivity is cyclic; it ceases some days after each mating. The advantage of this mechanism is that in the period before oviposition females are not so exposed, which decreases their vulnerability to predators while increasing their chances for survival and the opportunity to leave a greater number of offspring. According to Crews (1979), it is possible that female receptivity also decreases after a period of successive matings and ovipositions and that this behavior limits male activity, even though physiologically males are able to reproduce.

The existence of reproductive cycles means that in the majority of species several age cohorts synchronically occur. Adults predominate in different populations during the entire year. Juveniles and subadults can be present throughout the year; generally, however, their numbers are observed to increase during the final months of the wet season, to the extent that in some months they occur in larger proportions than do adults. In other cases, they appear only at the beginning and end of the year because reproduction ceases or notably decreases during the dry season.

According to Fitch (1973), tropical lizards from Costa Rica exhibit ecological adaptations in their reproductive behavior that lead to variations in population structure. Those whose reproduction is continuous throughout the year and whose populations include individuals of all sizes in constant proportion belong to type I. Those whose reproduction is continuous but whose levels of reproductive activity show small fluctuations in response to nonsevere seasonal climatic variations, and whose percentages of individuals of different sizes change, belong to type II. Populations whose reproduction is inhibited during the dry season (type III) have changing structures, with a majority of adults in one season and a majority of immatures in the other. Populations with very reduced reproductive stages have type IV structures.

In Costa Rica most species had type II population structures, which is the pattern that prevailed in the Caribbean lowlands. The cohort structure of Cuban lizards is known for four populations in species of the *lucius* series. *A. argenteolus* belongs to type II and *A. lucius, A. bartschi,* and *A. vermiculatus* to type III (Rodríguez Schettino and Martínez Reyes 1989a).

Communities

Some ecological relationships among iguanids in the Caribbean islands are fixed during the process of community settlement. These relationships are established not only intraspecifically but also interspecifically, which allows adequate use of environmental resources without considerable competitive interference. This has been demonstrated by Collette (1961), Schoener (1967, 1977), and Williams (1972, 1983), who have proposed that differential usage of microhabitat (type of substrate, height, and diameter of perch) constitutes an essential factor necessary for building a community.

In other words, species or intraspecific age groups of larger size are generally situated at greater heights on the trunks of trees and bushes. Moreover, adaptations to climate contribute to separation of the species or their cohorts within the microhabitat; there are thermoregulator and thermopassive species, which bask or not, and whose body temperatures are maintained within a very narrow range (stenotherms) or vary widely in relation to environmental temperatures (eurytherms). On the other hand, daily and seasonal activity cycles can differ among the members of a community, which is an important determinant of ecological segregation.

These differences imply that food resources (type and size of prey) can be secured by all members of a community in such a way that competitive relationships can be reduced and relative stability or ecological equilibrium achieved. All of this is closely related to morphological, physiological, and behavioral adaptations of the involved species, and to age and sex of the individuals that are in the community at any given time (Collette 1961; Schoener 1970a; Moermond 1979; Case 1983; Williams 1983; Estrada and Silva Rodríguez 1984; Berovides Álvarez et al. 1988). In this sense, studies of iguanids in the Antilles have demonstrated that, depending on the island area and the structural complexity of ecosystems, several species can coexist in sympatry and even in syntopy.

Species of the genus *Anolis* are the most dominant and abundant in the Caribbean islands. For example, in Hispaniola (Rand 1962; Rand and Williams 1969), Puerto Rico (Rand 1964; Williams 1972), Jamaica (Rand 1967b), and Grenada (Schoener and Gorman 1968), differences between the anolines that compose those communities in the usage of structural and climatic resources lead to segregation in obtaining available food. Few studies have been done on communities of Cuban iguanids and they comprise a small number of species, but the results obtained up to this moment agree in general with those mentioned before for other Caribbean islands.

Collette (1961) found six species of iguanids in Bosque de La Habana, on the western shore of the Almendares River, that differed in their perching behavior: *Leiocephalus cubensis* was found on the ground; *Anolis alutaceus* occupied holes of abandoned walls; *A. sagrei* was observed on the ground and on fence posts, at low heights and with the head down; *A. porcatus* was positioned on trunks of trees and on bushes, at great heights and with the head up; *A. angusticeps* occupied trunks and high twigs of trees and bushes; and *A. equestris* was found very high in trees. These differences are closely related to the respective species' morphology and thermoregulatory behavior.

Ruibal (1961) found that in Sierra de Cubitas the dominant species of the genus *Anolis* in a community of semideciduous forest behaved according to that pattern of distribution in space and climate. *A. allogus* and *A. jubar,* of similar morphology, occupied trunks of trees and bushes; the first species occurred in the more shaded zone of the forest and the second one at its border, with resultant differences in the environmental temperatures to which they were exposed. As a consequence, *A. allogus* had lower body temperatures than *A. jubar.* Two other species, *A. sagrei* and *A. allisoni,* occupied trunks and fence posts in open areas of pasture. They were under the sun and exposed to similar temperatures, but at different heights, the first one near the ground and the other farther up.

Lando and Williams (1969) detected ten species of iguanids in the area of the U.S. naval base at Guantánamo. In the coastal zone were found *Cyclura nubila* in caves of reefs; *Leiocephalus carinatus* on rocks; *L. raviceps* on sandy soil; *L. macropus* on the ground of the coastal shrubwood; and *A. argillaceus* on trunks of sea grapes (*Coccoloba uvifera*). Toward the forest zone were found *A. sagrei* on posts and trunks of bushes; *A. ophiolepis* among grasses; *A. homolechis* on trunks of trees, with head downward; *A. porcatus* on tree trunks; and *A. smallwoodi* on tree trunks, with head upward.

Sampedro Marín *et al.* (1979) found *L. macropus* and *L. raviceps* in sympatry on the southern coast of Guantánamo province, the first more abundant in dry forest under the shade, and the second in coastal shrubwood in shaded as well as sunny places.

Silva Rodríguez (1981) observed in the evergreen seasonal forest of Sierra del Rosario that *A. allogus* and *A. homolechis* used tree trunks and rocks as perches, males positioned higher than females and both species at similar heights. However, *A. allogus* was found in the wettest, most shaded zones of the forest, while *A. homolechis* occupied trees near the limits of the forest. Also in Sierra del Rosario, but in gallery forests, *A. allogus* was found on tree and bush trunks, *A. luteogularis* on high branches and trunks, and *A. vermiculatus* on trunks and branches of rose-apples (*Syzygium jambos*) near riversides (González Bermúdez and Rodríguez Schettino 1982).

Sampedro Marín *et al.* (1982) found *A. sagrei* and *A. homolechis* in a secondary shrub of La Habana province, where both species occupied tree and bush trunks and the ground at similar heights but *A. homolechis* occupied trunks of larger diameter. The researchers found differences in the proportions of prey types eaten by these lizards, which indicates that partitioning of trophic resources is what most separates them.

In every vegetation floor of Sierra del Turquino were found assemblages of iguanids in which spatial and climatic relationships among the species that composed them were observed. *A. argenteolus* and *A. porcatus* were the most ubiquitous species and occupied tree trunks, the first in shaded places and the second under the sun in open or half-shaded places. The remaining species in each community differed in their use of microhabitats. They reached a maximum of ten species on the submontane floor (Rodríguez Schettino 1985).

The same was observed in the three ecosystems studied at Península de Guanahacabibes by Rodríguez Schettino and Martínez Reyes (1985), where *L. stictigaster* and *A. quadriocellifer* are the dominant species: the first species occupied the ground, the second tree and bush trunks. The remaining species differed in perch, climatic adaptation, or both.

In the keys of the archipelagoes these relationships are also found. There, the most common species are *A. sagrei*, *A. porcatus*, *Cylura nubila*, and *L. carinatus*. The first two generally live in forests, *A. sagrei* near the ground and *A. porcatus* high on trees and bushes. *C. nubila* and *L. carinatus* inhabit the coastal zones, on sand and rocks under the sun, but they differ notably in their feeding (Garrido 1973*b, d;* Garrido and Schwartz 1969; Varona and Garrido 1970; Garrido *et al.* 1986; Estrada and Novo Rodríguez 1986*a*).

The most recent studies of iguanid communities in Cuba (García Rodríguez 1989; Alarcón Chávez *et al.* 1990; Hechevarría *et al.* 1990; Quesada Jacob *et al.* 1991; Ardines *et al.* 1992; Martínez Reyes 1995; Rodríguez Schettino *et al.* 1997) have demonstrated that one or two species are usually abundant; in these cited cases they are *A. homolechis* and *A. sagrei*. Depending on locality, season of the year, and habitat structure and complexity, the way and the degree to which environmental resources are used vary, but generally both species maintain the patterns that characterize them when they are living sympatrically: *A. homolechis* lives on tree and bush trunks under filtered sun, at the limits of shrubwoods and forests, *A. sagrei* on bush trunks and on the ground, under the sun in open places and in open zones of forests.

The remaining species that compose such communities are less abundant and differ in their use of environmental resources. *A. porcatus* perches high on trunks; *A. angusticeps* on high twigs; *A. equestris* or *A. luteogularis* on the crowns of the highest trees; *A. alutaceus* in grasses and bushes; *A. lucius* on rocks and tree trunks; species of the genus *Leiocephalus* on the ground and on small rocks. In communities where species of the genus *Chamaeleolis* live, they occupy the highest parts of tree trunks and branches.

All of these data indicate that communities of Cuban iguanids are composed of one or two dominant and up to eleven other sympatric species that differ in their use of environmental resources, not only intraspecifically but also interspecifically, in the dimensions of space, climate, and food. Although these different species have adaptations and specializations that allow them to use a given part of the environmental resources, they can show several degrees of amplitude in their utilization of such resources, depending on the community in which they are. That is, according to their ecological plasticity, some species vary their mode of using space and other habitat resources depending on the vegetation structure and in relation to the presence of other congeneric species.

4

Genetics

GEORGINA ESPINOSA LÓPEZ AND ADA R. CHAMIZO LARA

The genetic studies that have been done on Cuban iguanids deal mainly with biochemical genetics. There are some papers in which karyotypes are described and a few about morphological polymorphism in *Anolis porcatus*. Hass *et al.* (1993) described the DNA sequences for nine species of Cuban anoles and Losos *et al.* (1998) for eleven species.

A wide variety of laboratory assays is available for revealing molecular genetic markers. Among them are protein immunology, protein electrophoresis, DNA-DNA hybridization, restriction analysis, polymerase chain reaction, and sequencing of nuclear and mitochondrial DNA. Of these methods, only DNA sequencing can be applied at virtually any taxonomic level. Despite this advantage, protein electrophoresis is considered highly or marginally informative for comparing species of lower or intermediate taxonomic levels (Avise 1994).

The papers by Hunter and Market (1957), in which they presented the combination of electrophoresis and biochemical staining methods, and those by Harris (1966) and Lewontin and Hubby (1966), who measured the genic variation in natural populations using samples of proteins selected at random, became true events within population genetics.

Several excellent reviews about methodology, interpretation, results, and the meanings of these techniques for evolutionary biology can be consulted: Lewontin (1974), Ayala (1980), Nevo (1983), and Avise (1983, 1994).

Electrophoretic studies of the genus *Anolis*, which began in the 1960s, have been directed toward analysis of the effects of environmental factors on protein polymorphisms as well as toward systematics. On the one hand, such studies have been centered largely on the analysis of ecological variables that maintain and differentiate protein polymorphism and on the relationship of this with phenotypic diversity, which is supported by the selection theory (Lewontin 1974; Nevo 1983). On the other hand, partisans of the neutralist theory (Kimura 1968) argue for the use of these results in systematics.

The evolutionary causes of genetic variation clearly involve both neutrality and selection, but the problem is how little we know about the nature and importance of natural selection in shaping nucleotide variation. According to Kreitman (1991), the most serious obstacle to evaluation of the forces shaping patterns of variation and change may be the way we organize empirical population genetics: many groups working on many organisms. If we want to solve these problems, many groups must work on a small number of model species. Only in this way may we finally be able to understand the role of natural selection in maintaining genetic variation and put to rest some issues raised by the old neutralist-selectionist debate.

Protein Polymorphism in Iguanid Natural Populations

RESULTS IN THE GENUS *Anolis*

The relationships of protein polymorphisms with several factors such as geographic locality, season of the year, and sex have been studied in some species of *Anolis* using proteins that are characterized by their high degree of variation in most zoological groups. Despite the fact that many current authors do not accept the exactitude of such relationships, we would like to reveal here what we have found with Cuban species.

The esterase activity and its genetic variation are detected most frequently, and there are examples with a genetic basis in many species of animals (Manwell and Baker 1970). The genus *Anolis* is not an exception: some authors have studied the esterases of anolines (Buth *et al.* 1980; Mugica Valdés *et al.* 1982). Espinosa López *et al.* (1985), Espinosa López, González Suárez and Berovides Álvarez (1985), and Espinosa López (1989) reported that the genic frequencies of a system of muscle esterases of intermediate mobility are not influenced by sex, but are observed to change according to the season of the year in *A. sagrei, A. homolechis, A. porcatus,* and *A. allisoni* (fig. 9). That system has three alleles; one of them is more abundant, with approximately the same frequencies in the two seasons, while the other two alleles have alternating frequencies, which is an example of the action of seasonal diversificationary selection on genetic polymorphism maintenance.

The alkaline phosphatase was also studied by Espinosa López (1989), who indicated its possible genetic control and its usage in systematic stud-

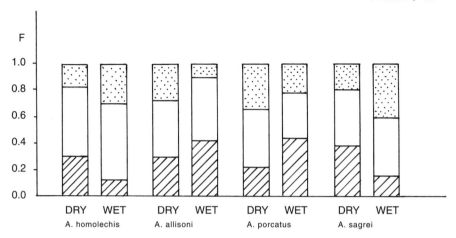

Fig. 9. Genic frequencies of esterases (F) in some species of *Anolis*, according to seasons of the year. *Striped columns*, pL (genic frequency of the slow allele); *white columns*, ql (genic frequency of the intermediate allele); *dotted columns*, rR (genic frequency of the fast allele).

ies of *Anolis*. The genic frequencies of this enzyme did not show a relationship with either sex or season of the year.

The amylase activity is related to sex. Espinosa López (1989) found that the active allele of this enzyme has the greatest frequency in males (fig. 10). Also, species that dwell in the most variable habitats (such as *A. homolechis*) are those of highest heterozygosity, while ecologically restricted species or

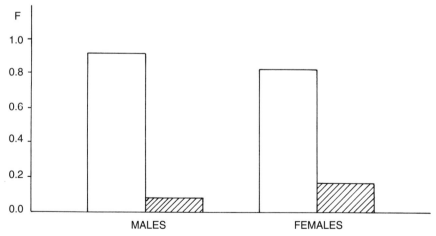

Fig. 10. Genic frequencies of amilase (F) in some species of *Anolis*, by sex. Data for *A. argenteolus, A. homolechis, A. jubar,* and *A. mestrei* were combined. *White columns,* p+ (genic frequency of the active allele); *black columns,* qo (genic frequency of the silent allele).

those with relatively stable habitats (*A. vermiculatus*, *A. lucius*) have the lowest polymorphism. Furthermore, this enzyme can be considered a good marker in population studies within the genus *Anolis* because of its interspecific polymorphism.

The genic frequencies of albumin are influenced by sex (Espinosa López 1989) and possibly interact in a sex-season-allele way that could not be demonstrated by this researcher due to the small size of her sample. This protein is a useful marker for differentiating populations, which is shown in the results obtained with populations of *A. porcatus* (fig. 11 and map 1). It is also appropriate for systematic studies, in comparing species of the same genus.

The zone detected by Schiff's reactive corresponding to glycoproteins is a useful marker that is related to sex and season; the genic frequencies characterize populations of *A. porcatus* and the interspecific polymorphism of that zone provides it with systematic value within the genus *Anolis*.

The pattern of plasmatic general proteins has been employed with the purpose of measuring genetic variability in relation to environmental stability. In this sense, Silva Rodríguez and Espinosa López (1983) found that *A. allogus*, whose habitat is less variable than that of *A. homolechis*, also displayed less variability in its electrophoretic patterns. Espinosa López *et al.* (1987) calculated the coefficient of variation of the number of bands of plasmatic general proteins as an indirect measure of genetic variability, and

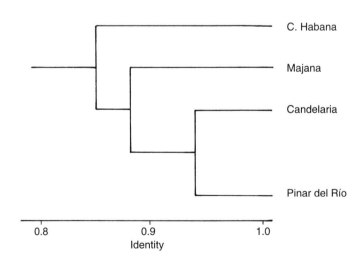

Fig. 11. Phenogram of similarities among four populations of *Anolis porcatus*, based on Hedrick's (1971) identity, calculated for loci of the muscle esterases and albumins.

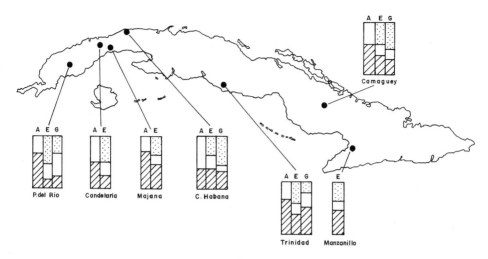

Map 1. Genic frequencies of three polymorphic proteins of *Anolis porcatus* and *A. allisoni* at several localities. Pinar del Río, Candelaria, Majana, and Ciudad de La Habana correspond to *A. porcatus* and the remaining localities to *A. allisoni. A,* albumin; *E,* esterases; *G,* glicoproteins; *striped columns,* pL (genic frequency of the slow allele); *white columns,* ql (genic frequency of the intermediate allele); *dotted columns,* rR (genic frequency of the fast allele).

niche width as a measure of environmental heterogeneity between sexes, seasons, and habitats, for *A. lucius, A. homolechis,* and *A. sagrei.* They found that genetic variability and environmental hetereogeneity were, in general, greater in females than in males, more pronounced in the dry season than in the wet season, and more prevalent in anthropic than in natural habitats. In addition, they observed that niche width and genetic variability tend to be positively correlated.

The general protein patterns are intraspecifically very stable (Espinosa López, Sosa Espinosa, and Berovides Álvarez 1990). The results obtained for twelve species of Cuban *Anolis* suggest that, as with fishes (Corzo *et al.* 1984), such patterns can characterize every species of *Anolis.*

The lactate dehydrogenase and the aspartate aminotransferase are monomorphic enzymes, but with interspecific polymorphism in several species of Cuban *Anolis* (Espinosa López, Sosa Espinosa, and Berovides Álvarez, 1990), which agrees with the results obtained for other species of the genus (Buth *et al.* 1980; Gorman, Buth *et al.* 1980).

Data about esterases, albumin, amylases, glycoproteins, and plasmatic general proteins may be the expression of the action of some selective mechanisms (equilibrating selection) in which one allele is favored or disfavored by sex or season; if so, this fact would support the selection theory. Furthermore, we have contrasted enzyme and nonenzyme proteins and found different behaviors in relation to natural selection among them.

Neither of the factors mentioned above seem to influence the genic frequencies of phosphatases. Therefore, it is possible that the changes that occur among species and populations for this enzyme are neutral with respect to those factors and that they have a historical origin in the populations. The constancy maintained by the genic frequencies of this protein support the neutrality theory. The same could be true for the monomorphic proteins already described.

Espinosa López *et al.* (1991) analyzed the estimates of heterozygosity in species of the *lucius* series (fig. 12) and found higher figures for nonenzyme proteins than for enzymes, similar to what Espinosa López (1989) found. When two populations of *A. vermiculatus* were compared, it was observed that for enzymes, the population from Soroa (more anthropic) was less heterozygotic; for nonenzyme proteins, this population was more heterozygotic than the one from Cayajabos (with a lesser degree of anthropism).

The calculated heterozygosities for enzymes and nonenzyme proteins are greater in the population of *A. lucius* collected at Jibacoa than in that collected at Jobo Rosado, the first locality more anthropized than the sec-

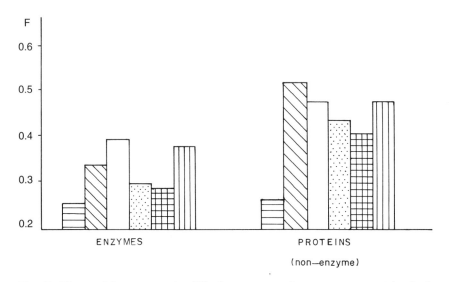

Fig. 12. Observed heterozygosity (H) of enzymes and non-enzyme proteins in the *lucius* series of *Anolis*. Horizontal stripes, *A. bartschi;* diagonal stripes, *A. vermiculatus* from Soroa; white columns, *A. vermiculatus* from Cayajabos; dotted columns, *A. lucius* from Jibacoa; crossed stripes, *A. lucius* from Jobo Rosado; vertical stripes, *A. argenteolus.*

ond one. However, these differences are smaller than those detected for *A. vermiculatus*. If the positive correlation between diversity of nonenzyme proteins and habitat width is taken into account (Nevo *et al.* 1984), the obtained results with nonenzyme proteins are expected, which indicates that they are more reflective of environmental changes than enzymes.

Analyzing the heterozygosity estimates for enzymes and nonenzyme proteins of the four species of the *lucius* series, it is observed that *A. argenteolus* and *A. vermiculatus* are the most variable, not only for enzymes, but also for nonenzyme proteins, followed by *A. lucius*; the least variable species is *A. bartschi*, which is also the one most unlike the others.

A. argenteolus is an arboricolous species that lives on small and large trees in natural habitats and even in trees and houses of the city of Santiago de Cuba (Ruibal 1964). This is a reflection of its high genetic variability. *A. vermiculatus* lives on tree trunks on riverbanks in Pinar del Río province (González Bermúdez and Rodríguez Schettino 1982), where it can dive under water and stay there, although both sexes differ in microhabitat use. This fact and the limited stability of its habitat indicate that although it is a geographically and ecologically restricted species, it has high estimates of heterozygosity.

A. lucius is the most widespread of the four species; therefore, it should be expected to have the most relative variation. However, it is one of the least variable. Its relatively uniform habitat (Rodríguez Schettino and Valderrama Puente 1986) seems to be responsible for this paradox. *A. bartschi* is the species with the least heterozygosity. This agrees with its limited ecogeographic distribution, since it is restricted to karstic *mogotes* of Pinar del Río province (Ruibal 1964). Furthermore, its microhabitat is very uniform and Estrada and Novo Rodríguez (1986c) have found no differences between the sexes.

All this seems to indicate that habitat stability is more important in estimating heterozygosity than geographic distribution of the species.

Regarding the factors that influence the degree of heterozygosity in a population, Espinosa López (1989) found that species that live in the most severe and unpredictable environments (anthropic localities, medium-high perch heights, and open habitats) show larger heterozygosity figures than those living in the most benign environments (natural localities, low perches, and close habitats; tables 4.1 and 4.2). Comparisons between populations of a single species (*A. homolechis*, *A. lucius*, and *A. vermiculatus*) demonstrated that conditions such as degree of anthropism (table 4.1) can influence the genetic variability of populations in such a way that they can face environmental changes. This offers a view of how much human activity affects natural populations, not only their ethoecology, obligating them to

Table 4.1. Observed heterozygosity figures from populations of *Anolis* with different degrees of anthropism. *d*, difference between more severe and more benign habitats; d > 0.10 is considered the significant level of difference.

Species	Locality	Enzymes	d	Proteins	d
	Anthropic	0.39		0.55	
A. homolechis			+0.07		+0.13
	Natural	0.32		0.42	
	Anthropic	0.33		0.47	
A. lucius			+0.03		+0.03
	Natural	0.30		0.44	

survive in habitats other than the original ones and to vary their habits, but also their genotypic frequencies.

The most heterozygotic genotypes will be better able to survive environmental changes; therefore, they will be more capable of expressing several phenotypes conditioned by different environments, more likely to tolerate habitat rebuilding, and more able to survive in other conditions where they are obligated to change ecologically and ethologically. On the other hand, individuals with less variation potential (the most homozygotic ones) cannot face environmental changes; they are eliminated from the population or do not leave descendants, and the population becomes stabilized with the most variable individuals.

USAGE IN SYSTEMATIC STUDIES

Classification of the genus *Anolis*, based on Etheridge (1960) and Williams (1976), has been widely discussed by Gorman, Buth, and Wyles (1980) and Shochat and Dessauer (1981), mainly regarding the division in alpha and beta sections proposed by Etheridge (1960). Wyles and Gorman (1980) and Gorman *et al.* (1984) have found contradictions among the genetic and osteological data, which led them to state that the alpha and beta dichotomy does not exist.

Table 4.2. Observed heterozygosity figures from populations of *Anolis* according to different ecologic variables (symbols as in table 4.1).

Ecologic variable	Type	Enzymes	d	Proteins	d
	Low	0.32		0.37	
Perch height			+0.04		+0.08
	Medium-high	0.36		0.45	
	Close	0.33		0.42	
Habitat			+0.04		+0.03
	Open	0.37		0.45	

Nevertheless, Guyer and Savage (1986) reanalyzed the osteological, karyological, and immunological data previously obtained by other authors and proposed to formalize the alpha and beta sections, with four genera in the first one and one in the second. The basis of this work has been criticized by Cannatella and de Queiroz (1989) and Williams (1989*b*). Burnell and Hedges (1990) concluded that the categories of section and subsection should be abandoned based on the analysis of twelve slow-evolving proteins in forty-nine Antillean species of *Anolis*. Hass *et al.* (1993) have provided evidence from two types of molecular data: albumin immunological distance and DNA sequencing information from a mithocondrial 16 S ribosomal RNA gene, implicating an origin by overwater dispersal for most of the Antillean vertebrate fauna, including *Anolis*, *Chamaeleolis*, and other related genera.

According to these statements, taxonomic discrepancies exist that indicate the need to continue studying the molecular genetics of the genus *Anolis* in order to increase evidence about the phylogenetic relationships among species and to help elucidate its classification.

To evaluate interspecific taxonomic relationships among Cuban species, Buth *et al.* (1980) electrophoretically compared *A. carolinensis* from Texas and Georgia, and *A. porcatus* from Cuba, using thirty-five genetic loci. The Cuban population differed from the North American ones in five of them. The Nei (1972) distance estimated between populations of Cuba and the United States is 0.165, which is slightly higher than the estimates of intraspecific distance established by Bezy *et al.* (1977) and Gorman, Buth, and Wyles (1980) for lizards. The latter authors accepted the hypothesis of the molecular clock, according to which the data obtained by Buth *et al.* (1980) allowed them to predict that *A. carolinensis* colonized North America starting from Cuba during Pliocene times.

Lieb *et al.* (1983) studied two populations of *A. sagrei* from the east and west coasts of Florida and compared them with one Cuban population. Assuming that Florida populations are derived from Cuban ones, they tried to determine the status of these mainland forms with respect to genetic equivalence, possible history of colonization, and role of genetic drift in the differentiation of populations. They scored thirty-two loci in every specimen, of which only nine were polymorphic. Comparisons among these populations demonstrated that the ones from Tampa and La Habana are more closely related, with a calculated genetic distance of 0.01, while both populations from Florida had a genetic distance of 0.06. In addition, the absence of alleles fixed in different ways supported the hypothesis of the recent origin of these populations as well as that of the independent colonization of the Tampa and Miami populations. The levels of heterozygosity were similar in the three populations; the observed mean heterozygosity was approximately 7%, and the polymorphic loci had a level of 18%. These high levels of variability suggest that a large number of individuals colonized the Florida peninsula.

Pérez-Beato and Berovides Álvarez (1982) studied the electrophoretic patterns in cellulose acetate of plasmatic proteins of one population of *A. equestris* from Ciudad de La Habana province and two populations of *A. porcatus*, one from Ciudad de La Habana and the other from Pinar del Río. The electrophoretic patterns allowed them to differentiate the two species and the electrophoretic mobility of the fraction 1 separated the specimens of both populations of *A. porcatus*.

Burnell and Hedges (1990) studied relationships among some species of *Anolis* using slow-evolving protein loci. These included the Cuban species *A. porcatus*, *A. smallwoodi*, *A. homolechis*, *A. ophiolepis*, *A. centralis*, and *A. sagrei*. As a result of their analysis, these species were grouped at a distance level of 0.8—not too high, considering their very distinct external morphological features—with the exception of *A. sagrei*, which was more separate from the others.

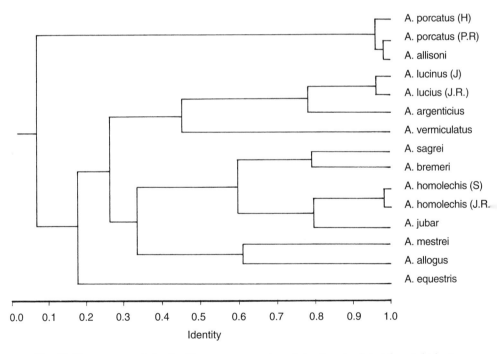

Fig. 13. Phenogram of similarities among species of *Anolis* calculated for eight loci, based on Nei's (1972) identity and UPGMA clustering. *H,* Ciudad de La Habana; *P.R.,* Pinar del Río; *J,* Jibacoa; *J.R.,* Jobo Rosado; *S,* Soroa.

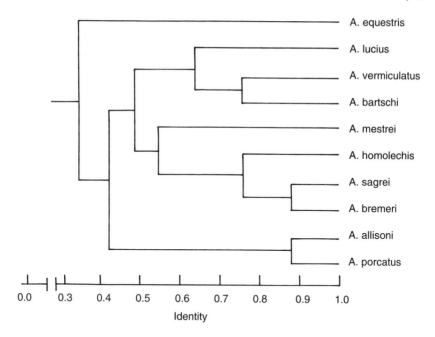

Fig. 14. UPGMA phenogram for species of *Anolis*, calculated for different mobilities of the miogen bands.

Espinosa López, Posada García *et al.* (1990), and Espinosa López, Sosa Espinosa, and Berovides Álvarez (1990) studied fourteen species of the genus *Anolis*. The results obtained by these authors, shown in figures 13 and 14, generally support the Williams (1976) division into superspecies, but there are two contradictory facts.

One of them is that the relationship between *A. allisoni* and *A. porcatus*, calculated by the Nei (1972) identity, is similar to those which Buth *et al.* (1980) and Lieb *et al.* (1983) estimated between conspecific populations of *Anolis*. Both species are also closely related in their morphology, to the extent that, for a long time, it was thought that in Cuba there was only one of these two members of the *carolinensis* group (*A. porcatus*). However, Ruibal and Williams (1961*a*) very well defined the existence of the two species and their localities, as well as the differentiation of *A. porcatus* in three forms: western, central, and eastern. There is evidence of additional differentiation in the western form of *A. porcatus*, not only morphologically but also in some proteins (Pérez-Beato and Berovides Álvarez 1979, 1982).

Pérez-Beato and Berovides Álvarez (1984) reported finding a population of *A. allisoni* whose individuals exhibited graduations of the blue color, which, together with the study of characteristics such as canthal ridges and ear opening shape in western populations of *A. porcatus*, led them to postulate either hybridization between these species or polymorphism of both; for these authors, the latter phenomenon is more likely. Such a statement is supported by the results of Espinosa López, González Suárez, and Berovides Álvarez (1985) and Espinosa López (1989) in relation to changes in genic frequencies of albumin, glycoproteins, and esterases in several populations of these two species (fig. 11 and map 1).

The other fact in which Espinosa López, Posada García *et al.* (1990) and Espinosa López, Sosa Espinosa, and Berovides Álvarez (1990) found contradictions is the location of *A. mestrei*, because of the separation of this species from *A. homolechis* and *A. jubar*. Ruibal (1964) analyzed this group and affirmed that *A. homolechis* and *A. mestrei* are similar species, differing only in color. However, the ecological characteristics of these three species are distinct: *A. homolechis* inhabits zones of forest borders, *A. jubar* dwells in semiarid coastal regions, and *A. mestrei* lives on karstic rocks in shaded places in Sierra del Rosario and Sierra de los Organos (Garrido 1982*b*). It is possible that although they are morphologically similar species, the biochemical characteristics have rapidly developed; this would allow stabilization of the differences detected in the ethological and ecological characteristics that separate them.

In relation to the species groups, a discrepancy exists. *A. vermiculatus* was once placed in the genus *Deiroptyx* with *A. bartschi*; later Etheridge (1960) considered the two species to be members of the genus *Anolis* based on their osteological characteristics. Later, because of the variation of its chromosomes (Gorman and Atkins 1968), its color, and its lack of dewlap, Garrido (1976*b*) again placed *A. vermiculatus* in the genus *Deiroptyx*. However, many authors do not recognize this placement. The results of Espinosa López, Posada García *et al.* (1990) and Espinosa López, Sosa Espinosa, and Berovides Álvarez (1990) support placement of this species within the *lucius* series and a closer relationship with *A. bartschi*.

The dendrogram obtained by Espinosa López *et al.* (1991) from the comparison of the species of the *lucius* series through the Nei (1972) distance is shown in figure 15. There is a slightly larger identity between populations of *A. lucius* than between those of *A. vermiculatus*. The similarity between *A. lucius* and *A. argenteolus*, which belong to the same superspecies, according to the classification of Williams (1976), is also remarkable. The identity estimates between those pairs of species are similar to those obtained by Yang *et al.* (1974) among species of *Anolis* of the *roquet* group (superspecies *richardi*, *luciae*, and *roquet*, whose genetic distances vary between 0.13 and 0.97), and to those observed in figure 13 for comparisons of species of the superspecies *homolechis* and *sagrei*.

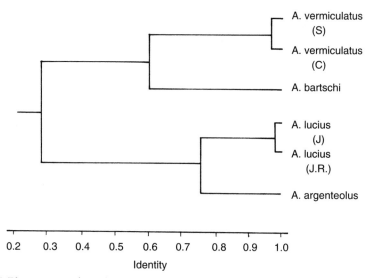

Fig. 15. Phenogram of similarities among species of the *lucius* series of *Anolis,* calculated for eight loci, based on Nei's (1972) identity and UPGMA clustering. C, Cayajabos; other abbreviations as in fig. 13.

These results corroborate the relationships between *A. vermiculatus* and *A. bartschi* discussed previously. The higher categories of series and section used by Williams (1976) are not supported by the data of Espinosa López, Posada García *et al.* (1990) and Espinosa López, Sosa Espinosa, and Berovides Álvarez (1990), but the existence of these categories has been the subject of much discussion; their presence will be confirmed when evidence explaining the relationships among different species of *Anolis* increases.

Results in the Genus *Leiocephalus*

If numerous studies have been done on protein polymorphism in the genus *Anolis* (Gorman and Kim 1976; Buth *et al.* 1980; Gorman, Buth *et al.* 1980; Gorman *et al.* 1983; Lieb *et al.* 1983; Espinosa López 1989), there are no papers that deal with this subject in the genus *Leiocephalus,* and the results shown here have been recently obtained (Chamizo Lara, Camacho, Rivalta *et al.* 1989, 1991; Chamizo Lara, Camacho, Torres *et al.* 1989; Chamizo Lara, Rivalta *et al.* 1994). The ten protein systems studied in three species of the genus (*L. carinatus, L. cubensis,* and *L. stictigaster,* collected in the Península de Zapata) are under genetic control of at least twenty loci, whose polymorphism is shown in table 4.3.

The tendency of a particular locus to be polymorphic or monomorphic goes beyond the limits of the population (Ayala 1975). Some hypotheses

Table 4.3. Allelic frequences of ten polymorphic loci from four populations belonging to three species of the genus *Leiocephalus.* **L. car.,** *L. carinatus;* **L. stic.,** *L. stictigaster;* **L. cub.,** *L. cubensis;* PG, **Playa Girón;** S, **Soplillar;** PL, **Playa Larga. Number in parentheses is number of analyzed specimens.**

Locus	Alleles	L. car. PG (7)	L. stic. PG (32)	L. cub. S (15)	L. cub. PL (10)
General	A	0	1.00	0.53	0.65
proteins (Gp III)	B	1.00	0	0.47	0.35
General	A	1.00	0	0	0
proteins (Gp V)	B	0	1.00	1.00	1.00
Phosphoglucose	A	1.00	0.33	1.00	1.00
isomerase (Pgi II)	B	0	0.67	0	0
Phosphoglucomutase	A	0	0.09	0.30	0.25
(Pgm)	B	1.00	0.91	0.70	0.75
Malic enzyme	A	1.00	0	0	0
(Me)	B	0	1.00	1.00	1.00
Malate	A	1.00	0	0	0
dehidrogenase (Mdh I)	B	0	1.00	1.00	1.00
Malate	A	1.00	1.00	0.85	0.95
dehidrogenase (Mdh II)	B	0	0	0.15	0.05
a-glicerophosphate	A	0.36	0.38	0	0
dehidrogenase (a-Gpdh)	B	0.64	0.62	1.00	1.00
Superoxid	A	1.00	0	0	0
dismutase	B	0	0	1.00	1.00
(Sod I)	C	0	1.00	0	0
Superoxid	A	1.00	0	0	0
dismutase	B	0	0	1.00	1.00
(Sod II)	C	0	1.00	0	0
Mean heterozygosity		0.024	0.054	0.061	0.049
±		0.025	0.039	0.034	0.030
% polymorphic loci		5	15	15	15

have been postulated in order to explain this general observation. One of them, by Johnson (1974), suggested that the regulating enzymes (those that modulate the lengthwise flow of the metabolic vias) are generally more polymorphic than the nonregulating enzymes. The data herein support such a hypothesis because three of the four enzymes in which polymorphism was detected are regulating ones, according to Zouros and Hertz (1984): alpha-glicerophosphate dehydrogenase, phosphoglucomutase, and phosphoglucoseisomerase.

The heterozygosity estimates obtained for *L. cubensis* and *L. stictigaster* (table 4.3) are within the range given by Selander and Johnson (1973) for four species of lizards and, in general, for the majority of vertebrates (0.05-0.07), unlike the estimate for *L. carinatus*, which was the lowest—almost of the same magnitude as those reported for genetically very homogeneous species such as the American alligator, *Alligator mississipiensis* (Adams *et al.* 1980).

Habitat anthropization in these populations (relatively similar in all) does not seem to influence observed differences in their genetic variability, as is reported in this chapter for some Cuban species of the genus *Anolis*. Little is known about the niche width for Cuban species of *Leiocephalus*, their generalist or specialist strategies in using the habitat, or the sizes and life spans of their populations. Consequently, it is difficult to make inferences regarding the interactions of these factors and the obtained estimates of heterozygosity. However, the smaller genetic variabilities calculated for the populations of *L. cubensis* from Playa Larga and *L. carinatus* could have been influenced by the sample sizes used for these populations, since degree of polymorphism is quite dependent, mainly on sample size (Gorman and Kim 1976).

USAGE IN SYSTEMATIC STUDIES

There are no great systematic complications in the genus *Leiocephalus* at the specific level, but many questions still need to be elucidated in determining subspecies, considering that a large number of them have been described based only on small morphological variations; moreover, some populations stay without being assigned to any subspecies.

The results of polymorphism for three species of the genus (*L. carinatus*, *L. cubensis*, and *L. stictigaster*) are shown in table 4.3. All loci of the enzymatic systems isocitrate dehydrogenase (Idh), lactate dehydrogenase (Ldh), and glutamate oxalate transaminase (Got) were monomorphic and fixed for the same allele in the three species. Five loci are divergent for those species: three of them are fixed for the same allele in *L. cubensis* and *L. stictigaster*, and for one alternating allele in *L. carinatus*. These are Gp-V, Me, and Mdh-I loci. The other two that have diagnostic characteristics (Sod-I and Sod-II) are fixed for different alleles in the three species.

With genic frequency data, the genetic distances among these populations were calculated using the coefficients of genetic distance of Nei (1972) and Rogers (1972), and a matrix of genetic identity was made (table 4.4), from which was obtained a dendrogram that clusters these four populations (fig. 16). As was expected, the largest degree of similarity was found between populations of *L. cubensis* from Soplillar and Playa Larga; therefore, they can be considered a single population.

Table 4.4. Genetic identity matrix among populations of *Leiocephalus*. Above the diagonal, Nei's (1972) genetic identity coefficients; below the diagonal, Rogers' (1972) similarity coefficients (symbols as in table 4.3).

Species	L. car.	L. stic.	L. cub. PL	L. cub. S
L. car.		0.6498	0.7107	0.7133
L. stic.	0.6402		0.8540	0.8477
L. cub. (PL)	0.6486	0.8149		1.0000
L. cub. (S)	0.6828	0.8064	0.9864	

Moreover, *L. cubensis* and *L. stictigaster* are more closely related to each other than to *L. carinatus*. The Rogers' similarity coefficient found between *L. cubensis* and *L. sticitgaster* (0.81) is almost at the end of the range reported by Webster *et al.* (1972) for conspecific populations of *Anolis* (0.69–0.82). These two species are also very closely related morphologically, such that for a long time it was thought there was only one of them: *L. cubensis*. Schwartz (1959*a*) reviewed the *cubensis* complex and described *L. stictigaster* as a new species. This close relationship between *L. cubensis* and *L. stictigaster* was confirmed by Pregill (1992), who proposed the phylogeny of the genus based mainly on osteological characteristics.

Of the studied populations, that of *L. carinatus*—the only species non-endemic to Cuba—is most separate genetically. Nevertheless, the genetic differences found between *L. carinatus* and the species *L. cubensis* and *L. stictigaster* are fewer than those reported by Avise and Aquadro (1982) for congeneric species of reptiles (0.51). Of the six Cuban species of the genus,

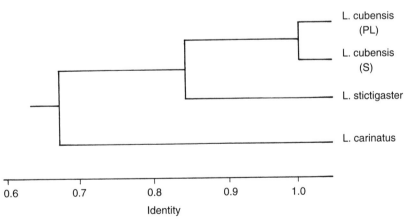

Fig. 16. Phenogram of similarities among populations of *Leiocephalus*, calculated for twenty loci, based on Nei's (1972) identity and UPGMA clustering. *PL*, Playa Larga; *S*, Soplillar.

L. carinatus is undoubtedly the most dissimilar, having notable morphological distinctions and, according to the biochemical results, the greatest genetic divergence.

Chromosomal Studies

Karyological studies have been done in some Cuban species of the genera *Anolis, Chamaeleolis,* and *Leiocephalus.* The cytological studies performed on *Anolis* supported and confirmed the classification of the genus. Gorman and Atkins (1968) stated that the majority of the *Anolis* from the alpha section that have been cytologically examined have six pairs of metacentric macrochromosomes and twelve pairs of microchromosomes (2n = 36), a number that seems to be primitive for the family Iguanidae *sensu lato* (Gorman *et al.* 1967). The beta *Anolis* have seven pairs of metacentric macrochromosomes and eight pairs of microchromosomes (2n = 30).

These authors cytologically examined twelve species of Cuban *Anolis* belonging to both alpha and beta sections. Among those of the alpha section, they studied *A. angusticeps, A. allisoni, A. argenteolus, A. bartschi, A. lucius, A. porcatus,* and *A. vermiculatus.* All of these species follow the general pattern described for the alpha section regarding the macrochromosomes; however, with respect to the twelve pairs of microchromosomes *A. vermiculatus* is an exception, since it has only eleven pairs. On the other hand, contradictory results have been found for *A. equestris:* the general pattern for the alpha section (Webster *et al.* 1972) and eleven pairs of macrochromosomes with seven pairs of microchromosomes (De Smet 1981).

The species from the beta section studied by Gorman and Atkins (1968) were *A. homolechis, A. mestrei, A. quadriocellifer, A. rubribarbus,* and *A. sagrei.* They have seven pairs of macrochromosomes and seven pairs of microchromosomes (2n = 28) without karyotypic variation among species, in contrast with what these researchers found for the beta section (eight pairs of microchromosomes) in their earlier studies (Gorman and Atkins 1966, 1967). Accordingly, Gorman and Atkins (1968) considered that radiation of the species groups *carolinensis* and *sagrei* has occurred in Cuba with little evolution of their chromosomes. Moreover, for *A. sagrei* De Smet (1981) reported ten pairs of macrochromosomes and four pairs of microchromosomes plus an extra X (2n = 29).

Porter *et al.* (1989) studied the karyotypes of *A. porcatus* and *A. homolechis,* and their results coincide with those described for both sections. Leal Díaz *et al.* (1991) corroborated the chromosomal number already described by Gorman *et al.* (1967) for *A. porcatus* and *A. allisoni.* In addition, they analyzed the C-banding pattern and the nucleolus organizer region (NOR) in these species.

Gorman *et al.* (1969) reported a 2n = 36 karyotype for the genus *Chamaeleolis*. This agrees with the position of Gorman *et al.* (1967) that such a number seems to be primitive for the family, since the genus *Chamaeleolis* is considered ancient and primitive (Williams 1969, 1989*b*). However, according to Hass *et al.* (1993), this genus is a recent addition to the Antillean fauna and a synonym of *Anolis*.

Paull *et al.* (1976) studied several species of lizards and found a chromosomal number of 2n = 36 for the genus *Cyclura* and of 2n = 32–36 for the genus *Leiocephalus*. Porter *et al.* (1989) found six pairs of macrochromosomes and eleven pairs of microchromosomes (2n = 34) for *L. cubensis*, and six pairs of macrochromosomes with nine pairs of microchromosomes (2n = 30) for *L. raviceps*. Leal Díaz and Morales Palmero (1991) reported 2n = 31 karyotype (twelve macrochromosomes and nineteen microchromosomes) for *L. cubensis*. As can be seen, despite the limited data for the genus *Leiocephalus*, there are discrepancies in the number of chromosomes reported by different authors. This indicates that only by studying other populations and species can new information be gained about the karyology of this interesting genus.

5

Parasites

ALBERTO COY OTERO

The parasitefauna of iguanids is quite typical. In general, they do not have the large number of parasitic species that characterizes other vertebrates, such as birds and mammals; nor do they experience, with some exceptions, a high degree of infestation. This is mainly due to the small size of almost all saurian species.

In Cuba, the endoparasites of these reptiles are well studied: almost 80% of hosts have been investigated and twenty-six forms, twenty of which are adults (table 5.1), have been determined. The ectoparasites, on the contrary, have been little studied; to date, seven species of hosts have been investigated and only five parasites are known (table 5.1), of which three are ticks. That is why in this chapter I will analyze both ecto- and endoparasites separately, taking into account some biological characteristics of those groups.

Parasites Found in Cuban Iguanids

Ectoparasites
  Order Acarina
    Suborder Metastigmata
    Family Ixodidae Murray, 1877
      Amblyomma albopictum Neumann, 1899
      Amblyomma torrei Pérez Vigueras, 1934
    Family Argasidae Canestrini, 1890
      Ornithodoros cyclurae Cruz, 1984
    Suborder Prostigmata
    Family Pterygosomidae Oudemans, 1910
      Cyclurobia javieri Cruz, 1984
    Family Trombiculidae Ewing, 1944
      Eutrombicula alfreddugesi Oudemans, 1910

Endoparasites
  Class Nematoda
  Order Oxyurida
    Family Atractidae Travassos, 1919
        Atractis opeatura Leidy, 1819
        Cyrtosomum scelopori Gedoelst, 1919
        Cyrtosomum longicaudatum Brennes and Bravo Hollis, 1960
    Family Ozolaimidae Pereira, 1935
        Ozolaimus monhystera (Linstow, 1902)
        Travassozolaimus travassosi Pérez Vigueras, 1938
        Paralaeuris cyclurae (Dosse, 1939)
    Family Oxyuridae Cobbold, 1864
        Parapharyngodon cubensis (Baruš and Coy Otero, 1969)
        Skrjabinodon anolis Chitwood, 1934
    Family Pharyngodonidae Travassos, 1919
        Spauligodon cubensis (Read and Amrein, 1953)
        Parathelandros sp. Baruš and Coy Otero, 1969
  Order Spirurida
    Family Physalopteridae (Leiper, 1908)
        Skrjabinoptera phrynosoma (Ortlepp, 1922)
        Physaloptera squamatae Harwood, 1932
        Abbreviata sp. larvae Baruš and Coy Otero, 1969
        Physalopteroides valdesi Coy Otero and Baruš, 1979
    Family Rhabdochonidae Skrjabin, 1946
        Trichospirura teixeirai (Baruš and Coy Otero, 1968)
    Family Spiruridae Oerley, 1885
        Cyrnea sp. larvae Coy Otero and Baruš, 1979
    Family Oswaldofilariidae Sonin, 1966
        Oswaldofilaria brevicaudata (Rodhain and Vuylsteke, 1937)
    Family Splendidofilariidae Sonin, 1966
        Piratuba digiticauda Lent and Freitas, 1941
  Order Strongylida
    Family Molineidae Durette-Desset and Chabaud, 1977
        Oswaldocruzia lenteixeirai Pérez Vigueras, 1938
  Order Ascaridida
    Family Anisakidae Skrjabin and Karokhin, 1945
        Porrocaecum sp. larvae (Pérez Vigueras, 1938)
  Class Digenea (Trematoda)
  Order Plagiorchiiformes
    Family Plagiorchiidae Ward, 1917
        Turquinia cubensis Coy Otero, 1989
    Family Mesocoeliidae Dolffus, 1933
        Mesocoelium monas Rudolphi, 1819
  Order Brachylaemiformes
    Family Urotrematidae Poche, 1926
        Urotrema scabridum Braun, 1900
        Urotrema wardi Pérez Vigueras, 1940

Class Cestoda
Order Pseudophyllidea
   Family Proteocephalidae Mola, 1928
      Proteocephalus sp. Coy Otero, 1970
   Family Dibothriocephalidae Luhe, 1902
      Sparganum sp. Pérez Vigueras, 1935

Ectoparasites

Of the seven species of examined iguanids, only the iguana (*Cyclura nubila nubila*) has four ectoparasites, two of which are specific to this host. Two ectoparasites have been found in two other species, *Leiocephalus carinatus* and *Anolis sagrei*; in the remaining species, only one (table 5.1).

The trombiculid acari, very frequent and diverse in reptiles, are present in Cuban iguanids, but they have not yet been studied. According to Cruz and Daniel (1991), only one species, *Eutrombicula alfreddugesi*, is currently known to have great significance for humans. This mite, known in Cuba as *abuje* or *abujo*, is a plague in humans and domestic animals, mainly preying on chicks. Its bite can cause intense itching, which in strong infestations causes anemia and, in some cases, death. It is supposed that abuje is a potential vector of some human diseases. Cruz and Abreu (1986) found it on *Anolis sagrei* from Cabo Cruz, Granma province.

Other mite species parasitize only reptiles, at least at the specific level of the parasite, although there is a doubtful report by Neumann (1899) of *Amblyomma albopictum* in porcupine (*Cercolabes villosus*) from Brazil. *A. albopictum* has a wide distribution all over the Cuban territory, including the archipelagoes; it seems to be associated with coastal forests and shrubwoods, the habitat of the iguana—one of the reptiles it parasitizes.

As in many such ticks, larvae and nymphs are less exigent for the host than adults. Adults are found on reptiles of large size, while immatures are found on smaller species. Adults have been found on snakes of the families Colubridae (*Alsophis cantherigerus*) and Boidae (*Epicrates angulifer*) and on iguana. Larvae and nymphs have been also found on the curly-tailed lizard *Leiocephalus carinatus*.

Little is known about *Amblyomma torrei*; it has been found in Puerto Rico, the Cayman Islands, and Cuba, where it occurs on the keys and coasts, always associated with coastal forests and shrubwoods occupied by the iguana. It is a species affecting three hosts. Immatures are less specific than adults and live on species of smaller size. Larvae and nymphs have been found to parasitize *Cyclura n. nubila, Leiocephalus macropus, L. cubensis, Anolis luteogularis*, and *A. sagrei*, while adults have been found only on the iguana. Currently it is known to parasitize only iguanids.

Table 5.1. Cuban iguanid parasites, geographic distribution and Cuban hosts.

Parasite	Distribution	Cuban hosts
A. albopictum	Cuba, Swan Islands, Hispaniola, Brazil(?), Guyana(?)	*C. nubila, L. carinatus*
A. torrei	Cuba, Puerto Rico, Cayman Islands	*C. nubila, L. macropus, L. carinatus, L. cubensis, A. luteogularis, A. sagrei*
O. cyclurae	Cuba	*C. nubila*
C. javieri	Cuba	*C. nubila*
E. alfreddugesi	Tropical America	*A. sagrei*
A. opeatura	Cuba, Haiti, Brazil	*C. nubila, A. sagrei*
C. longicaudatum	Cuba, Central America	*A. equestris, A. baracoae, A. luteogularis, L. carinatus, L. cubensis, L. stictigaster, C. chamaeleonides, C. porcus*
C. scelopori	Cuba, Central and North America	*A. equestris, A. baracoae, A. luteogularis, A. porcatus, A. allisoni, A. argillaceus, A. loysiana, A. centralis, A. lucius, A. argenteolus, A. bartschi, A. vermiculatus, A. cyanopleurus, A. sagrei, A. homolechis, A. bremeri, A. quadriocellifer, A. jubar, A. allogus, A. mestrei, A. rubribarbus, A. ophiolepis, L. carinatus, L. cubensis, L. macropus, L. raviceps, L. stictigaster, C. porcus*
P. cyclurae	Cuba, Bahama Isl.	*C. nubila*
O. monhistera	Cuba, Haiti, Central America	*C. nubila, A. sagrei*
T. travassosi	Cuba, Bahama Isl.	*C. nubila, C. chamaeleonides*
P. cubensis	Cuba	*A. luteogularis, A. porcatus, A. allisoni, A. lucius, A. vermiculatus, A. bartschi, A. sagrei, A. bremeri, A. homolechis, A. quadriocellifer, A. jubar, A. allogus, L. carinatus, L. cubensis, L. macropus*
S. anolis	Cuba, Puerto Rico	*A. alutaceus*
Parathelandros sp.	Cuba	*A. homolechis*
S. cubensis	Cuba	*A. bremeri*
Abbreviata sp.	Cuba	*A. luteogularis, A. baracoae, A. sagrei, A. homolechis, L. carinatus, L. cubensis, L. macropus*

Parasite	Distribution	Cuban hosts
P. squamatae	Cuba, USA	*A. equestris, A. baracoae, A. lucius, A. sagrei, A. bremeri, A. homolechis, A. allogus, L.carinatus, L. cubensis, L. macropus, L. stictigaster, L. raviceps*
P. valdesi	Cuba, South America	*A. argillaceus*
S. phrynosoma	Cuba, USA	*A. luteogularis, A.lucius, A. bartschi. A. alutaceus, A. sagrei, A. bremeri, A. homolechis, A. quadriocellifer, A. mestrei, L. carinatus, L. cubensis, L. raviceps*
T. teixeirai	Cuba	*A. baracoae, A. sagrei, A. homolechis, C. porcus, L. cubensis*
Cyrnea sp.	Cuba	*A. luteogularis, L. carinatus*
O. brevicaudata	Cuba, Central and South America	*C. nubila, A. baracoae*
P. digiticauda	Cuba, Brazil	*A. baracoae, A. vermiculatus, C. porcus*
O. lenteixeirai	Cuba, Puerto Rico	*A. equestris, A. baracoae, A. luteogularis, A. allisoni, A. loysiana, A. lucius, A. bartschi, A. sagrei, A. bremeri, A. homolechis, A. quadriocellifer, A. allogus, L. carinatus, L. cubensis, L. macropus, L. stictigaster, C. porcus, C. nubila*
Porrocaecum sp.	Cuba	*A. luteogularis, A. baracoae, A. smallwoodi, A. allisoni, A. loysiana, A. lucius, A. bartschi, A. alutaceus, A. sagrei, A. bremeri, A. homolechis, A. jubar, A. allogus, A. ophiolepis, L. carinatus, L. cubensis, L. macropus, L. stictigaster, L. raviceps, C. porcus*
T. cubensis	Cuba	*A. argenteolus*
M. monas	Cosmopolite	*C. porcus*
U. scabridum	Cuba, Central and South America	*A. porcatus, A. sagrei*
U. wardi	Cuba	*A. porcatus, A. sagrei*
Proteocephalus sp.	Cuba	*A. allogus*
Sparganum sp.	Cuba	*A. equestris*

Where *A. torrei* coincides with *A. albopictum* they form mixed infestations, although generally the second species is the more abundant. An exception was noted in one iguana collected in Playa Jaimanitas, Ciudad de La Habana province in 1964, on which were found three females, twenty-nine males, and six nymphs of *A. torrei* and two females and six nymphs of *A. albopictum* (Cerny 1966).

Ornithodoros (*Alectrobius*) *cyclurae* is the most surprising species of ectoparasite on Cuban reptiles. It is known based on only one larva, the unique specimen collected from the nasal cavity of an iguana from the cliffs near Cabo Cruz, Granma province (Cruz 1984*a*). *Cyclurobia javieri* is another unusual parasite on Cuban iguana; it was described based on sixteen males, fourteen nymphs, and three larvae (females are unknown) collected on a host from Cabo Cruz in March 1981. Little is known about this species, which seems to be a member of a primitive group parasitizing reptiles (Cruz 1984*b*).

The geographic distribution of ectoparasites (Cruz 1978, 1984*a, b*; Cruz and Abreu 1986) also suggests how little is known about them. Two species, *Ornithodoros cyclurae* and *Cyclurobia javieri*, have been found only in Cuba. One species, *Eutrombicula alfreddugesi*, is spread all over tropical America. *Amblyomma albopictum* was found in Cuba, Hispaniola, and the Swan Islands (Neumann 1899); finally, *A. torrei* lives in Cuba, Puerto Rico, and the Cayman Islands. It is quite possible that *A. torrei* also inhabits Hispaniola, but it has not been observed there to date.

The tick *A. albopictum* is vicariant with *A. antillarum* Kohls, 1969, living on the iguana from the Virgin Islands (Keiraus 1985), while *Ornithodoros cyclurae* seems to be associated with *O. elongatus* Kohls, Saxesline, and Clifford, 1965, from Hispaniola. The extinction of iguanas in some Antillean islands has contributed to a lack of understanding about the distributions of some mites, because as their hosts have disappeared so have the parasites, without being studied.

Endoparasites

Twenty-six endoparasites have been found to date in Cuban iguanids, of which twenty are nematodes, four are trematodes, and two are cestodes. None is harmful to the health of humans or domestic animals, and the majority are typical of this group of reptiles at the specific level. Exceptions are the nematodes *Oswaldocruzia lenteixeirai* and *Physalopteroides valdesi*, which also parasitize amphibians (Coy Otero and Baruš 1979*b*); *Spauligodon cubensis*, a geckonid (Coy Otero and Baruš 1979*b*); and a species of trematode, *Urotrema scabridum,* which also parasitizes bats (Groschaft and del Valle 1968).

Adults of twenty species have been determined. Of these, only five (*Parapharyngodon cubensis, Spauligodon cubensis, Trichospirura teixeirai, Urotrema wardi,* and *Turquinia cubensis*) are endemic to Cuba, while four others are spread throughout the Caribbean area and four in the southwestern region of North America. Only one (*Mesocoelium monas*) can be considered a cosmopolite because of its worldwide distribution.

Within Cuban iguanids, the subspecies *Cyclura nubila nubila* is noteworthy. Its colon is filled with nematodes, which number between 50,000 and 70,000 per host (Baruš *et al.* 1969). This intense level of invasion seems to be related to the Cuban iguana's digestive process, as is also the case in some herbivorous animals and other iguanas of the genus *Cyclura* (Iverson 1979). The parasites help to facilitate the digestion of cellulose from leaves, fruits, and stems, which constitute the greatest part of the diet in these animals. Furthermore, the fact that this interspecific relationship does not cause any type of disease in the host supports the idea that parasites take part in the digestion of the vegetable material. Such colic nematodes belong to the species *Travassozolaimus travassosi, Ozolaimus monhystera, Paralaeuris cyclurae,* and *Atractis opeatura,* which also occur in the big Neotropical iguanas of the genera *Ctenosaura, Iguana,* and *Cyclura,* indicating a relationship between hosts and parasites. In that regard, Baruš and Hubalek (1995), studying relationships among Neotropical iguanines based on their twenty-two nematode species, supported the opinions of de Queiroz (1987) and Norell and de Queiroz (1991) about a close relationship between *Iguana* and *Cyclura.*

Another species of nematode found in the Cuban iguana, but not as abundant as the previous ones, is *Oswaldofillaria brevicaudata,* which has also been found occasionally in *Anolis baracoae.*

Hosts belonging to the genera *Anolis, Chamaeleolis,* and *Leiocephalus* are parasitized, with some exceptions, by the same species of nematodes: *Parapharyngodon cubensis, Cyrtosomun sclepori, C. longicaudatum, Physaloptera sguamatae, Skrjabinoptera phrynosoma, Trichospirura teixeirai,* and *Oswaldocruzia lenteixeirai.* On the contrary, other species parasitize only one or two hosts. These are *Physalopteroides valdesi* in *Anolis argenteolus; Skrjabinodon anolis* in *A. alutaceus; Turquinia cubensis* in *A. argenteolus; Proteocephalus* sp. in *A. allogus;* and *Mesocoelium monas* in *Chamaeleolis porcus.*

The helminthofauna of hosts belonging to the genus *Chamaeleolis* has a close relationship with that of the giant *Anolis* (*A. equestris, A. luteogularis,* and *A. baracoae*). This is well observed in the nematode species *Piratuba digiticauda, Cyrtosomum scelopori,* and *C. longicaudatum.* The first parasitizes *Chamaeleolis porcus* and *Anolis baracoae. Cyrtosomum scelopori* parasitizes twenty-three species of the genus *Anolis,* while *C. longicaudatum* is found only in the previously mentioned hosts and in three species of the genus *Leiocephalus.* Mixed infestations of the two species of the genus *Cyrtosomum* have not yet been found in a single specimen.

Trichospirura teixeirai is undoubtedly one of the most interesting helminths of Cuban reptiles. The only member of the family Rhabdochonidae that parasitizes reptiles has been found in Cuba in five species belonging to the genera *Anolis, Leiocephalus,* and *Chamaeleolis* (*C. porcus*). From the genus *Trichospirura* three other species are known, all parasites of monkeys and bats from Brazil and Malaysia.

Baruš *et al.* (1996) studied coevolution patterns between the four genera of Cuban lizards of the family Iguanidae and the twenty species of nematodes parasitizing them. These researchers suggested that there is a shared similarity in the nematode fauna of species of the genera *Anolis, Chamaeleolis,* and *Leiocephalus* and a dissimilarity between these nematodes and those of the genus *Cyclura.* In addition, they considered that the relationship found between the nematodes of *Anolis* and *Leiocephalus* could be interpreted as ecological parasitism.

Among the helminth larvae found in Cuban iguanids, those that belong to the genus *Cyrnea* match imaginal forms parasitizing birds of the families Cuculidae and Phasianidae (Baruš 1966, 1969), whose life cycles have been completed. The larval forms of *Abbreviata,* which have been found in an advanced stage of development (IV larvae), are apparently adult forms found in the Cuban racer, *Alsophis cantherigerus* (Baruš and Coy Otero 1966, 1978). The finding of invasive larvae of *Abbreviata baracoa* in amphibians of the genus *Peltaphryne* (Baruš 1973) leads to the supposition that in this case saurians act as reservoir hosts; that is, larvae of the nematode are temporarily housed in them.

Larvae of the genus *Porrocaecum,* present in numerous species of iguanids, are imaginal forms of nematodes that parasitize birds of the orders Falconiiformes and Strigiiformes (Baruš 1966). Finally, the larvae of *Sparganum,* found in the subcutaneous tissue of *Anolis equestris,* are pleurocercoids, or juveniles of cestode species, that have been described in many studies of different hosts from other countries. In Cuba these larvae have also been found in the fishes *Lutjanus analis, Epinephelus striatus,* and *Cibium acervum,* as well as in the snake *Tretanorhinus variabilis* (Pérez Vigueras 1936). These are forms that match adult cestodes of the genus *Diphyllobothrium,* which parasitize dogs, cats, and humans. In this case saurians are accidentally parasitized; the normal intermediary hosts are crustaceans and freshwater fishes.

Of the twenty-six species of helminths found in Cuba—including both larval and adult forms—eight nematodes have widespread geographic distributions. They are found all over the Cuban archipelago, which also agrees with the great variety of hosts they parasitize. These are the same species mentioned earlier in the analysis of the specificity of parasitism. The five species of nematodes that parasitize the iguana are distributed in the areas the iguana occupies; that is, in the coastal areas and keys.

Two species, *Piratuba digiticauda* and *Trichospirura teixeirai,* show interesting distributional patterns. Despite the fact that they have been found in different regions of the country and in several different hosts, they have been collected abundantly only in one species and only at one locality. The first species was found in *Anolis vermiculatus* from the region of Mil Cumbres (Sierra del Rosario, Pinar del Río province) and the second in *A. sagrei* from Cayo Cantiles and Cayo el Rosario, Archipiélago de los Canarreos. The eleven remaining species of helminths have local distributions and parasitize few hosts, and, in general, have been found in low numbers.

6

Biogeography

LOURDES RODRÍGUEZ SCHETTINO

Historical biogeography deals with the origin and evolution of biota in relation to the geological history of a given region of the world. For the Caribbean area in particular, numerous hypotheses have been proposed since the beginning of the twentieth century because of the particular interest in clearing up the enigmas of a depauperate Antillean biota—compared to that of the adjacent mainland—and of an extremely complex geology.

Ecological biogeography, on the other hand, deals with the current distribution of organisms in a given region, their interrelationships, and the environmental factors of greatest importance in determining and maintaining the distribution patterns observed at a given time. In conducting such investigations it is necessary to consider the distribution of the habitats occupied by the species and the distribution of the species within those habitats. For instance, Lee (1980) found two groups of environmental factors that greatly influenced the distribution of amphibians and reptiles of the Península de Yucatán: the amount and seasonality of precipitation and the structural heterogeneity of vegetation.

There have been few studies dedicated to the ecological biogeography of the Antillean herpetofauna. Research on the herpetofauna of other regions, such as that of Duellman (1978, 1987) on the communities of amphibians and reptiles from Amazonian forests, and of that of Lee (1980), previously mentioned, is of great interest.

In this chapter I will discuss some aspects of the historical and ecological biogeography of iguanians in the Caribbean region in general and in Cuba in particular, based on papers published by several authorities in this field.

Historical Biogeography of the Caribbean Region

There are two classic but opposite arguments for explaining the poverty of the fauna of the Caribbean islands: (1) the islands arose without any biota and were colonized by overwater dispersal; and (2) the island faunas are fragments of the mainland ones, and their poverty is a consequence of extinctions that occurred after separation from the mainland.

Those who endorse the first argument consider overwater dispersal the only way for species to have colonized the islands. In contrast, those who support the second propose a former connection of the islands to the mainland; thus, they assert, the fauna colonized directly by land.

Since the first years of the twentieth century this controversy has existed among the followers of the overwater dispersal hypothesis (rafters) and those who advocate the hypothesis of faunistic interchange through terrestrial connections (bridge builders)—either by the union of islands to the mainland or by island or promontory chains currently submerged. In general, members of the first group (Aguayo 1949; Simpson 1940, 1956; Darlington 1957; Etheridge 1960; Williams 1969) as well as those of the second (Barbour 1914; Corral 1939) based their conclusions on a stable geology for the Antilles; that is, the Caribbean islands were always in their current place.

However, the revolution in the earth sciences that occurred in the 1960s radically changed the mode of thinking: the evident truth of plate tectonics and of continental drift notably influenced biogeographic theories, to the extent that even such contemporary dispersalists as Pregill (1981a) no longer accept a stabilist hypothesis.

These new ideas led to the formulation of new geological models for the Caribbean area and, thereafter, of new biogeographic models. Thus the theory of vicariance appeared, which interprets the current patterns of biotic distribution as a result of the division of ancestral biotas in response to changes in geography climatically, physiographically, or tectonically induced (Rosen 1976), in opposition to the theory of overwater dispersal. Nevertheless, the two opposite ways of thinking have been maintained to date, with only a few modifications (Crother and Guyer [1996] and Hedges [1996a] are two recent examples).

Perfit and Williams (1989) took into account the most accepted geological models for the Caribbean region, proposed by Malfait and Dinkelman (1972), Ladd (1976), Perfit and Heezen (1978), Coney (1982), Pindell and Dewey (1982), Wadge and Burke (1983), and Donnelly (1985), and concluded that the history before the Cretaceous cannot be reconstructed with certainty because of the paucity of rocks of pre-Cretaceous age in the Caribbean region. Moreover, they summarized the information and recognized the existence of four significant periods in Caribbean paleogeography.

The first period lasted from the Cretaceous to the mid-Eocene. The Caribbean Plate was present in the Pacific Ocean at the end of the Cretaceous, west of México and between México and South America, as a magmatic arc composed of the Leeward proto-Antilles, the Nicaraguan Rise, the Cayman Ridge, and the Aves Ridge. Then it began to migrate to the east and was progressively breaking, forming islands from west to east.

The second period extended from the mid-Eocene to the Miocene, during which time the Caribbean Sea invaded the islands. Nevertheless, there is evidence that some sections of Cuba, Hispaniola, Puerto Rico, and even Jamaica were never submerged.

The third period ran from the Miocene to the mid-Pliocene. This was a rising period for the majority of islands. It is generally accepted that the Greater Antilles acquired their current configuration and position approximately during the Upper Miocene (Furrazola *et al.* 1964; Judoley and Meyerhoff 1971; Iturralde-Vinent 1975, 1981, 1982, 1988; Coney 1982; Donnelly 1985; MacPhee and Iturralde-Vinent 1994; Iturralde-Vinent *et al.* 1996). The Lesser Antilles, which had arisen as a volcanic arc since the previous period, finished their development.

The fourth period endured from the Pliocene to the present. Periodic changes in sea level during the Pleistocene modified the extension of the emergent lands; thus the area and configuration of islands varied from time to time.

Bearing in mind that the generalized geological model of a proto-Antillean archipelago between North and South America at the end of the Mesozoic is accepted by most geologists and biogeographers, the question is whether that archipelago was continuous and connected to both mainlands, like an isthmus, or whether it was composed of isolated islands. If the first theory is correct, separation of the archipelago from the mainland and its fragmentation thereafter had a vicariant effect on the fauna (Rosen 1976; Hedges 1982; Savage 1982; Guyer and Savage 1986; Roughgarden 1995; Crother and Guyer 1996). If the second one is true, overwater dispersal is the event that allowed the arrival of faunas on the proto-Antillean archipelago (Pregill 1981*a;* Williams 1989*a, b;* Hass 1991; Hass *et al.* 1993; Hedges 1996*a, b, c*).

There is not enough geological evidence to support either of these two arguments. Nevertheless, the second one is more accepted; in addition, there is the theory of the bolide impact in the Yucatán peninsula (at 65 my BP, Cretaceous-Tertiary), used by Hedges *et al.* (1992) to explain the almost total absence of an ancient Antillean fauna. That impact probably caused huge waves and tsunamis that should have devastated the biota living there. However, the discovery of some very ancient fossils in the Antilles that are related to extant fauna, although insufficient so far, indicates the possible existence of a terrestrial biota established very early during the formation of the islands (for example, in the proto-Antilles, before the bolide impact) that did not disappear at all. Williams (1989a) compiled paleontological information of the greatest age and geographic interest and reported that the following biota are known: one primate (20,000 to 60,000 years), one rodent (100,000 to 180,000 years), and one sirenian (from the Eocene), from Jamaica; plants from the Cretaceous to the Pleistocene, and a fish from the Miocene of Haití; *Anolis, Sphaerodactylus, Eleutherodactylus,* mammalian hair, bird feathers, ants, and spiders in amber from the Oligocene or Miocene of the Dominican Republic (according to Lambert *et al.* [1985], the Dominican amber was dated as being from the Eocene, but Hedges [1996c], based on other authors, concluded that it is from the Upper Oligocene to Lower Middle Miocene); and an echimyd, an iguanid, turtles, and a sirenian from the Oligocene or Miocene of Puerto Rico.

Those species, or their ancestors, could have been living in the proto-Antilles before they separated from the mainland and traveled with them; however, dispersal could have happened during the Eocene or earlier, especially since distances between the islands and the mainland should have been less than they are now (Williams 1989a). Moreover, the Caribbean islands have been joined or separated, their area has increased and diminished, and their place and position have changed; all of these events have allowed vicariance of the faunas and dispersal of faunistic elements as well (Perfit and Williams 1989).

Another possible path for the arrival of South American fauna in the Antilles was postulated by MacPhee and Iturralde-Vinent (1994): terrestrial communication by continuous land or groups of very close islands through the Aves Ridge from South America to western Cuba, during the Oligocene, due to the uplifting of the Greater Antilles Ridge and the Aves Ridge, and a drop in sea level of about 160 m. The extensive emergence of land at that time could have allowed the passage of some species from South America to the Greater Antilles. In addition, MacPhee and Iturralde-Vinent (1994) and Iturralde-Vinent *et al.* (1996) think that the Equatorial Atlantic Current could not have flowed to the Caribbean Sea at that time, precluding overwater dispersal of some elements of the fauna from eastern South America until the transition between the Oligocene and Miocene.

Hedges (1996b) did not recognize the significance of that path because it endured only about 3 million years. Thus he argued that almost the only way for herpetofauna to reach the Antilles was by overwater dispersal on flotsam from the large northeastern South American rivers, carried by marine currents that, according to Hedges, flowed in the same clockwise direction as they do today. Although Iturralde-Vinent et al. (1996) recognized such a direction of the marine currents during the Oligocene, they argued that the Greater Antilles Ridge, together with the Aves Ridge, probably compelled the Equatorial Current to flow northward and not enter the Caribbean Sea. During the Miocene the fragmentation of the Antilles Ridge and the subsidence of the Aves Ridge allowed the currents to flow through the Caribbean Sea.

As has been postulated by various authors (MacFadden 1980, 1981; Hedges 1982; Savage 1982; Rodríguez Schettino and González Alonso 1984; Guyer and Savage 1986; Burgess and Franz 1989; Hedges 1989; Joglar 1989; Thomas 1989; Williams 1989a, b; Burnell and Hedges 1990), the origin of the Caribbean fauna is complex and should not be interpreted from one point of view alone. Although overwater dispersal can occur at any time, it depends on existent barriers, and successful colonization depends on the distances that must be traversed, the ecological conditions in the territory to be colonized, and the vagility and genotypic characteristics of the taxon that allow it to adapt and reproduce in such a new territory. In fact, different species are passively transported today over floating rafts of vegetation and others are actively transported, either by air (fliers) or by water (swimmers). This means that dispersal is feasible and not too rare (Perfit and Williams 1989).

On the other hand, vicariance is also a reality in the Caribbean region, in the sense of fragmentation and aggregation of islands with their respective faunas, but concordance in time and place in the separation of the islands and of the taxa has not been proved, although it should have happened during the earlier stages of the proto-Antilles formation (Perfit and Williams 1989).

In reviewing many articles about the geology and biogeography of the Caribbean region, I realized that all of them present evidence that is more or less supported. This leads to the conclusion that it is not possible, given the current geological and biological evidence, to consider one biogeographic model more valid than another, or one excluding the other. However, just as the geology of the Caribbean region is complex and not yet completely elucidated, many factors have been involved in the origin of its biota. Vicariance and overwater dispersal, colonizations and extinctions, speciation and adaptive radiation—all have occurred and still occur, in a continuous process of earth and organic evolution.

Zoogeography of the Cuban Iguanids

The stabilist concept of geology and the few fossils of iguanids found in Cuba led Barbour (1914) and Aguayo (1949) to believe that such species reached the island in relatively recent times from Central and South America. After studying the Antillean anolines, Etheridge (1960) and Williams (1969) concluded that overwater dispersal should have been the only possible way for species to colonize Cuba, and that two big invasions from tropical America were necessary for successful colonization. From the first one, which occurred before the great Caribbean Miocene inundation, only *Chamaeleolis* survived in Cuba. During the second invasion, the remaining species arrived.

The most primitive iguanids have been found in the Middle to Late Jurassic period of the Gondwana Plate (Estes 1983). There were also iguanids in the Late Cretaceous and in the Paleocene from Brazil (Pregill 1981*a*), and in the Miocene they had a wide distribution in South America; thus the extant genera were probably settled in the Antilles in the middle of the Cenozoic. Nevertheless, Pregill and Olson (1981) considered the climatic and sea-level changes that occurred during the Pleistocene to be so drastic that survival of any pre-Pleistocenic species would have been precluded; consequently, the current Cuban fauna is the result of post-Pleistocenic colonizations by overwater dispersal (but see MacPhee and Iturralde-Vinent 1994).

According to Hedges (1982) and Savage (1982), the extant species had their origins in Central America and colonized Cuba not only by dispersal but also by vicariance between the island and the mainland, and between the island and other islands of the Greater Antilles. Rodríguez Schettino and González Alonso (1984) asserted that some characteristics of Cuban species cannot be explained if these species arrived only recently; therefore they stated, although not conclusively, that there was probably a connection between the Central American and Cuban territories that predated their present configuration.

Vanzolini and Heyer (1985) considered that anolines are a South American group with a big radiation in Central America, since there is much evidence that their passage from South to Central America occurred during the Cretaceous or early Tertiary, before the formation of the Isthmus of Panama in the Pliocene. The biogeographic model proposed by Guyer and Savage (1986) is based on some vicariant events that occurred at the end of the Mesozoic in the present Central American territory. These events allowed subsequent differentiation in situ of the ancestral species that stayed with the fragments from which the Antillean islands originated. The conclusions of Guyer and Savage (1986) have been criticized by Williams (1989*a, b*), who argued that dispersal from the mainland and between islands served as a filter barrier and that this better explains the impoverished and unbalanced nature of the Antillean fauna.

The genus *Cyclura* is found nowadays in Cuba, the Cayman Islands, the Bahama Islands, Jamaica, Hispaniola, and Anegada, and the known fossils are consistent with extant species of Cuba, the Cayman Islands, and Anegada (Morgan 1977; Pregill 1981*b;* Morgan *et al.* 1993). According to Schwartz and Carey (1977), *Cyclura* is closely related to the genus *Ctenosaura* from Central America, but de Queiroz (1987) found that it shared a more recent common ancestor with *Iguana,* from Central and South America, than with *Ctenosaura,* based on his phylogenetic analysis of seventy-four skeletal and twenty-one nonskeletal characteristics. Analyzing the nematodes prasitizing the Neotropical iguanines, Baruš and Hubalek (1995) found a close relationship among *Ctenosaura, Iguana,* and *Cyclura,* stronger between the last two. In addition, Hedges (1996*b*) stated that *Cyclura* is a close relative of *Iguana* and that its origin was by overwater dispersal from Central or South America during the Miocene, based on the immunological distance of 20 between the two genera.

The genus *Leiocephalus* lives in Cuba, the Cayman Islands, the Bahama Islands, Hispaniola, and Martinique. According to Etheridge (1966*b*), its characteristics were fixed in the Antilles during the mid-Cenozoic, from South American tropidurine ancestors; this was corroborated by Pregill (1992), based on phylogenetic analysis of thirty-nine morphological characteristics (osteology, scalation, and coloration). However, Hedges (1996*b*) said that the immunological distances and the DNA sequencing data indicate an origin by overwater dispersal from North America during the mid-Cenozoic, from *Crotaphytus* and *Sceloporus.*

The genus *Chamaeleolis* is endemic to Cuba; fossils are not known, but its features, considered primitive by Etheridge (1960) and Williams (1969, 1989*b*) indicated to these researchers that it originated from tropical American ancestors at the end of the Mesozoic, and that it could have survived the big Caribbean inundation of the mid-Tertiary. However, using the immunological distances and the sequences of DNA, Hass *et al.* (1993) found that *Chamaeleolis* is a relatively recent addition to the anoline fauna of the Antilles.

The genus *Anolis* is found in the Antilles and in North, Central, and South America. Skin fragments exist from the Mexican Oligocene, and there is a Dominican species from the mid-Cenozoic in amber (Rieppel 1980). Williams (1969) considered that the beta *Anolis* were endemic to México and Central America, from which they traveled to Jamaica, Cuba, and the Bahamas. The alpha *Anolis* were typically South American fauna that arrived in the Lesser Antilles, Greater Antilles, and North America by overwater dispersal, and in Central America by land.

Guyer and Savage (1986) revealed that the ancestors of extant anoles lived in tropical South America and the proto-Antilles, and that, at moving the last, the alpha *Anolis* remained isolated in South America (genus *Dactyloa*) and in the proto-Antilles (genera *Semiurus, Ctenonotus,* and *Anolis*), while the beta *Anolis* (genus *Norops*) stayed in Central America, later traveling to Jamaica and Cuba. Williams (1989*b*) considered that the proposition of Guyer and Savage's genera is based on inadequate data, which makes their recognition uncertain, and ratified his hypothesis of overwater dispersal for the arrival of alpha and beta *Anolis* in Cuba.

In contrast, Burnell and Hedges (1990) stated that the data on albumin immunological distances between Antillean and North American species suggest their separation after the breakage of the proto-Antilles, so that dispersal between islands at the middle of the Cenozoic and later speciation are what probably occurred. Nevertheless, some aspects of this theory are still unclear.

Although Hedges *et al.* (1992) recognized the possibility of the occurrence of vicariance for the Antilles with respect to North and South American faunas in the Cretaceous, they presented evidence on the evolution of albumin in different groups of amphibians and reptiles that does not prove the ancient origin of such fauna. They suggested that dispersal was the principal means of colonization for the Antilles and attributed the present absence of an ancient biota to the collision of a bolide in the western Caribbean region at the end of the Cretaceous, which had fatal consequences for the fauna living at that time, especially those inhabiting areas of lower elevation.

Hass *et al.* (1993) argued that the molecular data they reported do not support the alpha and beta dichotomy for *Anolis*, which indicates a mid-Cenozoic origin from Central America for some lineages and from South America for others, by overwater dispersal. However, based on the few anole fossils known, the combination of skeletal features of each anole island fauna of the Greater Antilles, the high diversity of species, and the exceptional independent adaptive radiation on each island that has produced remarkably similar sets of species led Losos and de Queiroz (1997) to believe that such lizards have been there for a long time and that the evolutionary convergence seen to date is an early occurrence, suggesting that all four sets of species descended from a common ancestor.

Summarizing everything I have revealed here, the most recent evidence of geology and biology suggests that the main ancestors of Cuban iguanids came principally from Central and South America, through one means or another (in other words, by overwater dispersal, vicariance, or both). Once on the lands that through geological evolution were forming the present territory of Cuba, they underwent a great adaptive radiation by means of allopatric, parapatric, and sympatric speciation, convergence, and other diverse evolutive mechanisms that have resulted in the present faunal diversity and endemicity.

On the other hand, according to their current geographic distribution (table 6.1), several Cuban species live in every geographic region represented in map 2. Zones with extreme climatic conditions—very wet (Península de Zapata) or very dry (southern coast of Guantánamo province and the archipelagoes)—have fewer species than zones corresponding to mountain ranges (Cordillera de Guaniguanico, Macizo de Guamuhaya, Sierra Maestra, and Macizo de Sagua-Baracoa; map 3), where a greater number of regional or local endemic species live.

The Macizo de Sagua-Baracoa (zone 8) accommodates not only the largest number of species but also the highest number of Cuban species and species endemic to the zone. This agrees with the high diversity of vegetation zones that can be found there, which, combined with the complex and heterogeneous structure of the vegetation and the relatively well-conserved territories, allows many different kinds of lizards to find the microhabitats they require for surviving in sympatry. In addition, the

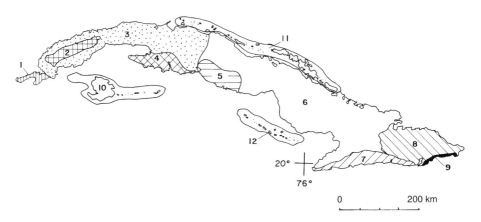

Map 2. Geographic regions of Cuba in which iguanids are distributed. *1*, Península de Guanahacabibes; *2*, Guaniguanico Range; *3*, western lowlands; *4*, Península de Zapata; *5*, Guamuhaya Range; *6*, central eastern lowlands; *7*, Sierra Maestra; *8*, Sagua-Baracoa Range; *9*, southern coast of Guantánamo province; *10*, Canarreos Archipelago; *11*, Sabana-Camagüey Archipelago; *12*, Jardines de la Reina Archipelago.

Table 6.1. Geographic distribution of Cuban iguanids in the twelve zones of map 2. *GDP*, geographic distribution pattern; *PC*, pancuban; *CPC*, cuasipancuban; *R*, regional; *P*, population; *L*, local. Endemic species are marked with an asterisk (*).

Species	Zones												GDP
	1	2	3	4	5	6	7	8	9	10	11	12	
L. carinatus	X	X	X	X	X	X	X	X	X	X	X	X	PC
L. cubensis*			X	X	X	X	X	X		X	X	X	CPC
L. macropus*	X	X	X		X	X	X	X	X				P
L. raviceps*			X			X		X	X				P
L. stictigaster*	X	X	X	X	X	X	X			X	X		CPC
L. onaneyi*									X				L
C. nubila	X	X	X		X	X	X	X	X	X	X	X	CPC
Ch. chamaeleonides*	X	X	X	X		X	X			X			CPC
Ch. porcus*						X	X	X					R
Ch. barbatus*		X											R
Ch. guamuhaya*					X								R
A. equestris*		X	X		X	X					X		R
A. luteogularis*	X	X	X	X						X			R
A. noblei*							X	X					R
A. smallwoodi*							X	X	X				R
A. baracoae*								X					R
A. pigmaequestris*											X		L
A. porcatus*	X	X	X	X	X	X	X	X	X	X	X	X	PC
A. isolepis*						X	X	X	X				P
A. allisoni			X	X	X	X							R
A. angusticeps	X	X	X	X	X	X	X	X		X	X		CPC
A. paternus*			X							X			R
A. guazuma*						X							L
A. alayoni*								X					R
A. garridoi*					X								L
A. loysiana*	X	X	X	X	X	X	X	X					CPC
A. argillaceus*						X	X	X	X				R
A. centralis*						X	X	X	X				R
A. pumilus*	X	X	X		X	X				X	X		R
A. lucius*			X		X	X							R
A. argenteolus*						X	X	X	X				R
A. vermiculatus*		X											R
A. bartschi*		X											R
A. alutaceus*	X	X	X	X	X	X	X	X		X			CPC
A. clivicola*						X							P
A. anfiloquioi*								X					P
A. inexpectata*								X					L
A. macilentus*								X					L

(continued on next page)

Table 6.1. (*continued*)

Species	1	2	3	4	5	6	7	8	9	10	11	12	GDP
A. vescus*								X					L
A. alfaroi*								X					L
A. cyanopleurus*								X					L
A. cupeyalensis*						X		X					P
A. mimus*							X						R
A. fugitivus*								X					L
A. juangundlachi*			X										L
A. spectrum*		X	X			X							P
A. vanidicus*					X								R
A. sagrei	X	X	X	X	X	X	X	X	X	X	X	X	PC
A. bremeri*			X							X			R
A. homolechis*	X	X	X	X	X	X	X	X		X			CPC
A. quadriocellifer*	X												R
A. jubar*						X	X	X	X		X		R
A. guafe*							X						L
A. confusus*							X						L
A. mestrei*		X											R
A. allogus*	X	X	X			X	X	X					CPC
A. ahli*					X								R
A. delafuentei*					X								L
A. rubribarbus*							X						R
A. imias*									X				L
A. birama*						X							L
A. ophiolepis*		X	X	X		X			X				CPC
Number of species	15	21	24	13	21	28	26	31	13	15	11	5	
Cuban endemics	11	17	19	9	16	23	22	27	10	11	7	2	
Species endemic to the zone	1	4	1	0	4	1	5	10	2	0	1	0	

lowlands (zones 3 and 6) lodge many species, mainly those of wide geographic distribution that are adapted to places modified by humans, such as cultivated and urban areas; however, the level of endemicity there is low, with only one species in each zone.

Only five species, *Leiocephalus carinatus, Cyclura nubila, Anolis angusticeps*, *A. allisoni*, and *A. sagrei*, are not endemic of Cuba, which implies 91.6% endemicity. The first three species are endemic to the Antilles, while *A. allisoni* also inhabits the Islas de la Bahía, Honduras. *A. sagrei* also lives on the mainland, although only along the moist Atlantic coast of the Gulf of México and the Caribbean Sea.

0   200 km

Map 3. Principal mountain ranges of Cuba (small dots) and the highest elevation in each range: *1,* Cordillera de Guaniguanico (Pan de Guajaibón, 699 m); *2,* Alturas de La Habana-Matanzas (Pan de Matanzas, 381 m); *3,* Alturas de Bejucal-Madruga-Coliseo (Sierra de Camarones, 358 m); *4,* Alturas del Nordeste (Sierra de Jatibonico, 443 m); *5,* Macizo de Guamuhaya (Pico de San Juan, 1,140 m); *6,* Sierra de Cubitas (Cerro de Tuabaquey, 330 m); *7,* Sierra de Najasa (Sierra del Chorrillo, 301 m); *8,* Grupo de Maniabón (Sierra Verde, 347 m); *9,* Sierra Maestra (Pico Turquino, 1,972 m); *10,* Macizo de Sagua-Baracoa (Pico del Cristal, 1,231 m); *11,* Sierra de Caballos (295 m).

According to these recent data on geographic distribution in the Cuban territory (Schwartz and Henderson 1988, 1991; Rodríguez Schettino 1989, 1993; table 6.1), it is possible to corroborate the five geographic distribution patterns reported by Rodríguez Schettino (1986*a*), although with some modifications in the number of species that each pattern involves. Three species, *Leiocephalus carinatus, Anolis porcatus,* and *A. sagrei,* show a Pan-Cuban distribution: they can be found in every zone of map 2.

The species that inhabit almost all of the country, with the exception of one or a few zones, have a quasi–Pan-Cuban distribution; these include *Leiocephalus cubensis, L. stictigaster, A. alutaceus, A. homolechis, A. allogus, A. ophiolepis, A. angusticeps, Chamaeleolis chamaeleonides,* and *Cyclura nubila.*

Twenty-six species have regional distribution; they are found in one or more adjacent zones. Of these, *Chamaeleolis barbatus, Anolis paternus, A. vermiculatus, A. bartschi, A. bremeri, A. quadriocellifer,* and *A. mestrei* live in the western region; *A. equestris, A. luteogularis,* and *A. pumilus* are in the western and central regions; *C. guamuhaya, A. vanidicus,* and *A. ahli* inhabit the central region (Macizo de Guamuhaya), while *A. allisoni* and *A. lucius* also live in the central region but are somewhat more extended toward the west and east; *A. argillaceus, A. centralis, A. argenteolus,* and *A. jubar* live in the central eastern region; and *C. porcus, A. noblei, A. smallwoodi, A. baracoae, A. alayoni, A. mimus,* and *A. rubribarbus* dwell in the eastern region.

Eight species have an allopatric distribution; that is, they live in distinct localities that can be in one or more zones. Of them, *L. raviceps, L. macropus,* and *Anolis loysiana* are the most widespread, with populations in different places throughout the island of Cuba. *A. isolepis* has populations in the central and eastern mountain ranges; *A. clivicola* and *A. anfiloquioi,* in Sierra Maestra and in Sagua-Baracoa, respectively; *A. spectrum,* in zones 2, 3, and 6; and *A. cupeyalensis,* in zones 6 and 8.

Each local species has a restricted distribution in only one zone: *A. juangundlachi* in zone 3; *A. delafuentei* and *A. garridoi* in zone 5; *A. birama* in zone 6; *A. guazuma, A. guafe,* and *A. confusus* in zone 7; *A. inexpextata, A. macilentus, A. vescus, A. alfaroi, A. cyanopleurus,* and *A. fugitivus* in zone 8; *L. onaneyi* and *A. imias* in zone 9; and *A. pigmaequestris* in zone 11.

Because iguanids comprise the majority of the Cuban terrestrial reptiles, their patterns of geographic distribution are consistent with the faunistic areas proposed by Rodríguez Schettino (1993), determined by reptile assemblages of one or more zones. Thus regional, allopatric, and local iguanids typify the different faunistic areas: in western Cuba there are seven species; in central Cuba, eight; in eastern Cuba, sixteen; in the archipelagoes, six; in Macizo de Guamuhaya, four; and on the southern coast of Guantánamo province, eight.

Elevation above sea level is a limiting factor, mainly because of changes in climate and in vegetation physiognomy that take place when the altitude increases. Although in Cuba the mountain ranges are not very high, there is an altitudinal distribution of iguanids (Rodríguez Schettino 1985; Moreno and Valdés 1991). Of the sixty-two species, twelve subsist exclusively in the lowlands (group I) and twenty-three live only at some height above sea level (group III). The remaining species (group II) can be found in the lowlands and uplands. Only four species reach the highest altitude in Sierra del Turquino: *A. angusticeps, A. homolechis, A. isolepis,* and *A. clivicola.* The first two are distributed from the coast to Pico Cuba; the last two are typically altitudinal.

Of the species in groups II and III, sixteen live in Cordillera de Guaniguanico, but only one (*Chamaeleolis barbatus*) is restricted to the highlands and endemic to this mountain range. In Macizo de Guamuhaya sixteen species are found, six of which belong to group III (*A. isolepis, A. garridoi, A. vanidicus, A. ahli, A. delafuentei,* and *Chamaeleolis guamuhaya*) and are endemic to this region, with the exception of *A. isolepis.* There are sixteen species living in Sierra Maestra and La Gran Piedra, five of them altitudinal: *A. noblei, A. isolepis, A. guazuma, A. clivicola,* and *A. mimus;* the last three are endemic to this zone. In the remaining eastern mountains, twenty-four species are found; among them are fifteen species of group III (*Leiocephalus onaneyi, Anolis noblei, A. smallwoodi, A. isolepis, A. baracoae, A. alayoni, A. anfiloquioi, A. inexpectata, A. macilentus, A. vescus, A. alfaroi, A. cupeyalensis, A. cyanopleurus, A. fugitivus,* and *A. rubribarbus*), of which the last eleven are endemic to Macizo de Sagua-Baracoa.

Ecological Biogeography of Cuban Iguanids

Environmental conditions such as macro- and microhabitat, climate, and altitude determine the ecogeographic distribution of iguanids in Cuba. In general, almost all of the sixty-two species are endemic and territorial, with little ability to translate over great distances, which allows them to be considered good zoogeographic indicators (Rodríguez Schettino 1993).

Not even Pan-Cuban species are uniformly distributed; they are limited by the boundaries of the vegetation zones (macrohabitats) in which they live. Taking into account the typical macrohabitats of these species, including their geographic ranges, iguanids are represented in the map of terrestrial reptiles (Rodríguez Schettino 1989), with the exception of the eight last described species.

The iguanid distribution in fourteen of the vegetation zones, adapted and generalized after Capote López *et al.* (1989; plate 58), indicates that the majority of the species are generally found in forests (see table 3.1). These are mainly mesophyllous evergreen and semideciduous forests, not only primary but also secondary ones. Highly homogeneous vegetation zones, such as swamp and mangrove forests, shelter very few species. It can also be observed that the cloud and montane rain forests, found only in Sierra Maestra, are occupied by the same four species; two of them (*A. clivicola* and *A. isolepis*) are actually altitudinal species; the other two (*A. angusticeps* and *A. homolechis*) are among the more ecologically plastic species, occurring in a wide variety of vegetation zones.

On the other hand, sixteen species can be found in places modified by humans—in urban zones or agroecosystems—especially *L. cubensis, A. equestris, A. luteogularis, A. porcatus, A. sagrei,* and *A. homolechis,* because of their presence and abundance. Where they live in sympatry these six species divide the space vertically, so they must coexist with less interference. Thus *L. cubensis* is found on the ground; *A. sagrei* and *A. homolechis* are trunk-ground dwellers, the second one higher and more shaded than the first; *A. porcatus* lives on trunks; and *A. equestris* or *A. luteogularis* (depending on their geographic distribution) inhabit the crowns. This is one of the most valuable evolutive mechanisms of adaptation: it allows several species to coexist without a large degree of competitive interference (Schoener 1977, 1983).

As to the composition of iguanid lizard assemblages, the more similar groups are those living in cloud and montane forests (fig. 17) and also those inhabiting cultivated and urban zones. Figure 17 shows that there are three groups of lizard assemlages. One is composed of the four species living in high mountain vegetation zones; the second is formed by the species that live in the most complex and heterogeneous vegetation zones; in the third are clustered the few species dwelling in vegetation zones with extreme conditions not favorable for reptile life.

The extension of the vegetation zones has changed with time and this has modified the ecogeographic distribution of lizards. During the Pleistocene, dry and wet periods alternated, although the last Wisconsian glaciation imposed the driest and coolest climate (Pregill and Olson 1981). Consequently, species not requiring moist forests should have been dominant

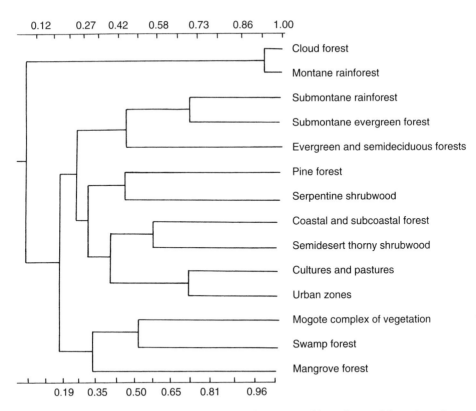

Fig. 17. Dendrogram of similarities among Cuban iguanid lizard assemblages based on the Sorensen coefficient of similarity.

during that time. Afterward, the Antilles became wetter, with most of their surfaces covered by forests; some relictual arid habitats remained, now inhabited by species that had a wider distribution during the Pleistocene glaciations. The genera *Leiocephalus* and *Cyclura*, both endemic to the Antilles, are now found mainly in the driest places, such as coastal and subcoastal microphyllous shrubwoods. The known fossils demonstrate that these two genera had a wider distribution during the Pleistocene. For instance, on the semidesertic southern coast of Guantánamo province is found a community of lizards that is very adapted to dry conditions, including *Cyclura nubila*, *L. carinatus*, *L. macropus*, *L. raviceps*, *L. onaneyi*, and eight species of *Anolis*, Of these species, *L. onaneyi* and *A. imias* are endemic to the zone, living in quite arid habitats—the first at the top of a hill with limestone, almost barren of vegetated ground, and the second among and inside boulders of the xeric coastal belt. In addition, in the several keys of the northern archipelagoes are dry habitats occupied by species that do not require moist forests, such as *Cyclura*, *L. carinatus*, *A. porcatus*, *A. angusticeps*, *A. pumilus*, *A. sagrei*, and *A. jubar*.

Forests formerly made up the dominant landscape in Cuba, occupying more than 90% of the land surface at the time the Spaniards arrived—not only in the uplands but also in the lowlands. With the uncontrolled development of society that followed, forests began to diminish in size: in 1812 they covered 90% of Cuban territory; in 1900, 54%; and in 1959, 14%. Since 1960 forests have undergone a slight recovery due to the implementation of reforestation plans and the protected areas system; thus in 1975 they occupied 18% of Cuba (Ayala Castro 1978), 19% in 1985 (Arcia Rodríguez 1989). Now forests can be found mainly in mountainous areas and in some protected places such as Península de Guanahacabibes and Península de Zapata. As a consequence, lizards able to tolerate only deep-forest conditions have become extinct or remained in very restricted localities at high elevations, like some of the components of the *alutaceus* group in the eastern mountains. More generalized species such as *A. porcatus*, *A. angusticeps*, *A. sagrei*, and *A. homolechis* tolerate many conditions and have occupied almost all of the new habitats left by the forests.

Places for basking and air and substrate temperatures are other factors that influence the spatial distribution of iguanids. In general, species inhabiting forests have few available sites for basking; to find basking sites to increase their body temperatures would require too much energy expenditure. Thermopassivity is the evolutive mechanism by which these species adapt to such environments. On the contrary, species that are effective thermoregulators can survive in open places, where the cost of this behavior is not too great. Examples of the first type are species of the *lucius* series of the genus *Anolis;* the second type includes *L. carinatus, L. stictigaster, C. nubila,* and *A. sagrei.*

Species of the *sagrei* group provide another interesting example of the influence of climate on the ecological distribution of lizards. Even being of the same ecomorph (the trunk-ground type), some of these species occur in sympatry in many places. This is possible because they have different ecological requirements such as perch height, heat, and light. In general, *A. sagrei* lives in open places with high light intensity, near and in human constructions; *A. jubar* and *A. quadriocellifer* can also inhabit open areas. *A. homolechis* is a typical edge-dweller in forests and ecotones. The other species are more adapted to deep forests.

At Península de Guanahacabibes lives *A. sagrei,* only in the few towns there and near roads and paths; *A. homolechis* lives on the borders of the forest and in the boundaries between semideciduous and coastal forest; and *A. quadriocellifer,* the endemic and most abundant species of the zone, occurs in both coastal xeric vegetation and semideciduous forest (Rodríguez Schettino and Martínez Reyes 1985). At Soroa, in the Sierra del Rosario, Pinar del Río province, live four species of this group: *A. sagrei* in open areas of the forest and in human constructions, where there is a great deal of sunlight; *A. homolechis* at the edge of the forest, which receives filtered sun; and *A. allogus* and *A. mestrei* in deep shade within the forest, the first species occupying tree trunks more than rocks and the second, rocks more than tree trunks; of the last two species, *A. mestrei* is found on higher perches than *A. allogus* (Rodríguez Schettino et al. 1997).

At Topes de Collantes, Macizo de Guamuaya, the situations of *A. sagrei, A. homolechis,* and *A. allogus* are similar (Rodríguez Schettino *et al.* 1979), but two other species, *A. ahli* and *A. delafuentei,* live in deep forest with *A. allogus.* The difference in resource usage between *A. ahli* and *A. allogus* must be examined in more detail, taking into account other dimensions such as feeding, because apparently the species do not differ in the kind of substrate or light intensity required. With *A. delafuentei,* such differences are not understood because the species is only known on the basis of one male. At La Mula, a forest on the southeastern coast of the Santiago de Cuba province, the dominant species is *A. jubar; A. homolechis* dwells at the edge of the forest and *A. sagrei* is restricted to an adjacent town (Rodríguez Schettino 1985).

Another environmental resource, type of food, does not seem to be what determines whether a given species has a wider or narrower distribution, since in all Cuban ecosystems there are numerous species of invertebrates,

vertebrates, and plants that may be eaten by iguanids. To date no studies have proved the existence of an obligatory relationship between any Cuban iguanid and a given type of food. With the exception of the iguana, all species are insectivorous and some may include small vertebrates and even vegetable material in their diets. However, none of them uses a given taxon as basic food. Instead, like members of the family from other countries, they direct their efforts toward prey that will provide the richest source of energy at the least expense; that is, toward the most abundant, most easily captured prey in the places they inhabit.

7

Systematic Accounts of the Species

Lourdes Rodríguez Schettino

In this chapter I have presented whatever information I was able to gather, published or not, about the taxonomy, geographic distribution, morphology, and natural history of Cuban iguanids. In compiling this data I consulted more than 500 bibliographic references published from 1820 to 1998, numerous works by students, master's theses and doctoral dissertations, oral presentations, posters of many symposia, and some papers in press. In addition, I spoke with various researchers, who offered their valuable knowledge to enrich this chapter. Another important source of information was the herpetological collection of the Institute of Ecology and Systematics, Havana. The specimens there are preserved in 70% ethanol, kept in glass jars in air-conditioned rooms.

As explained in chapter 1 of this book, the order followed in the presentation of genera and species of the family Iguanidae in Cuba is based on Etheridge (1960) and Williams (1976), while observing the chronology in the original descriptions of species and subspecies. In preparing synonymies I considered the synonyms that are most important for understanding the history of each taxon and, in the case of genera, only the ones involving Cuban species.

In this chapter I have considered the date of publication of the vast work of Ramón de la Sagra to be 1838, based on the enlightening paper by Smith and Grant (1958). This matter is complex and confusing because Sagra's work was published in parts, from 1838 to 1843, in both Spanish and French editions. Chapter 4, which deals with fishes and reptiles, was written by T. Cocteau and G. Bibron. These researchers described various species, some of which had already been published by Duméril and Bibron (1837); that is why the latter work takes precedence for the original descriptions of such species. Seemingly, this happened because Bibron collaborated in both works and because the work by Duméril and Bibron was published a year before the first edition of Sagra's work was published.

I have provided type localities of each taxon according to the way they were originally published, that is, with the geographic names that were used at the time of publication. Under each subhead titled "Localities" I have included in parentheses the first published reference for that locality, the name of the individual who offered a new datum, the CZACC abbreviation if the information comes from that collection, or the initials of the author, always in chronological order by province (for names of provinces and of some cities and towns, see map 4, I–XV), beneath each species. In each of these subheads I have used the current geographic names, taken from the National Atlas of Cuba (I.C.G.C. 1978). Maps for each species were prepared taking all of these data into consideration

In preparing the subhead "Description" I took the characteristics given in the original descriptions, considering external morphological characteristics that are not doubtful or ambiguous, and corroborated or modified them from live individuals in most of the species and subspecies. To indicate the size of each taxon, snout-vent length, head length, and tail length were measured for the greatest number of specimens preserved for the longest time, in the same way described in chapter 2 of this book. Each species was then classified according to one of the following intervals of snout-vent length:

Size Interval

Very small   from 30.0 to 49.9 mm

Small   from 50.0 to 69.9 mm

Medium   from 70.0 to 99.9 mm

Intermediate   from 100.0 to 139.0 mm

Large   from 140.0 to 199.9 mm

Very large   200.0 mm or more

The color illustrations were drawn using live individuals as models, except for those of *Leiocephalus onaneyi*, *Anolis pigmaequestris*, *A. fugitivus*, and *A. cupeyalensis*, which were made from descriptions and photographs. Only

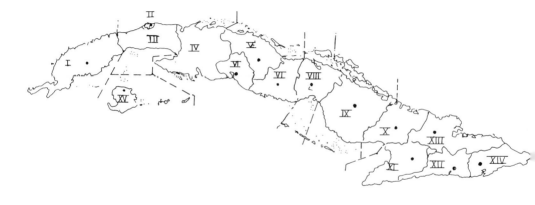

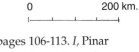

Map 4. Provinces of Cuba (capitals above marked with dots), pages 106-113. *I*, Pinar del Río (Pinar del Río); *II*, Ciudad de La Habana (La Habana); *III*, La Habana (La Habana); *IV*, Matanzas (Matanzas); *V*, Villa Clara (Santa Clara); *VI*, Cienfuegos (Cienfuegos); *VII*, Sancti Spíritus (Sancti Spíritus); *VIII*, Ciego de Ávila (Ciego de Ávila); *IX*, Camagüey (Camagüey); *X*, Las Tunas (Las Tunas); *XI*, Granma (Bayamo); *XII*, Santiago de Cuba (Santiago de Cuba); *XIII*, Holguín (Holguín); *XIV*, Guantánamo (Guantánamo); *XV*, Isla de la Juventud (Nueva Gerona).

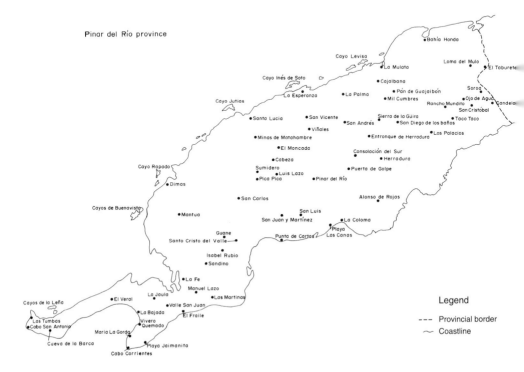

Ciudad de La Habana province

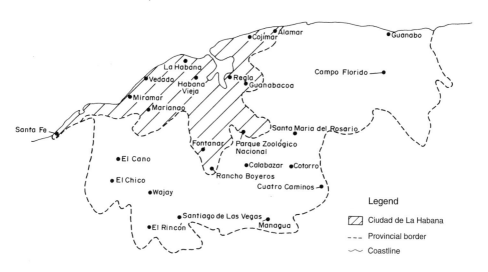

Alamar
•Cojimar
•Guanabo

La Habana
•Vedado
Habana
•Regla
Vieja
•Guanabacoa
•Miramar
•Marianao

Campo Florido —•

Santa Fe•

•Santa Maria del Rosario

•Fontanar
•Parque Zoológico
Nacional

•El Cano

•Calabazar •Cotorro
•Rancho Boyeros
•El Chico
Cuatro Caminos•

•Wajay

•Santiago de Las Vegas
•El Rincón
Managua•

Legend

▨ Ciudad de La Habana

--- Provincial border

⌒ Coastline

La Habana province

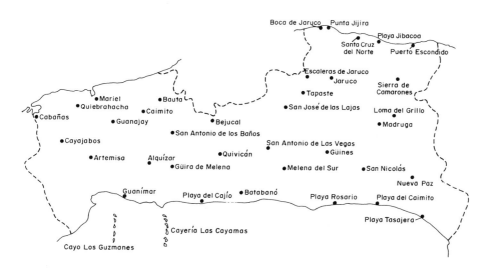

Boca de Jaruco Punta Jijira
Playa Jibacoa•
Santa Cruz
del Norte•
Puerto Escondido•

•Escaleras de Jaruco
•Jaruco
Sierra de
Camarones
•Tapaste
Loma del Grillo•
•Mariel
•Bauta
•San José de las Lajas
•Quiebrahacha
•Caimito
•Madruga
•Cabañas
•Guanajay
•Bejucal
•Cayajabos
•San Antonio de los Baños
San Antonio de Las Vegas
•Artemisa
•Quivicán
•Güines
Alquízar
•Güira de Melena
•Melena del Sur
•San Nicolás
Nueva Paz•
Guanímar•
Playa del Cajío•
•Batabanó
Playa Rosario•
Playa del Caimito•
Playa Tasajera•
Cayería Las Cayamas
Cayo Los Guzmanes

Legend

--- Provincial border

⌒ Coastline

Matanzas province

Cayo Cruz del Padre
Cayo Galindo
Punta Hicacos
Cayos Jutías
Varadero
Cayos de las Cinco
Bacunayagua
Leguas
Punta de La Maya
Salinas Bidos
Matanzas
Cárdenas
Pan de
Matanzas
Limonar
Martí
Unión de Reyes
Jovellanos
Perico
Los Arabos
Bolondrón
Colón
Alacranes
Agramonte
Jagüey Grande
Calimete
Maneadero
Santo Tomás
Pálpite
Playa Larga
Caleta del Rosario
Legend
Cayos de Diego Pérez
La Salina
El Cenote
Caleta Buena
- - - Provincial border
Playa Girón
⁓ Coastline

Villa Clara province

Cayo Blanquizal
Cayo Sotavento
Cayo Esquivel del Sur
Cayo Las Picúas
Cayo Lanzanillo
Corralillo
Isabela de
Sagua
Sierra Moreno
Rancho Veloz
Cayos del Pajonal
Sagua La Grande
Cayo Fragoso
Quemado de Güines
Cayo Santa María
Cayo Francés
Cifuentes
Cayo Cobos
Manacas
Encrucijada
Santo Domingo
Cayo Conuco
Caibarién
Cayos de La Herradura
Remedios
Camajuaní
Santa Clara
Ranchuelo
Placetas
Legend
Manicaragua
Guinía de Miranda
- - - Provincial border
Jibacoa
⁓ Coastline
Guanayara

Cienfuegos province

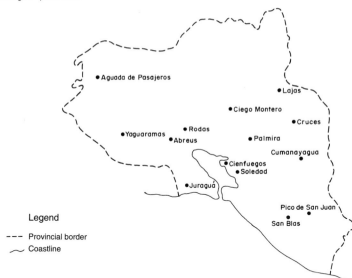

● Aguada de Pasajeros

● Lajas

● Ciego Montero

● Cruces

● Rodas
● Yaguaramas
● Abreus

● Palmira

Cumanayagua ●

● Cienfuegos
● Soledad

●Juraguá

Pico de San Juan ●

San Blas ●

Legend

--- Provincial border
～ Coastline

Sancti Spíritus province

Cayo Lucas
Cayo Salinas

● Yaguajay

● Mayajigua
Jobo Rosado ●

●Fomento

● Cabaiguán

Arroyo Blanco ●

●Topes de Collantes

● Taguasco

● Jatibonico
● Sancti Spíritus

● Trinidad
● Banao

La Sierpe ●
●Casilda

Tunas de Zaza ●

● El Jíbaro

Legend

--- Provincial border
～ Coastline

Ciego de Ávila province

Cayo Guillermo

Cayo Coco

Cayo Paredón Grande

Cayo

Cayo Cruz

Isla de Turiguanó

Romano

● Chambas

Loma de Cunagua

● Florencia

● Morón

● Bolivia

● Velazco

● Majagua

● Ciego de Ávila

● Venezuela

● Baraguá

Júcaro

Cayo de Ana María

Camagüey province

Cayo Romano

● Esmeralda

Cayo Guajaba

Cayo Sabinal

● Sierra de Cubitas

● Florida

Nuevitas ●

Playa Santa Lucía

● Minas

● Camagüey

● Jimaguayú

● Vertientes

● Sibanicú

● Cascorro

● Martí

Najasa →

● Sierra del Chorrillo

Guáimaro

Santa Cruz del Sur

Las Tunas province

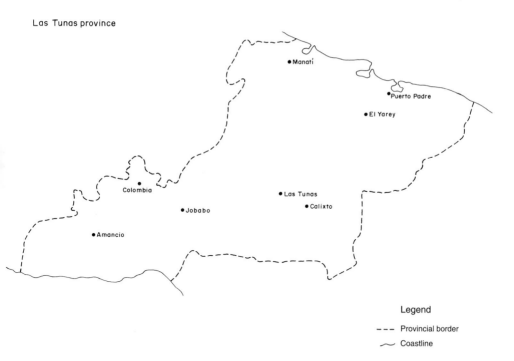

- Manatí
- Puerto Padre
- El Yarey
- Colombia
- Las Tunas
- Calixto
- Jobabo
- Amancio

Granma province

- Birama
- Río Cauto
- Cauto Cristo
- Bayamo
- Jiguaní
- Manzanillo
- Yara
- Bueycitos
- Guisa
- Campechuela
- Buey Arriba
- Media Luna
- Pico Bayamesa
- Niquero
- Belic
- Pilón
- Portillo
- Vereón
- Pesquero de la Alegría
- Cabo Cruz

Santiago de Cuba province

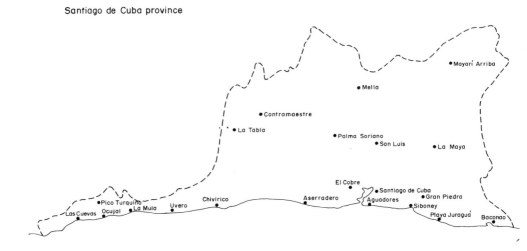

- Mayarí Arriba
- Mella
- Contramaestre
- La Tabla
- Palma Soriano
- San Luis
- La Maya
- El Cobre
- Aserradero
- Santiago de Cuba
- Aguadores
- Gran Piedra
- Siboney
- Chivirico
- Uvero
- La Mula
- Ocujal
- Pico Turquino
- Las Cuevas
- Playa Juraguá
- Baconao

Legend
- - - Provincial border
~ Coastline

Holguín province

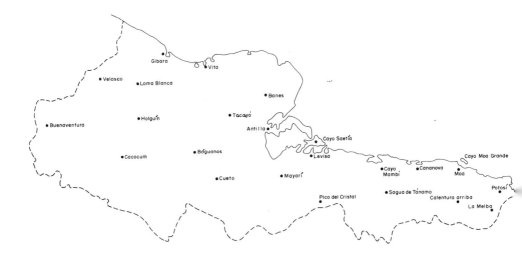

- Gibara
- Vita
- Velasco
- Loma Blanca
- Banes
- Buenaventura
- Holguín
- Tacayó
- Antilla
- Cayo Saetía
- Cayo Moa Grande
- Cacocum
- Báguanos
- Levisa
- Cananova
- Moa
- Cueto
- Mayarí
- Cayo Mambí
- Sagua de Tánamo
- Potosí
- Pico del Cristal
- Calentura arriba
- La Melba

Legend
- - - Provincial border
~ Coastline

Guantánamo province

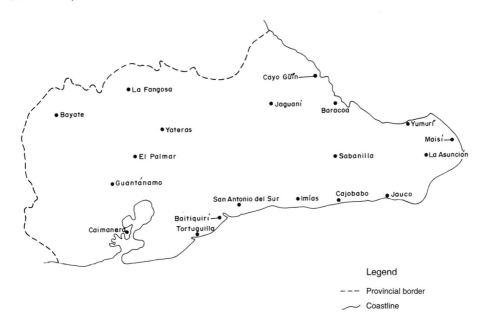

- Cayo Güin
- La Fangosa
- Jaguaní
- Baracoa
- Bayate
- Yumurí
- Maisí
- Yateras
- El Palmar
- Sabanilla
- La Asunción
- Guantánamo
- San Antonio del Sur
- Imías
- Cajobabo
- Jauco
- Baitiquirí
- Tortuguilla
- Caimanera

Legend

- - - Provincial border
~~ Coastline

Isla de la Juventud

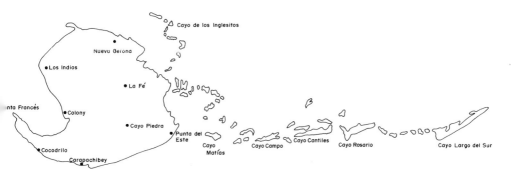

- Cayo de los Inglesitos
- Nueva Gerona
- Los Indios
- La Fé
- nta Francés
- Colony
- Cayo Piedra
- Punta del Este
- Cocodrilo
- Carapachibey
- Cayo Matías
- Cayo Campo
- Cayo Cantiles
- Cayo Rosario
- Cayo Largo del Sur

males were illustrated, with the exception of species with very marked sexual dichromatism or great age variation; in these cases, females or juveniles were also illustrated.

All of the references consulted for each species have been grouped under the subhead "Bibliographic Sources" to avoid intercalated quotations and to facilitate reading and understanding. I have provided known information only for Cuban populations of nonendemic species; where it was necessary to provide some data about non-Cuban populations, I have explained them in the text.

Family Iguanidae (iguanas, *chipojos, bayoyas, lagartijas*; iguanas, giant anoles, anoles, lizards)

There are four genera and sixty-two living species recognized for Cuba, distributed all over the island of Cuba, Isla de la Juventud, the archipelagoes of Los Colorados, Los Canarreos, Sabana-Camagüey, and Jardines de la Reina.

Genus *Leiocephalus* (*bayoyas, perritos de costa*; ground-dwelling lizards, curly-tailed lizards)

Leiocephalus Gray, 1827, *Phil. Mag.* 2(2):208. Type species: *Leiocephalus carinatus. Holotropis* Cocteau, in Duméril and Bibron, 1837, *Erp. Gén.* 4:264. Type species: *Leiocephalus carinatus.* Geographic range: Greater Antilles (Cuba; Hispaniola); Cayman Islands; Navassa Island; Bahamas Islands; Lesser Antilles (Martinique).

CONTENT AND DESCRIPTION

There are twenty-one recognized species. The Cuban members of the genus (six species, forty subspecies) are diagnosed as follows:

Small, medium, or intermediate size. Mean snout-vent length from 60.9 to 104.3 mm in males and from 50.9 to 96.3 mm in females. Robust, strong appearance; limbs adapted to running and digging, with long, slender digits. Long tail, with the final region bent laterally or upward, or coiled. Caudal autotomy. Lateral longitudinal skin fold in most species.

Zones or longitudinal body lines, alternating brown, gray, and white in the majority of species (fig. 18). Keeled, imbricate dorsal scales, sometimes multicarinate, lanceolate, or denticulate; generally smooth, circular ventral scales. Longitudinal middorsal crest of large, very keeled scales. Three or four large, preauricular scales in males.

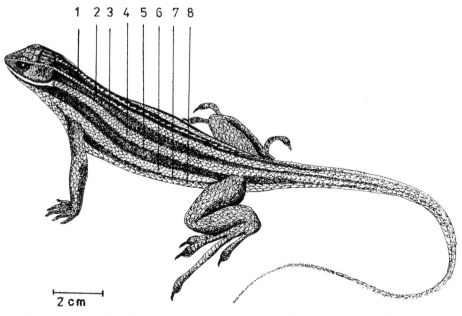

Fig. 18. Longitudinal lines or zones in a specimen of the genus *Leiocephalus*. The first (represented by the number 1) is at the middorsal line. The remaining ones, numbered consecutively in ascending order, are counted from the middorsal line to the ventral region.

Leiocephalus carinatus ("*perrito de costa*"; curly-tailed lizard; plate 1)

Leiocephalus carinatus Gray, 1827, *Phil. Mag.* 2(2):208. Holotype BMNH 1946.8.29.75. Type locality: Cuba, restricted to La Habana (Schwartz and Ogren 1956). *Holotropis microlophus* Cocteau, in Duméril and Bibron, 1837, *Erp. Gén.* 4:264. Lectotype MNHN 2392. Type locality: Cuba. *Leiocephalus macleayi* Gray, 1845, *Cat. Lizards Brit. Mus.*:218. Syntypes BMNH 1946.8.10.58, BMNH 1946.8.11.82. Type locality: Cuba.

Geographic Range

Isla de Cuba; Isla de la Juventud; Archipiélago de los Jardines de la Reina; Cayo Conuco, Archipiélago de Sabana-Camagüey; Bahamas Islands; Cayman Islands.

Localities (map 5)

Pinar del Río: San Vicente, Viñales (Cochran 1934); Península de Guanahacabibes; Cueva del Cable, Viñales (Schwartz 1959b). *La Habana and Ciudad de La Habana:* northern coast of both provinces (Cocteau and Bibron 1838); Jardín Zoológico de La Habana (Silva Lee 1985); Escaleras de Jaruco (González González 1989). *Matanzas:* Camarioca (Gundlach 1880); Playa Larga, Península de Zapata (Garrido 1980*b*); Playa Girón; Caleta Buena (Martínez Reyes 1994*c*); Bacunayagua (Soto Ramírez 1994); El Cenote (L.R.S.). *Cienfuegos:* Cienfuegos (Schwartz and Ogren 1956). *Sancti Spíritus:* Casilda; Sierra de Trinidad (Schwartz 1959*b*). *Villa Clara:* Cayo Conuco (Garrido 1975*a*); Cayo Fragoso (Martínez Reyes 1998). *Ciego de Ávila:* Archipiélago de los Jardines de la Reina (Cayo Caballones; Cayo Grande) (Cochran 1934); Cayo Coco (Martínez Reyes 1998). *Camagüey:* Archipiélago de los Jardines de la Reina (Cayo Cachiboca); Bahía de Nuevitas (Schwartz 1959*b*); Playa Santa Lucía (Garrido and Jaume 1984); Cayo Sabinal; Cayo Romano (Martínez Reyes 1998). *Las Tunas:* Las Tunas (Díaz Castillo *et al.* 1991). *Holguín:* Banes (Schwartz 1959*b*); Gibara (Garrido and Jaume 1984); Guardalavaca (L.R.S.). *Granma:* Río Bayamo (Gundlach 1867); Cabo Cruz (Cochran 1934); Jiguaní (Garrido and Jaume 1984). *Santiago de Cuba:* Santiago de Cuba (Barbour 1914); Las Cuevas; La Mula (Rodríguez Schettino 1985); Siboney (Fong and del Castillo 1997); Parque Baconao (L.R.S.). *Guantánamo:* Guantánamo (Cochran 1934); Maisí (Schwartz and Ogren 1956); Tortuguilla; Baitiquirí; Imías (CZACC). *Isla de la Juventud:* Caleta Grande; Punta del Este; Puerto Francés; Sierra de Casas (Schwartz 1959*b*); Cayo Inglés (Silva Lee 1985).

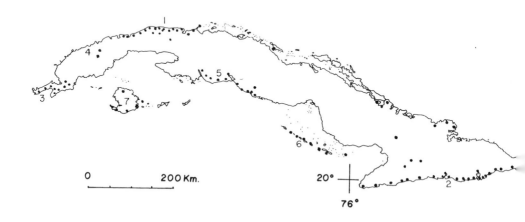

Map 5. Geographic distribution of *Leiocephalus carinatus* in Cuba. *1, L. c. carinatus; 2, L. c. aquarius; 3, L. c. zayasi; 4, L. c. mogotensis; 5, L. c. labrossytus; 6, L. c. cayensis; 7, L. c. microcyon.*

CONTENT

Polytypic species; thirteen subspecies are recognized, of which seven live in Cuba:

Leiocephalus carinatus carinatus

Leiocephalus carinatus Gray, 1827, *Phil. Mag.* 2(2):208. Type locality: Cuba, restricted to La Habana (Schwartz and Ogren 1956). *Leiocephalus carinatus carinatus:* Barbour and Shreve, 1935, *Proc. Boston Soc. Nat. Hist.* 40(5):360. Geographic range: northern coast of Cuba, from Cabañas to Varadero.

Leiocephalus carinatus aquarius

Leiocephalus carinatus aquarius Schwartz and Ogren, 1956, *Herpetologica* 12(2):100; holotype ChM 55.1.62. Type locality: Aguadores, near of Santiago de Cuba, Oriente province, Cuba. Geographic range: southern coast of Granma, Santiago de Cuba and Guantánamo provinces, from Cabo Cruz to Punta de Maisí.

Leiocephalus carinatus zayasi

Leiocephalus carinatus zayasi Schwartz, 1959, *Reading Public Mus. Art. Gallery Sci. Publ.* 10:9; holotype AMNH 77756. Type locality: northern shore of Ensenada de Cortés. Geographic range: Península de Guanahacabibes, from Ensenada de Corrientes to 10 km *SE* of Cayuco (nowadays Manuel Lazo).

Leiocephalus carinatus mogotensis

Leiocephalus carinatus mogotensis Schwartz, 1959, *Reading Public Mus. Art Gallery Sci. Publ.* 10:12; holotype AMNH 77755. Type locality: Cueva del Cable, San Vicente, Pinar del Río province, Cuba. Geographic range: San Vicente; Valle del Ancón; Valle de las Dos Hermanas.

Leiocephalus carinatus labrossytus

Leiocephalus carinatus labrossytus Schwartz, 1959, *Reading Public Mus. Art Gallery Sci. Publ.* 10:33; holotype AMNH 77757. Type locality: 5 km *SE* of Pasa Caballo, Las Villas province, Cuba. Geographic range: southern region of Matanzas, Cienfuegos and Sancti Spíritus provinces, from Península de Zapata to Casilda, including Sierra de Trinidad.

Leiocephalus carinatus cayensis

Leiocephalus carinatus cayensis Schwartz, 1959, *Reading Public Mus. Art Gallery Sci. Publ.* 10:38; holotype AMNH 77758. Type locality: Faro de Cayo Cachiboca, Jardines de la Reina, Camagüey province, Cuba. Geographic range: Cayos Cachiboca, Caballones, Grande, and others in Archipiélago de los Jardines de la Reina.

Leiocephalus carinatus microcyon

Leiocephalus carinatus microcyon Schwartz, 1959, *Reading Public Mus. Art Gallery Sci. Publ.* 10:43; holotype AMNH 81271. Type locality: Caleta Grande, Isla de Pinos. Geographic range: Isla de la Juventud.

DESCRIPTION

The species is characterized in Cuban territory as follows:

Medium or intermediate size. Mean snout-vent length from 98.5 to 109.1 mm in males and from 90.9 to 100.9 mm in females.

Pale or dark brown color, with or without transverse yellow or brown bands, ill defined, on the body and limbs. Without zones or longitudinal lines. Generally with yellow and brown transverse bands. Pale yellow or gray gular region, with very dark brown or black streaks. Yellow or gray ventral region, with brown or black spots. Green iris.

Without lateral longitudinal skin fold. Vertically oval and large ear opening. Very keeled, imbricate, lanceolate dorsal scales; smooth or weakly keeled, imbricate, circular, ventral scales, near the same size as dorsals. Three or four large, preauricular scales. Dorsal and caudal crest with very keeled and large scales. Postcloacal scales similar to the others.

Six pairs of macrochromosomes and eleven pairs of microchromosomes (2n = 34).

BIBLIOGRAPHIC SOURCES

Gray (1827): original description. Cocteau and Bibron (1838): detailed description; coloration; habitat data; behavior. Gundlach (1867, 1880): detailed description of the coloration in life; habitat data; behavior; abundance. Barbour (1914): habitat data; abundance. Barbour and Ramsden (1916a, b): habitat data. Barbour and Ramsden (1919): description of an adult; drawing of the dorsal head scales (pl. 14, fig. 11); habitat data; behavior; abundance; description of juveniles. Dunn (1920): scalation. Barbour (1930b, 1935, 1937): habitat data; behavior. Conant and Hudson (1949): longevity. Alayo Dalmau (1951): habitat data; abundance. Alayo Dalmau (1955): coloration; drawings of the dorsal head scales (pl. 5, fig. 1, no. 61), of the dorsal scales (pl. 5, fig. 1, no. 62); habitat data; behavior; abundance. Schwartz and Ogren (1956): detailed description; habitat data; behavior; abundance. Schwartz (1959b): detailed description; photographs of the dorsal head surface (figs. 4, 7, 10, and 13), of the gular region scales (figs. 5, 8, 11, and 14), of the dorsal region (figs. 6, 9, and 12); habitat data; behavior; description of juveniles; sexual variation. Petzold (1962): reproduction in captivity. Buide (1966, 1967): habitat data. Etheridge (1966a): drawing of a caudal vertebra (fig. 7A), of the second digit of the hindlimb (fig. 8A), of a caudal scale (fig. 9A); taxonomy. Lando and Williams (1969): coloration. Schwartz (1969): synonyms. Coy Otero (1970): nematodes. Coy Otero and Baruš (1979a): nematodes. Coy Otero and Lorenzo (1982): helminths. Buide (1985): two photographs; habitat data; behavior; abundance; feeding. Rodríguez Schettino (1985): habitat data. Rodríguez

Schettino and Martínez Reyes (1985): habitat data; abundance; thermic relations. Schwartz and Henderson (1985): photograph of a specimen (fig. 72). Silva Lee (1985): two photographs; behavior; reproduction; description of the juveniles. Armas (1987): feeding. Martínez Reyes and Rodríguez Schettino (1987): cannibalism. Chamizo Lara, Camacho, Rivalta *et al.* (1989): electrophoretic patterns of nine enzymatic systems. Chamizo Lara, Camacho, Torres *et al.* (1989): electrophoretic patterns of seven proteic systems. Porter *et al.* (1989): karyotype; photograph of the ideogram (fig. 1A). Sexto (1989): three photographs of several specimens; feeding; reproduction. Alarcón Chávez *et al.* (1990): abundance. Chamizo Lara *et al.* (1991): electrophoretic patterns of eleven proteic systems of muscle and liver; photograph of a specimen. Schwartz and Henderson (1991): feeding; reproduction. Martínez Reyes (1994a): habitat data; feeding. Martínez Reyes (1994c): abundance. Chamizo Lara and Rodríguez Schettino (1996): habitat data. Cubillas Hernández and Polo (1997): abundance. Fong and del Castillo (1997): habitat data.

MORPHOLOGICAL VARIATION

Without metachromatism. Coloration is individual and geographically quite variable, as are the arrangement and intensity of transverse bands on body and tail, and gular streaks.

Geographic Variation

The seven Cuban subspecies are characterized as follows:

Leiocephalus carinatus zayasi (Guanahacabibes): dark brown color, with yellow and pale brown isolated scales. Brown and dark brown bands on tail, not well defined. Pale yellow gular region, with very dark streaks. Mean snout-vent length of 106.6 mm in males and of 90.9 mm in females.

L. c. mogotensis (Pinar del Río): bluish gray color, with blue isolated scales. Brown and yellowish brown transverse bands on tail. Yellow gular region with black spots. Mean snout-vent length of 98.5 mm in males and of 99.5 mm in females.

L. c. carinatus (La Habana, Ciudad de La Habana, and Matanzas provinces): yellowish brown color, with dark brown transverse bands on body, limbs, and tail. Pale yellow gular region, with gray streaks and spots. Mean snout-vent length of 108.9 mm in males and of 100.9 mm in females.

L. c. labrossytus (central region): yellowish brown color, with yellow isolated scales and brown transverse bands. Dark brown and ochraceous transverse bands on tail. Gray gular region, with black dashes and spots. Mean snout-vent length of 103.2 mm in males and of 93.8 mm in females.

L. c. cayensis (Archipiélago de Jardines de la Reina): gray color, with green isolated scales and greenish yellow transverse bands on tail. Pale yellow gular region, with gray spots. Mean snout-vent length of 109.9 mm in males and of 97.4 mm in females.

L. c. aquarius (southern region of eastern provinces): grayish yellow color, with longitudinally aligned black dots on dorsum and dorsal crest. Post-auricular and suprascapular black spots. Black and yellowish brown transverse bands on tail. Gray gular region, with dark gray spots. Mean snout-vent length of 99.8 mm in males and of 91.8 mm in females.

L. c. microcyon (Isla de la Juventud): dark brown color, with transversely aligned yellow scales. Brown and yellow transverse bands on tail. Yellow gular region, with dark brown streaks. Snout-vent length of the holotype (male) of 125.0 mm and shorter in females (Schwartz 1959*b*).

The populations that live at Cayo Conuco, Playa Santa Lucía, and Gibara have not been assigned to any subspecies since they have not been taxonomically studied.

Sexual Variation

There are no sexual differences in color patterns in any of Cuban subspecies; males do not have differentiated postcloacal scales. That is why sexual variation is not easily detected. Only in size can some differences be seen (table 7.1).

Age Variation

The coloration of juveniles is similar to that of adults. Juveniles can have different colors in the gular region, but with the same patterns, and the

Table 7.1. Comparison of the snout-vent length *(SVL)*, head length *(HL)*, and tail length *(TL)* between male and female adults of *Leiocephalus carinatus. N*, number of specimens measured; *X*, arithmetic mean; *CV*, coefficient of variation; *m*, minimum; *M*, maximum; *t*, estimated Student test; *p*, probability of error. All measures in mm.

Sex	N	X	CV	m	M	t	p
				SVL			
Males	30	104.3	13.56	47.3	133.2	2.78	<0.01
Females	34	96.3	8.92	78.4	115.8		
				HL			
Males	29	28.5	3.50	15.2	32.6	5.01	<0.01
Females	34	26.2	8.84	21.2	31.4		
				TL			
Males	21	121.1	29.69	67.2	185.2	0.53	n.s.
Females	26	125.4	14.66	64.0	154.8		

change takes place in individuals with 50.0 mm of snout-vent length. Their mean snout-vent length is 39.0 mm, which is about one-third of the adult size. They are subspecifically alike.

Natural History

This species lives in coastal rocky zones of *"diente de perro,"* where it seeks shelter in the interstices, crevices, and cavities of rocks. It also inhabits sandy coasts with an abundance of sea grapes (*Coccoloba uvifera*), where it uses dead leaves for refuge and as food-searching sites. It is found not only in coastal places but also in *"mogotes"* and low uplands with rocks, out-croppings, *"casimbas,"* and cavities that can be used as shelters. Apparently, elevation above sea level is not what constrains its geographic distribution, but type of substrate.

It also uses human constructions such as old buildings and roadsides. Generally, it takes refuge for sleeping under stones, fallen logs, or dead leaves, inside hollows of rocks or tree trunks, or by burying itself in sand. If a refuge is uncovered between 8:00 A.M. and 10:00 A.M., some of these liz-ards can be seen sleeping, crowded together (L. Moreno, pers. comm.).

It usually basks on stones, rocks, and even tree trunks, on which it can easily climb. Moreover, it can walk on walls and cliffs. The mean height above ground at which individuals at Península de Guanahacabibes were observed was 0.8 m. However, in other places of the country, this species has been seen at heights of up to 6 m above ground, on tree trunks or rocks. It is a diurnal heliothermic species that spends many hours basking near shelters, on stones or logs, along roadsides, on roads, or in other open areas. Body temperatures of two individuals captured at Península de Guana-hacabibes during the wet season were 33.2 and 35.6°C, while air and sub-strate temperatures were 29.4 and 31.4°C, respectively, which indicates that attainment of corporal heat from environment is efficient.

The greatest number of active individuals can be observed during hours of intense heat, either basking on the most exposed rocks or foraging, as well as during the last hours of the afternoon. Adults usually stay buried in sand or hidden in their refuges in the dry season, and only active juveniles are usually found during the warmest hours. It is a very wary species, with agile movements, which quickly escapes. It stays near shelters and at the first signal of danger hides for a while, appearing again afterward. If it is surprised far from the refuge, it runs short distances before stopping, looks around, and runs again. Young individuals spend little time in selecting refuges; they can hide under stones or logs and in rock, wall, or cliff cavities. When they are surprised, they try to remain unnoticed by partially burying themselves in sand or staying motionless.

In captivity this species spends almost all its time under stones or buried in sand when there is not enough light and heat in the terrarium; however, when such conditions are available, it sprawls over stones and logs.

It was called *roquet* in the last century because of its fearless attitude when attacked and its habit of bearing its tail curled over it back, like the French dogs of this name. Later it was given the name *perrito de costa* for the same reason.

The species feeds on arthropods. Ants seem to be the most commonly eaten insects; it also can eat cooked meals. In the stomach contents of twenty-four specimens from *L. c. cayensis* were found sixteen types of arthropods, one "perrito de costa," and abundant vegetable material, especially fruits. At Playa Jibacoa, La Habana province, the stomach contents were composed of 95.9% ants and 3.4% coleopterans; at Playa Girón, Matanzas province, of 73.8% ants, 13.1% vegetable material, and the remainder other arthropods. According to one curious but factual account, several "perritos" went to eat cooked hens' eggs, which were offered at a house in Gibara, Holguín province. The lizards that inhabit the Jardín Zoológico de La Habana eat sweet-meats and remnants of food offered other species, at the same time capturing the insects they find; cannibalism has occurred on at least one occasion. In terraria they have accepted cockroaches (*Periplaneta americana*) and other large insects, guava (*Psidium guajaba*), "*cundeamor*" (*Momordica charantia*), and pieces of horseflesh. These data indicate that *L. carinatus* is a facultative omnivorous species, with a wide spectrum of consumption possibilities.

There is little known about reproduction in this species. Formerly, young were collected only since the second half of July during the summer of 1957, for it was believed that eclosion began in this month. However, based on bimonthly observations conducted over a two-year period (from May 1984 to April 1986) at the Jardín Zoológico de La Habana, juveniles are present throughout the year, although the greatest numbers were observed in January, August, and September. This seems to indicate that under those conditions, reproduction is continuous, increasing in some months.

The eggs are yellowish white with leathery shells; they have been observed in May and July. Mean lengths and widths of three eggs measured after several years of preservation in 70% ethanol were 19.8 and 10.9 mm, respectively. One egg measuring 25.1 x 11.5 mm was found in the oviduct of a female with a snout-vent length of 102.0 mm collected in July. M. Martínez Reyes (pers. comm.) has observed up to nine oviductal eggs of similar size, which indicates that clutches may be multiple. Eggs are buried in earth or sand. Those observed in captivity measured 16.3–17.5 mm x 10.5–11.8 mm;

incubation lasted from 91 to 101 days. The mean snout-vent length of hatchlings was 35.0 mm.

It is a long-lived species that has lived in captivity, according to known information, up to three years and nine months.

This is an abundant species in natural populations. It was found with a relative abundance of 1.9 individuals/hour in June 1990, in the coastal shrubwood of El Narigón on the northeastern coast of La Habana province. This is similar to what was observed in the cliffs of Península de Guanahacabibes (1.8) and greater than what was found in the coastal shrubwood of the latter locality (0.6) in the same season of the year. At Península de Zapata, the relative abundance varied between 1.0 and 3.0 individuals/hour, depending on the locality (Playa Larga, Playa Girón, or Caleta Buena). Eight individuals were introduced into the National Zoological Park, where the calculated density was 238.1 ind./ha after five years; thus the species shows a great capacity for adaptation and propagation.

Currently ten species of parasitic nematodes are known for *Leiocephalus carinatus: Parapharyngodon cubensis, Cyrtosomum scelopori, C. longicaudatum, Physaloptera squamatae, Skrjabinoptera (Didelphysoma) prhynosoma, Oswaldocruzia lenteixeirai, Porrocaecum sp.* larvae, *Abbreviata* sp. larvae, *Cyrnea* sp. larvae, and Physalopteridae gen. sp. In addition, there are two species of ectoparasite acari: the ticks *Amblyomma albopictum* and *A. torrei*.

Leiocephalus cubensis (bayoya, *iguanita;* little iguana; plate 2)

Tropidurus (Leiolaemus) cubensis Gray, 1840, *Ann. Mag. Nat. Hist.* 5:110. Holotype BMNH 98 a. Type locality: Cuba, restricted to the vicinity of Guanabacoa, La Habana province, Cuba (Schwartz 1959a). *Holotropis microlophus* Cocteau and Bibron, 1838 (*part.*), *Hist. Fís. Pol. Nat. Isla de Cuba, Reptiles* 4:56. Lectotype MNHN 239. Type locality: Cuba. *Holotropis vittatus* Hallowell, 1857, *Proc. Acad. Sci. Philadelphia* 8:151. Holotype unlocated. Type locality: Cuba. *Leiocephalus vittatus:* Boulenger, 1855 (*part.*), *Cat. Lizards Brit. Mus.* ed. 2, 2:163. *Leiocephalus cubensis:* Stejneger, 1917, *Proc. U.S. Natl. Mus.* 53(2205):273. *Leiocephalus raviceps* Cochran, 1934 (*part.*), *Smithsonian Misc. Coll.* 92(7):39.

Geographic Range

Isla de Cuba (from Artemisa to Sagua de Tánamo); Isla de la Juventud; Archipiélago de Sabana-Camagüey and Jardines de la Reina.

LOCALITIES (MAP 6)

Pinar del Río: Cayo Juan García (Varona and Garrido 1970). *La Habana and Ciudad de La Habana:* La Habana; Madruga (Barbour 1914); Guanabacoa; Boca de Jaruco; Caimito (Schwartz 1959*a*); Artemisa (Schwartz and Thomas 1975); Escaleras de Jaruco (González González 1989); E Quibú River (Martínez Reyes 1994*b*); Playa Jibacoa; Vedado (CZACC). *Matanzas:* Cárdenas (Gundlach 1880); Matanzas (Barbour 1914); Ciénaga de Zapata (Schwartz 1959*a*); Península de Hicacos (Buide 1966); Soplillar; Playa Larga (Garrido 1980*b*). *Cienfuegos:* Cienfuegos (Barbour 1914); Ciego Montero; Soledad; Cruces; Rodas; Cumanayagua (Schwartz 1959*a*). *Sancti Spíritus:* Trinidad; Casilda (Schwartz 1959*a*). *Ciego de Ávila:* Salto de la Tinaga (Cochran 1934); Morón; Ciego de Ávila; Loma de Cunagua (Schwartz 1959*a*); Cayo Coco (Garrido 1976*a*). *Camagüey:* Camagüey (Barbour 1914); Cayo Cachiboca (Cochran 1934); Sierra de Cubitas (Schwartz 1959*a*). Holguín: Sagua de Tánamo (Schwartz 1959*a*). *Santiago de Cuba:* Santiago de Cuba (Alayo Dalmau 1951). *Granma:* Bayamo (Barbour 1914); Jiguaní; Birama (Schwartz 1959*a*). *Isla de la Juventud:* Caleta Grande; Paso de Piedras; Puerto Francés; Jacksonville; Punta del Este; Cayos Campo and Hicacos (Schwartz 1959*a*).

CONTENT

Polytypic endemic species; five subspecies are recognized:

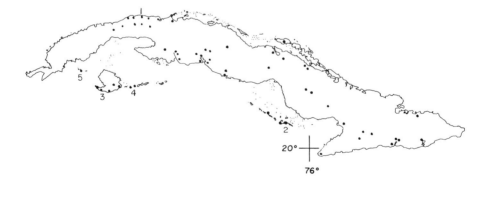

Map 6. Geographic distribution of *Leiocephalus cubensis.* 1, *L. c. cubensis;* 2, *L. c. paraphrus;* 3, *L. c. gigas;* 4, *L. c. pambasileus;* 5, *L. c. minor.*

Leiocephalus cubensis cubensis
Leiocephalus cubensis Gray, 1840, *Ann. Mag. Nat. Hist.* 5:110. Type locality: Cuba, restricted to the vicinity of Guanabacoa, La Habana province, Cuba (Schwartz 1959a). *Leiocephalus cubensis cubensis:* Schwartz, 1959, *Bull. Florida State Mus.* 4(4):107. Geographic range: from Artemisa to Sagua de Tánamo.

Leiocephalus cubensis paraphrus
Leiocephalus cubensis paraphrus Schwartz, 1959, *Bull. Florida State Mus.* 4(4):111; holotype AMNH 78005. Type locality: southernmost point of large, unnamed key 3 km W of Cayo Cachiboca lighthouse, Jardines de la Reina, Camagüey province, Cuba. Geographic range: Archipiélago de los Jardines de la Reina, Camagüey province.

Leiocephalus cubensis gigas
Leiocephalus cubensis gigas Schwartz, 1959, *Bull. Florida State Mus.* 4(4):113; holotype AMNH 81056. Type locality: Caleta Grande, Isla de Pinos. Geographic range: Isla de la Juventud (south of Ciénaga de Lanier).

Leiocephalus cubensis pambasileus
Leiocephalus cubensis pambasileus Schwartz, 1959, *Bull. Florida State Mus.* 4(4):118; holotype AMNH 81068. Type locality: Cayo Hicacos, Archipiélago de los Canarreos, La Habana province, Cuba. Geographic range: Archipiélago de los Canarreos (Cayos Campo and Hicacos).

Leiocephalus cubensis minor
Leiocephalus cubensis minor Garrido, in Varona and Garrido, 1970, *Poeyana* 75:18; holotype IZ 2754. Type locality: Cayo Juan García, Cayos de San Felipe, Archipiélago de los Canarreos, Pinar del Río province, Cuba. Geographic range: known only from the type locality.

DESCRIPTION

Medium size. Snout-vent length from 60.5 to 87.5 mm in males and from 54.3 to 74.5 mm in females.

Brown dorsal region and limbs, with zones 1, 2, and 3 generally of equal coloration; some dark brown scales arranged as ill-defined transverse bands on the dorsal region of the tail. Yellowish white or grayish labials. Dark brown postorbital stripe, which discontinuously extends to the lateral region of the body (zone 5), where dark and pale brown alternate. Very narrow zones 4 and 6, ivory in color. Zone 7 (lateral inferior) narrow and dark brown. Laterally and below zone 7, an orange or reddish region with brown or green vertical dashes that reach the ventral region. Gular and ventral regions grayish yellow. Some subspecies with a gular pattern of dark brown streaks, spots, and dots. Brown iris.

With longitudinal lateral skin fold. Vertically oval ear opening. Keeled, imbricate, and lanceolate dorsal scales; smooth, imbricate, circular ventral scales; both scales near the same size. Three large, triangular, preauricular scales. Enlarged postcloacal scales in males.

Six pairs of macrochromosomes and ten pairs of microchromosomes (2n = 32).

Bibliographic Sources

Gray (1840): original description. Gundlach (1867): habitat data; altitudinal distribution. Gundlach (1880): description of the coloration of a live individual; habitat data; altitudinal distribution. Barbour (1914): habitat data; abundance. Barbour (1916): habitat data; coloration. Barbour and Ramsden (1916b): abundance. Barbour and Ramsden (1919): description of the coloration of a live adult; drawing of dorsal head scales (pl. 14, fig. 12); habitat data; abundance. Dunn (1920): scalation. Barbour (1930b, 1935, 1937): habitat data. Alayo Dalmau (1955): drawings of the dorsal head scales (pl. 4, fig. 4, no. 63; pl. 7, fig. 4, no. 12), of the lateral head scales (pl. 4, fig. 4, no. 64), of the dorsal scales in head length (pl. 4, fig. 4, no. 65); abundance. Schwartz (1959a): drawing of the hemipenis (fig. 3); photographs of a specimen in dorsal view (figs. 4 and 5); detailed description; sexual variation; habitat data; evolution. Collette (1961): habitat data. Buide (1966): habitat data; abundance. Etheridge (1966c): morphology. Baruš and Coy Otero (1969a): parasites. Coy Otero (1970): nematodes. Varona and Garrido (1970): habitat data. Coy Otero and Baruš (1979a, b): nematodes. Garrido (1980b): dispersion. Coy Otero and Lorenzo (1982): helminths. Milera (1984): shelters. Buide (1985): photograph of a specimen; coloration; habitat data. Polo and Moreno (1986): feeding, behavior, and reproduction in captivity. Martínez Reyes and Fernández García (1988): feeding; habitat data. Chamizo Lara, Camacho, Rivalta et al. (1989): electrophoretic patterns of nine enzymatic systems. Chamizo Lara, Camacho, Torres et al.(1989): electrophoretic patterns of seven proteic systems. Chamizo Lara et al. (1991): electrophoretic patterns of eleven proteic systems of muscle and liver; photograph of a specimen. Leal Díaz and Morales Palmero (1991): karyotype; photographs of diakinesis, metaphase I, and the mitotic metaphase. Chamizo Lara and Moreno (1994): morphometry. Garrido (1994): distribution. Martínez Reyes (1994a): habitat data; feeding. Martínez Reyes (1994b): reproduction. Martínez Reyes (1994c): abundance. Martínez Reyes and Fernández García (1994): habitat data; feeding. Chamizo Lara and Rodríguez Schettino (1996): habitat data.

MORPHOLOGIC VARIATION

Without metachromatism. Color patterns are individually and geographically variable, as are length of the postorbital stripe and body color; the scalation varies little.

GEOGRAPHIC VARIATION

The five subspecies are characterized as follows:

Leiocephalus cubensis cubensis (Isla de Cuba): brown color with greenish sheen. Very dark brown postorbital stripe, which is laterally discontinuous, with dark and pale brown alternating on a reddish ground with green and reddish brown dots. Blue gular region with black streaks and spots. Yellow ventral region. Mean snout-vent length of 87.5 mm in males and of 74.5 mm in females.

L. c. minor (San Felipe keys): brown color with greenish sheen. Dark brown postorbital stripe, paler onto the lateral region. An irregular black blotch up and behind the ear opening. Pale green gular region, with black dots. Yellowish white ventral region. Mean snout-vent length of 60.5 mm in males and of 54.3 mm in females.

L. c. gigas (Isla de la Juventud, southern region): greenish brown color. Black postorbital stripe to the groin. Yellow gular region, with dark gray longitudinal discontinuous streaks. Blue in the anterior part of the ventral region and yellow in the posterior part. Mean snout-vent length of 83.3 mm in males and of 72.7 mm in females.

L. c. pambasileus (Cayo Hicacos): pale greenish brown color. Black postorbital stripe to the groin. Pale yellow gular region, with black spots. Yellow ventral region. Mean snout-vent length of 83.4 mm in males and of 65.4 mm in females.

L. c. paraphrus (Archipiélago de los Jardines de la Reina): greenish brown color, with yellow and orange dots. Black postorbital stripe to the scapular region. Blue gular region without spots. Yellow ventral region with blue sheen. Mean snout-vent length of 79.5 mm in males and of 56.0 mm in females.

Populations living at Península de Zapata and Cayo Coco have not been assigned to any subspecies since they have not been taxonomically studied. Chamizo Lara and Moreno (1994) found that each of the thirty-one morphometric measurements they analyzed are not sufficient for sustaining the status of any subspecies.

Sexual Variation

The pattern of coloration is similar in both sexes. Females are generally distinguished by the presence of brown longitudinal zones—instead of black or dark brown ones, as males have—and a less-defined postorbital stripe. Males are distinguished by enlarged postcloacal scales. Differences in size are shown in table 7.2.

Age Variation

Juveniles have coloration similar to that of adult females. Males have enlarged postcloacal scales.

NATURAL HISTORY

This species lives in lowlands with little vegetation, such as sandy coasts and some keys, sugarcane fields, sisal fields, and others. It was also observed in the mountains in the last century, but nowadays it is not found in uplands. Usually it does not enter forests but lives at the edges of forests and roadways, and on roads that cross them. It also lives in urban zones, gardens, house yards, and cemeteries. It is found in keys, under crawling vegetation that grows over the sand, in open areas, and in ecotones with coastal shrubwood or in such vegetation types.

It is a quite terrestrial species that usually stays on ground or sand, on dead leaves, or on a little high rock. The ground is used more often than stony terrains. It seeks shelter by burying itself in earth or sand, within crawling plants, or under dead leaves or stones. It also searches for refuge inside crevices and hollows of paved roads in urban places.

Table 7.2. Comparison of snout-vent length *(SVL)*, head length *(HL)*, and tail length *(TL)* between male and female adults of *Leiocephalus cubensis* (symbols as in table 7.1).

Sex	N	X	CV	m	M	t	p
				SVL			
Males	21	78.2	19.52	42.3	106.0	2.93	<0.01
Females	21	64.8	21.74	42.2	106.3		
				HL			
Males	21	21.3	17.53	12.2	27.7	3.81	<0.01
Females	21	17.6	13.54	12.2	21.8		
				TL			
Males	20	125.4	21.36	72.6	155.3	2.76	<0.01
Females	16	101.9	23.06	52.5	139.0		

A diurnal heliothermic species, it spends many hours basking; however, it leaves the refuge only when the air temperature is relatively high. On cloudy days and days with low temperatures, it stays hidden in its usual refuge until noon approaches. In captivity it begins to come out of its shelter after about 8:00 A.M., if the terrarium begins to receive enough light and heat. Most of the time it stays on perches, and after 5:00 P.M. it buries itself, passing through the soil head first. After October and during the dry season, it leaves the refuge only every three or four days, remaining outside for a little while to eat.

It is an easily frightened species that runs away quickly. To escape it runs short distances to reach any shelter or buries itself rapidly in sand. This has led some researchers to believe that it does not have the ability to orient itself and find its own refuge. When it is guided to a place with obstacles, it does not find the exit and can be easily captured (L. Moreno, pers. comm.). It carries the last portion of the tail curled slightly sideward when running.

This species feeds on insects, although it can eat remnants of human food. Formicids made up 74.6% of the food contents in samples obtained during six months in Ciudad de La Habana, while 15.8% of the specimens had vegetable material, mainly seeds, in their digestive tracts. Furthermore, there were four cases of saurophagy, three of them of cannibalism. The species has eaten lepidopterans, orthopterans, blatopterans, *Tenebrio molitor* larvae, arachnids, annelids, oligochaetes, small lizards, and ground beef in captivity but has rejected young tree frogs (*Osteopilus septentrionalis*) and chilopods. These data indicate that *L. cubensis* is a facultative omnivorous species, exhibiting a greater consumption of insects and rejection of some kinds of animals.

The greatest reproductive activity has been reported in April and July, based on specimens captured from February to November 1985 in Ciudad de La Habana. Adult females had from one to four developing ovules, and ovulation occurred when one of these ovules reached between 11.5 and 13.5 mm in length. Of the females with oviductal eggs, 66.6% had two and 33.3% three, plus nonoviductal eggs of different sizes; thus it is believed that this species can lay several eggs at once. The length of the oviductal eggs varied from 13.5 to 18.0 mm (x = 16.1 mm).

Measurements and laying dates of four eggs deposited in the herpetological collection of the Instituto de Ecología y Sistemática were as follows: two eggs, collected in August 1950 in a flowerpot, measuring 18.0 x 10.9 and 16.3 x 9.4 mm; one egg, measuring 19.0 x 13.0 mm immediately before eclosion in September 1950; one egg laid by a captive female in July 1965, measuring 20.0 x 9.0 mm. The first two were measured after several years of preservation in 70% ethanol, which explains why they were smaller than the latter two eggs; however, their measurements correspond with those of oviductal eggs.

The reproductive behavior of a captive pair was as follows: the female began courtship with quick, vertical head bobs, to which the male responded in the same way; then the female crawled and the male began to come closer and pursue her. When the male succeeded in positioning himself on the female, he bit her at the side of the nape until she stopped; then he began slow body movements from back to front, placing his tail perpendicular and below the female, the position in which copulation was accomplished. When they separated, the female moved away rapidly and dragged her cloaca over the ground of the terrarium, while the male slowly retreated in the opposite direction.

In general, every population contains numerous individuals. Males are more conspicuous, maybe because of their greater size and vivid colors. The proportion of immature individuals is larger than that of adults in the first and final months of the year; during the wet season adults are dominant as a consequence of annual changes in reproductive activity in the locality of La Habana. The relative abundance at Playa Larga, Matanzas province, was 24.0 individuals/hour.

To date, nine species of parasitic nematodes are known for *Leiocephalus cubensis: Parapharyngodon cubensis, Cyrtosomum scelopori, C. longicaudatum, Physaloptera squamatae, Skrjabinoptera (Didelphysoma) phrynosoma, Trichospirura teixeirai, Oswaldocruzia lenteixeirai, Porrocaecum* sp., and *Abbreviata* sp. Moreover, there is an ectoparasitic mite, the tick *Amblyomma torrei*.

Leiocephalus raviceps (bayoya; plate 3)

Leiocephalus raviceps Cope, 1862, *Proc. Acad. Nat. Sci. Philadelphia* 14:183. Syntypes ANSP 8601-03, MCZ 10928, USNM 4162. Type locality: eastern Cuba, restricted to mountains near Guantánamo, Oriente province, Cuba (Gundlach 1880).

GEOGRAPHIC RANGE

Isla de Cuba (Cabo Cortés; Península de Hicacos; Gibara; Santiago de Cuba; Guantánamo).

LOCALITIES (MAP 7)

Pinar del Río: San Waldo (Schwartz and Garrido 1968*a*). *Matanzas:* Península de Hicacos (Schwartz 1960*a*). *Villa Clara:* Cayo Lanzanillo, Archipiélago de Sabana (Garrido 1973*d*). *Holguín:* Los Cocos, Gibara (Garrido 1973*d*). *Las Tunas:* W Puerto Padre (Armas 1987). *Santiago de Cuba:* La Socapa, Santiago de Cuba; Laguna de Baconao (Alayo Dalmau 1955); 18.2 km *E* of Siboney (Schwartz 1960*a*). *Guantánamo:* mountains of Guantánamo (Gundlach 1880); Monte Verde, Sierra de Yateras (Barbour and Ramsden 1919); north of Guantánamo Bay; Boquerón (Cochran 1934); Caimanera; Baracoa (Schwartz 1960*a*); Imías (Sampedro Marín *et al.* 1979); Tortuguilla (CZACC).

CONTENT

Polytypic endemic species; five subspecies are recognized:
Leiocephalus raviceps raviceps
Leiocephalus raviceps Cope, 1862, *Proc. Acad. Nat. Sci. Philadelphia* 14:183. Type locality: eastern Cuba, restricted to mountains near Guantánamo (Gundlach 1880). Geographic range: from Guantánamo Bay to the area north of Cajobabo. *Leiocephalus raviceps raviceps:* Schwartz, 1960, *Proc. Biol. Soc. Washington* 73:74.
Leiocephalus raviceps uzzelli
Leiocephalus raviceps uzzelli Schwartz, 1960, *Proc. Biol. Soc. Washington* 73:70; holotype AMNH 79321. Type locality: 18.2 km *E* of Siboney, Oriente province, Cuba. Geographic range: from La Socapa to Guantánamo Bay.
Leiocephalus raviceps kilinikowskii
Leiocephalus raviceps kilinikowskii Schwartz, 1960, *Proc. Biol. Soc. Washington* 73:77; holotype AMNH 83326. Type locality: 4.5 km *SW* of Varadero, Matanzas province, Cuba. Geographic range: Península de Hicacos.

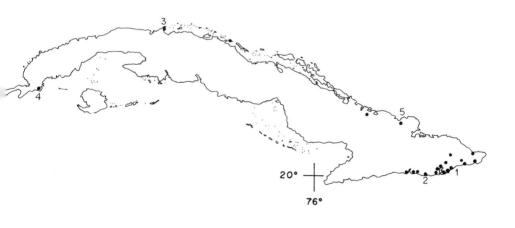

Map 7. Geographic distribution of *Leiocephalus raviceps. 1, L. r. raviceps; 2, L. r. uzzelli; 3, L. r. klinikowskii; 4, L. r. jaumei; 5, L. r. delavarai.*

Leiocephalus raviceps jaumei

Leiocephalus raviceps jaumei Schwartz and Garrido, 1967, *Proc. Biol. Soc. Washington* 82:24; holotype IZ 349. Type locality: San Waldo, 4 km N of Cortés, on the road from Cortés to Isabel Rubio, Pinar del Río province, Cuba. Geographic range: known only from the vicinity of the type locality.

Leiocephalus raviceps delavarai

Leiocephalus raviceps delavarai Garrido, 1973, *Torreia* 30:4; holotype IZ 2774. Type locality: Los Cocos, 6 km of Gibara, Oriente province, Cuba. Geographic range: known only from the type locality.

DESCRIPTION

Small size. Mean snout-vent length from 56.2 to 66.7 mm in males and from 45.8 to 53.8 mm in females.

Pale yellowish brown color on dorsum and limbs, with black dots and dashes and isolated yellow scales. Zones 1, 2, and 3 weakly defined. Zones 4 and 6 ivory-colored or pale brown. Black postorbital stripe to the ear opening; it continues (zone 5) brown with yellow dots and red dashes. Lateral region (zone 7) yellowish white. Pale brown iris.

With longitudinal lateral skin fold. Vertically oval ear opening. Keeled, imbricate, lanceolate, and small dorsal scales; smooth, circular ventral scales, larger than dorsals. Three or four large, triangular, preauricular scales. Enlarged postcloacal scales in males.

Six pairs of macrochromosomes and nine pairs of microchromosomes (2n = 30).

BIBLIOGRAPHIC SOURCES

Cope (1862): detailed description (original). Gundlach (1880): translation of the original description. Stejneger (1917): synonymy; phylogenetic relationships. Barbour and Ramsden (1919): drawing of the dorsal head scales (pl. 10, fig. 4); description of an adult. Dunn (1920): scalation. Cochran (1934): scalation. Alayo Dalmau (1955): drawing of the dorsal head scales (pl. 3, fig. 9, no. 4). Schwartz and Ogren (1956): habitat data. Schwartz (1960*a*): drawings of the body dorsal region (figs. 1–3), of the dorsal head scales (figs. 4–10); detailed description; sexual variation; habitat data. Buide (1966): habitat data; dispersion. Schwartz (1967): morphology. Schwartz and Garrido (1968*a*): detailed description. Lando and Williams (1969): description of the coloration in life of *L. r. uzzelli;* habitat data; behavior; feeding in captivity. Coy Otero (1970): nematodes. Garrido (1973*e*): detailed description; habitat data. Coy Otero and Baruš (1979*a*): nematodes. Sampedro Marín *et al.* (1979): habitat data; behavior; feeding. Buide (1985): photograph of a specimen. Silva Lee (1985): photograph of a specimen; habitat data; behavior; feeding; coloration. Porter *et al.* (1989): karyotype; photographs of the ideogram (figs. 1B and 1C), diakinesis (fig. 2A), and the secondary spermatocyte (fig. 2B).

Morphological Variation

Without metachromatism. Coloration is individually and geographically variable, as are distribution and color of the longitudinal lines; scalation is not too variable.

Geographic Variation

The five subspecies are characterized as follows:

Leiocephalus raviceps jaumei (Cortés, Pinar del Río province): pale brown color, with yellow dots. Black postorbital stripe with a vertical yellow bar. Lateral region (zone 5) brown with black dots. Pale gray gular region with some gray dots. White immaculate ventral region. Mean snout-vent length of 56.2 mm in males and of 45.8 mm in females.

L. r. klinikowskii (Península de Hicacos): yellowish white color, with black dorsal spots. Undefined longitudinal lines. Black postorbital stripe to the ear opening; it continues (zone 5) very pale brown with black dots. Gray ventral region. Mean snout-vent length of 58.6 mm in males and of 53.0 mm in females.

L. r. delavarai (Gibara): grayish brown color, with black dorsal spots. Black postorbital stripe to the ear opening. Dorsal part of the head pale yellow with black dots. Reddish lateral region (zone 5), with black and yellow dots. White or pale gray ventral region. Hindlimbs with reddish brown dots. Mean snout-vent length of 61.9 mm in males and of 50.9 mm in females.

L. r. uzzelli (W Guantánamo): pale brown color, with black or dark brown paired spots, disposed as weakly defined transverse bands. Gray postorbital stripe with brown spots. Dark brown lateral region (zone 5) with red and yellow scales. Pale gray ventral region. Mean snout-vent length of 63.2 mm in males and of 48.2 mm in females.

L. r. raviceps (E Guantánamo): yellowish brown color, with very small brown dots. Pale brown postorbital stripe. Red lateral region (zone 5) with isolated green scales. Pale gray ventral region. Mean snout-vent length of 66.7 mm in males and of 53.8 mm in females.

Populations living at Cayo Lanzanillo (Archipiélago de Sabana) and Baracoa have not been assigned to any subspecies since they have not been taxonomically studied.

Sexual Variation

Coloration is similar in both sexes; generally, females have more dorsal discontinuous longitudinal lines of dark brown color and have black dots in the gular region in some subspecies. Males are distinguished by their enlarged postcloacal scales. Size differences are shown in table 7.3.

Age Variation

Coloration of juveniles is similar to that of females. Males have enlarged postcloacal scales.

NATURAL HISTORY

The species lives in sandy coastal zones where the vegetation is scarce and xerophylic, with a predominance of grasses, cacti, spiny bushes, and some palms. It is not accustomed to entering the adjacent forests. The subspecies of Punta Hicacos is found only on loose sand and not on the stones that arise from it. At Imías, *L. r. raviceps* lives on the sandy ground of the coastal shrubwood, without reaching the contiguous sea grape forest. It takes refuge in burrows it makes in the sand or in existing burrows, under stones, fallen trunks, and even in anthropic places such as gardens and roadside ditches. It is a very elusive, fugitive species that escapes by running short distances before stopping, then running again until it reaches the refuge or buries itself in sand.

It is a diurnal heliothermic species that is found outside shelters during the intense heat of the day. Preferences for sunny or shaded places were

Table 7.3. Comparison of snout-vent length (*SVL*), head length (*HL*), and tail length (*TL*) between male and female adults of *Leiocephalus raviceps* (symbols as in table 7.1).

Sex	N	X	CV	m	M	*t*	p
			SVL				
Males	23	60.9	10.31	48.7	74.1	5.66	<0.01
Females	19	50.9	9.68	40.0	62.7		
			HL				
Males	23	15.9	10.61	12.3	19.7	5.84	<0.01
Females	19	13.4	6.16	12.0	15.8		
			TL				
Males	17	93.9	14.97	51.5	115.5	3.52	<0.01
Females	14	78.2	12.53	43.2	98.6		

similar at Imías: 45.2% of the individuals were in sunny places and 58.8% in shaded ones. The number of active individuals was larger between 10:00 A.M. and 11:00 A.M. and between 1:00 P.M. and 2:00 P.M. in August 1977. In general, this species begins to leave its refuges after 10:00 A.M. and returns at sunset. At Guantánamo it was observed exploring the area near the refuges, searching for insects among grasses and under small things, and foraging for food in sunny places during the morning and at noon.

The specimens captured at Imías had large amounts of vegetable material in their stomachs, mainly flowers, in addition to arthropods, principally insects of the orders Hymenoptera, Coleoptera, Lepidoptera, Hemiptera, and Homoptera but also scorpions, spiders, and mollusks. In captivity the species has eaten cockroaches, insect larvae, ants, and nocturnal butterflies. All of these data indicate that it is an omnivorous species, with a wide spectrum of consumption possibilities.

Very few isolated facts are known about reproduction. One pair was observed copulating on 6 March 1967, and a female captured at that time had an oviductal egg apparently ready for oviposition. An egg laid in June 1965, measuring 18.0 x 8.0 mm, was deposited in the herpetological collection of the Instituto de Ecología y Sistemática.

Populations of this species are scattered all over Cuban territory, with great distances between them, except for the subspecies *raviceps* and *uzzelli*, which are parapatric. However, each population in its geographic range is composed of numerous individuals living in relatively small areas.

To date, six species of parasitic nematodes are known for *Leiocephalus raviceps*: *Cirtosomum sclepori, Physaloptera squamatae, Skrjabinoptera (Didelphysoma) phrynosoma, Oswaldocruzia lenteixeirai, Porrocaecum* sp. larvae, and *Abbreviata* sp. larvae.

Leiocephalus macropus (*bayoya de la sierra*; plate 4)

Leiocephalus macropus Cope, 1862, *Proc. Acad. Nat. Sci. Philadelphia* 14:184. Syntypes USNM 25819–23, USNM 25825–26, USNM 12254 A-D, MCZ 10930; lectotype USNM 25919. Type locality: eastern Cuba, restricted to Monte Verde, Oriente province, Cuba (Stejneger 1917).

GEOGRAPHIC RANGE

Isla de Cuba (Península de Guanahacabibes; Bacunayagua; San Miguel de los Baños; Sierra de Trinidad; Loma de Cunagua; Playa Santa Lucía; Gibara; Baracoa; from Cabo Cruz to Maisí.

LOCALITIES (MAP 8)

Pinar del Río: Península de Guanahacabibes (Zug 1959); Rangel (Schwartz and Garrido 1967); El Mulo, Sierra del Rosario (Martínez Reyes 1995). *La Habana:* El Narigón, Puerto Escondido (L.R.S.) *Matanzas:* San Miguel de los Baños; Bacunayagua; Cueva del Negro (Garrido 1979). *Sancti Spíritus:* Finca La Pastora, Trinidad (Zug 1959). *Villa Clara:* Cayo Santa María (Martínez Reyes 1998). *Ciego de Ávila:* Loma de Cunagua (Zug 1959); Morón (Schwartz and Henderson 1985). *Camagüey:* Playa Santa Lucía (Schwartz and Garrido 1967). *Holguín:* Sagua de Tánamo (Barbour and Ramsden 1919); Miranda; Banes (Schwartz and Garrido 1967); Los Cocos; Gibara; Guardalavaca; Bahía de Naranjo (Garrido 1973d); Mayarí; Cayo Saetía (Garrido and Jaume 1984); Cupeyal del Norte; Farallones de Moa; Calentura Arriba; La Melba (Estrada *et al.* 1987). *Granma:* Guisa (Gundlach 1880); Cabo Cruz, Río Puerco; Punta Hicacos (Cochran 1934). *Santiago de Cuba:* Santiago de Cuba (Barbour 1914); Río Magdalena (Cochran 1934); Siboney (Alayo Dalmau 1951); Ocujal; southern slope of Pico Turquino (Hardy 1958b); Aserradero (Schwartz and Garrido 1967); Cayo Damas (Garrido 1975a); La Mula (Rodríguez Schettino 1985). *Guantánamo:* Monte Líbano; Baracoa (Barbour and Ramsden 1919); La Casimba, Maisí (Garrido 1967); southern slope of Sierra del Purial; Cayo Güín; Toa; Río Duaba; Bahía de Miel; Boca del Río Yumurí (Schwartz and Garrido

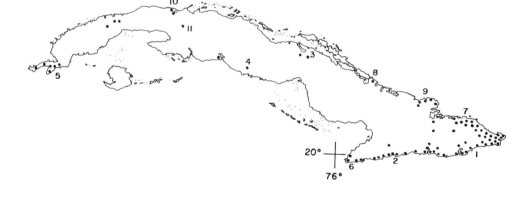

Map 8. Geographic distribution of *Leiocephalus macropus. 1, L. m. macropus; 2, L. m. inmaculatus; 3, L. m. hoplites; 4, L. m. hyacinthurus; 5, L. m. koopmani; 6, L. m. phylax; 7, L. m. asbolomus; 8, L. m. aegialus; 9, L. m. lenticulatus; 10, L. m. felinoi; 11, L. m. torrei.*

1967); Boca de Jauco (Sampedro Marín *et al.* 1979); Cupeyal (Garrido and Jaume 1984); Cayo Probado, springs of Jaguaní River; Ojito de Agua; springs of Yarey River; Cayo Fortuna; Alto del Yarey (Estrada *et al.* 1987); Tortuguilla; Cajobabo (CZACC); Imías (L.R.S.).

CONTENT

Polytypic endemic species; eleven subspecies are recognized:

Leiocephalus macropus macropus

Leiocephalus macropus Cope, 1862, *Proc. Acad. Nat. Sci. Philadelphia* 14:184. Type locality: eastern Cuba, restricted to Monte Verde, Oriente province, Cuba (Stejneger 1917). Geographic range: southern coast of the Santiago de Cuba and Guantánamo provinces, from Bahía de Santiago de Cuba to Cajobabo. *Leiocephalus macropus macropus:* Zug, 1959, *Proc. Biol. Soc. Washington* 72:144.

Leiocephalus macropus inmaculatus

Leiocephalus macropus inmaculatus Hardy, 1958, *J. Washington Acad. Sci.* 48(9): 294; holotype USNM 138412. Type locality: vicinity of Ocujal, Oriente province, Cuba. Geographic range: southern coast of Santiago de Cuba province, from Magdalena River to Bahía de Santiago de Cuba.

Leiocephalus macropus hoplites

Leiocephalus macropus hoplites Zug, 1959, *Proc. Biol. Soc. Washington* 72:140; holotype AMNH 78020. Type locality: 12 mi. *E* Morón, Loma de Cunagua, Camagüey province, Cuba. Geographic range: known only from the type locality.

Leiocephalus macropus hyacinthurus

Leiocephalus macropus hyacinthurus Zug, 1959, *Proc. Biol. Soc. Washington* 72: 145; holotype AMNH 78015. Type locality: Finca La Pastora, 2 km *NW* of Trinidad, Las Villas province, Cuba. Geographic range: known only from the type locality.

Leiocephalus macropus koopmani

Leiocephalus macropus koopmani Zug, 1959, *Proc. Biol. Soc. Washington* 72:146; holotype MCZ 55541. Type locality: near the base of Cabo Corrientes, Pinar del Río province, Cuba. Geographic range: Península de Guanahacabibes to the vicinity of Manuel Lazo.

Leiocephalus macropus aegialus

Leiocephalus macropus aegialus Schwartz and Garrido, 1967, *Reading Public Mus. Art Gallery Sci. Publ.* 14:15; holotype AMNH 83255. Type locality: Playa Santa Lucía, Camagüey province, Cuba. Geographic range: known only from the type locality.

Leiocephalus macropus phylax

Leiocephalus macropus phylax Schwartz and Garrido, 1967, *Reading Public Mus. Art Gallery Sci. Publ.* 14:17; holotype IZ 556. Type locality: Verreón, near Cabo Cruz, Oriente province, Cuba. Geographic range: southern coast of Granma province, from Vereón and Cabo Cruz to Punta Hicacos.

Leiocephalus macropus asbolomus

Leiocephalus macropus asbolomus Schwartz and Garrido, 1967, *Reading Public Mus. Art Gallery Sci. Publ.* 14:30; holotype IZ 568. Type locality: El Guayabo, Baracoa, Oriente province, Cuba. Geographic range: northern coast of Holguín and Guantánamo provinces, from Guardalavaca to Yumurí River, in Baracoa.

Leiocephalus macropus lenticulatus

Leiocephalus macropus lenticulatus Garrido, 1973, *Torreia* 30:10; holotype IZ 2782. Type locality: Los Cocos, 6 km of Gibara, Oriente province, Cuba. Geographic range: known only from the type locality.

Leiocephalus macropus felinoi

Leiocephalus macropus felinoi Garrido, 1979, *Poeyana* 188:2; holotype IZ 4751. Type locality: shore of the Bacunayagua River, Matanzas province, Cuba. Geographic range: known only from the type locality.

Leiocephalus macropus torrei

Leiocephalus macropus torrei Garrido, 1979, *Poeyana* 188:7; holotype IZ 3725. Type locality: slopes at the Paredones River, San Miguel de los Baños, Matanzas province, Cuba. Geographic range: vicinity of the type locality.

DESCRIPTION

Medium size. Mean snout-vent length from 60.2 to 88.0 mm in males and from 48.3 to 64.2 mm in females.

Grayish brown color in the dorsal region and limbs, with some very pale brown isolated scales. Zones 1, 2, and 3 generally of equal coloration. Dark brown or black postorbital stripe (zone 5); this becomes very dark at the scapular region, as a black patch with a yellow vertical bar, and continues to the hindlimbs; it is bordered by an upper longitudinal pale yellow stripe (zone 6). Black or dark brown postnuchal band of variable shape and intensity according to the subspecies. Two black or dark brown sacral dots. Gular and ventral region gray, pale yellow or pink according to the subspecies. Black paramedian and diagonal streaks on the gular region in some subspecies. Brown iris.

With longitudinal lateral skin fold. Vertically oval and large ear opening. Keeled, imbricate, lanceolate, dorsal scales; smooth, imbricate, circular ventral scales, larger than dorsals. Three large, triangular, preauricular scales. Enlarged postcloacal scales in males.

BIBLIOGRAPHIC SOURCES

Cope (1862): detailed description (original). Gundlach (1880): coloration in a free-living adult. Barbour and Ramsden (1919): description of a juvenile; drawing of the dorsal head scales (pl. 10, fig. 5); behavior. Dunn (1920): scalation. Barbour (1930b, 1935, 1937): abundance. Alayo Dalmau (1951): abundance. Alayo Dalmau (1955): drawings of the dorsal head scales (pl. 3, fig. 9, no. 5; pl. 7, fig. 7, no. 66), of the dorsal scales in the head length (pl. 7, fig. 7, no. 67); abundance. Schwartz and Ogren (1956): habitat data. Hardy (1958a): drawing of the mental region (figs. 1A and B); scalation. Hardy (1958b): drawings of the dorsal head scales (fig. 5 A-L), of the postorbital stripe (fig. 2, 1–5); detailed description; coloration. Zug (1959): drawing of the gular region (figs. 1, 3, 5 and 7), of the postorbital stripe (figs. 2, 4, 6 and 8); detailed description; coloration. Schwartz (1967): morphology. Schwartz and Garrido (1967): drawings of the first half of the body in lateral view (figs. 2 A-D, 4 A-D), of the gular region (fig. 3 A-D); detailed description; morphology; evolution; phylogenetic relationships. Garrido and Schwartz (1968): habitat data. Cerny (1969a, b): ectoparasites. Lando and Williams (1969): coloration of a free-living adult; habitat data; behavior; abundance; feeding in free life and captivity. Coy Otero (1970): nematodes. Garrido (1973c): habitat data. Garrido (1973e): detailed description. Coy Otero and Baruš (1979a): nematodes. Garrido (1979): photographs of the gular region (fig. 1 A-B), of the postorbital stripe (fig. 2 A-B), of the dorsal pattern (fig. 3 A-B); detailed description; habitat data. Sampedro Marín et al. (1979): habitat data; behavior; abundance; feeding. Coy Otero and Lorenzo (1982): helminths. Buide (1985): photograph of a specimen. Rodríguez Schettino (1985): habitat data; altitudinal distribution. Rodríguez Schettino and Martínez Reyes (1985): habitat data; behavior; thermic relations; abundance. Schwartz and Henderson (1985): photograph of a specimen (fig. 73). Silva Lee (1985): photograph of a specimen; feeding; coloration. Fong and del Castillo (1997): habitat data.

MORPHOLOGICAL VARIATION

Without metachromatism. Coloration pattern is individually and geographically variable, as are the presence of the postorbital stripe, the postnuchal band, and the gular region pattern. Scalation is not too variable.

Geographic Variation

The eleven subspecies are characterized as follows:
  Leiocephalus macropus koopmani (Guanahacabibes): greenish gray color. Brown postorbital stripe, poorly defined, with a greenish yellow vertical

bar in the scapular patch. Dark brown postnuchal band; two brown sacral dots. Brown head. Greenish yellow gular region, with black paramedian lines. Mean snout-vent length of 73.7 mm in males and of 64.2 mm in females.

L. m. felinoi (Bacunayagua): pale brown color, with dark brown spots. Black postorbital stripe with a yellow L-shaped bar in the scapular patch. Black oval postnuchal band; two black sacral dots. Terra-cotta-colored head. Yellow gular region with wide black paramedian lines. Ochraceous ventral region; rosaceous with yellow dots under the tail. Mean snout-vent length of 79.7 mm in males and of 62.6 mm in females.

L. m. torrei (San Miguel de los Baños): rosaceous brown color. Black postorbital stripe, with a yellow L-shaped bar in the scapular patch. Black postnuchal band, joined at its ends with the postorbital stripe, resembling a ring; two black sacral dots. Dark brown head. Yellow gular region with black paramedian lines. Orange ventral region with black dots in the thoracic region. Mean snout-vent length of 76.6 mm in males and of 54.8 mm in females.

L. m. hyacinthurus (Trinidad): pale brown color with black scale tips. Black postorbital stripe to the vertical yellow bar. Without postnuchal band or sacral dots. Purplish gular region with black paramedian and diagonal lines. Holotype (male) snout-vent length of 88.0 mm; females are not known.

L. m. hoplites (Loma de Cunagua): brown color with greenish and purplish sheens. Black postorbital stripe to the vertical yellow scapular bar. Black postnuchal band joined to the postorbital stripe as a ring; two black sacral dots. Yellow gular region with black paramedian and diagonal lines. Mean snout-vent length of 71.9 mm in males and of 63.0 mm in females.

L. m. aegialus (Playa Santa Lucía): brown color with greenish sheen and isolated yellow scales. Brown postorbital stripe with a vertical yellow bar over the weakly defined black scapular patch. Black postnuchal band divided into two parts; two black sacral dots. Gray gular region, without streaks, with some yellow scales. Holotype (male) snout-vent length of 70.0 mm.

L. m. lenticulatus (Gibara): pale brown color. Terra-cotta-colored postorbital stripe that is vanishing to the scapular region and has no vertical bar. Dark brown postnuchal band; two black sacral dots. Gray gular region with some black, scattered spots. Mean snout-vent length of 61.5 mm in males and of 48.6 mm in females.

L. m. phylax (Cabo Cruz): pale brown color with greenish sheen; some individuals with dark brown transverse lines. Terra-cotta-colored postorbital stripe to the scapular region, with a small vertical yellow bar. Absent or weakly defined brown postnuchal band; two black sacral dots only in females. Pink gular region with ill-defined black paramedian lines. Mean snout-vent length of 60.2 mm in males and of 48.3 mm in females.

L. m. inmaculatus (Ocujal): pale brown color. Without postorbital stripe; an ill-defined black scapular patch with a vertical yellow bar. Weakly defined black postnuchal band divided into two parts; without sacral dots. Gray gular region without lines. Mean snout-vent length of 65.3 mm in males and of 56.4 mm in females.

L. m. macropus (Guantánamo): pale greenish brown color. Black postorbital stripe with a yellow vertical bar over the scapular patch. Absent or small brown postnuchal band; two dark brown sacral small dots. Pale gray gular region without lines. Mean snout-vent length of 74.6 mm in males and of 59.9 mm in females.

L. m. asbolomus (Baracoa): grayish brown dorsal color and dark brown at sides, with some greenish yellow vertical lateral lines. Black postorbital stripe with a small yellow vertical bar over the scapular patch. Absent or small brown postnuchal band; two brown sacral dots. Pale gray gular region, without lines and with isolated pale yellow dots. Mean snout-vent length of 76.1 mm in males and of 57.3 mm in females.

Populations living in Sierra del Rosario, Cayo Saetía, Mayarí, and Cupeyal del Norte have not been assigned to any subspecies since they have not been taxonomically studied.

Sexual Variation

Color pattern is similar in both sexes; in eastern populations females lack the postorbital stripe. Generally, females have dark brown transverse lines in the dorsal region and the postorbital stripe is ill defined. Males are distinguished by enlarged postcloacal scales and their larger size (table 7.4).

Age Variation

Coloration of juveniles is similar to that of adults. Their snout-vent length ranges from 29.0 to 31.0 mm. Males have enlarged postcloacal scales.

NATURAL HISTORY

The species lives in semideciduous forests, generally alongside small streams and rivers or in coastal shrubwoods with a predominance of sea grapes (*Coccoloba uvifera*), on dead leaves. It perches on rocks, stones, and fallen logs and can even climb tree trunks up to 1 m above ground. Nevertheless, sandy, stony, ground or ground covered by dead leaves are the most

Table 7.4. Comparison of snout-vent length *(SVL)*, head length *(HL)*, and tail length *(TL)* between male and female adults of *Leiocephalus macropus* (symbols as in table 7.1).

Sex	N	X	CV	m	M	t	p
			SVL				
Males	40	71.2	14.54	53.0	90.5	7.00	<0.01
Females	39	56.8	13.44	39.0	67.3		
			HL				
Males	40	20.0	13.19	11.6	25.0	7.77	<0.01
Females	39	16.0	11.87	12.1	19.5		
			TL				
Males	36	116.8	21.73	55.3	161.2	4.38	<0.01
Females	31	92.4	20.86	45.8	123.1		

frequently used substrates. It is a fugitive species and is accustomed to running rapidly, bending the tail. It uses crevices of rocks as refuges, hides under dead leaves, or buries in sand or loose earth.

It is a diurnal nonheliothermic species. Most of the individuals of *L. m. macropus* at Boca de Jauco were found in shaded places, and this is usually observed in other populations as well. On the other hand, the greatest relative abundance was found between 10:00 A.M. and 11:00 A.M., which indicates that it takes refuge during hours of intense heat.

At Península de Guanahacabibes *L. m. koopmani* was found only on the rocky limestone floor of the semideciduous forest, during the wet season, after 10:00 A.M., in places where mean air and substrate temperatures were 29.0 and 27.4°C, respectively. Most individuals were observed in shaded places (54%), although the difference in the percentage of individuals under the sun (46%) was not very great.

During the night and on exceptionally cool mornings, it is found under loose coral pieces, stones, or fallen logs, but only in places where there was shade during the day.

It uses stones and trunks as sites of exposition in terraria; it takes refuge under stones, where it spends the night and the first hours of the morning.

This species forages during the day among dead leaves; its diet is based mainly on small cockroaches, larvae, moths, and nocturnal butterflies sheltered under dead leaves. On examination of the stomach contents of a sample of *L. m. macropus* from Boca de Jauco, a large amount of vegetable material was found, mainly flowers. Arthropods were also found, among which insects were widely represented; these were principally hymenopterans, coleopterans, lepidopterans, heteropterans, isopterans, and blatopterans, although acari, scorpions, spiders, chilopods, and mollusks were also present. The species has been fed in captivity with cockroaches, ants, grasshoppers, moths, and insect larvae; despite the great interest it has shown in earthworms, it has never eaten them even when deprived of other kinds of food; it procured water from small pots. These data indicate that this species is omnivorous, with a wide spectrum of consumption possibilities, although with a notable preference for insects.

Little is known about reproduction in this species. At La Mula, on the southern coast of Santiago de Cuba province, *L. m. inmaculatus* was collected in February, March, May, and September, and only in the first three months were gravid females found. The nonoviductal eggs of three females varied from 3.1 to 10.2 mm in length (x = 6.6 mm), with oviductal eggs measuring from 12.0 to 20.3 mm (x = 15.2 mm). All females had two oviductal eggs (M. Martínez Reyes, pers. comm.).

The different subspecies are geographically separated and cover all Cuban territory. There are numerous individuals in the places where they are found, with the exception of *L. m. felinoi, L. m. hyacinthurus,* and *L. m. aegialus,* from which only a few specimens have been found.

To date, six species of parasitic nematodes are known for *Leiocephalus macropus: Parapharyngodon cubensis, Cyrtosomum scelopori, Physaloptera squamatae, Oswaldocruzia lenteixeirai, Porrocaecum* sp. larvae, and *Abbreviata* sp. larvae. Moreover, there is a species of parasitic mite: the tick *Amblyomma torrei.*

Leiocephalus stictigaster (bayoya; plate 5)

Leiocephalus stictigaster Schwartz, 1959, *Bull. Florida State Mus.* 4(4):121. Holotype AMNH 77864. Type locality: Beach on Cabo Corrientes, Pinar del Río province, Cuba. *Holotropis microlophus* Cocteau and Bibron, 1838 (*part.*), *Hist. Fís. Pol. Nat. Isla de Cuba, Reptiles* 4:56. Lectotype MNHN 2392. Type locality: Cuba. *Leiocephalus vittatus* Boulenger, 1885 (*part.*), *Cat. Lizards Brit. Mus.* ed. 2, 2:163. *Leiocephalus cubensis* Barbour, 1916 (*part.*), *Ann. Carnegie Mus.* 10(2):304.

Geographic Range

Isla de Cuba (Cabo Corrientes; Sierras de los Órganos and del Rosario; Cienfuegos; Trinidad; Santa Clara; Camagüey; Santa Lucía; Gibara; Contramaestre; Sierra de la Gran Piedra); Isla de la Juventud; Archipiélago de Sabana-Camagüey.

LOCALITIES (MAP 9)

Pinar del Río: Península de Guanahacabibes; El Cayuco (Manuel Lazo); Herradura; Las Pozas; San Vicente (Schwartz 1959a). *Matanzas:* eastern Ciénaga de Zapata (Garrido 1980b); Playa Larga; Playa Girón; Caleta Buena (Martínez Reyes 1994c). *Cienfuegos:* Juraguá; Bahía de Cienfuegos (Schwartz and Garrido 1968b). *Sancti Spíritus:* Casilda (Schwartz and Garrido 1968b). *Villa Clara:* Cayo Santa María; Cayo Francés (Garrido 1973d). *Ciego de Ávila:* Cayo Coco (Schwartz *et al.* 1978); Cayo Guillermo (Martínez Reyes 1998). *Camagüey:* Playa Santa Lucía; south of the Sierra de Cubitas; Cayo Sabinal (Schwartz 1964b); Cayo Guajaba (Garrido *et al.* 1986); Cayo Paredón Grande (Martínez Reyes 1998). *Las Tunas:* Puerto Manatí; Bahía de Malagueta (Garrido and Jaume 1984). *Holguín:* Gibara (Schwartz and Garrido 1968b); Punta de Mulas (Garrido and Jaume 1984). *Granma:* Bayamo (Barbour 1914). *Santiago de Cuba:* Contramaestre; Sierra de la Gran Piedra (Schwartz and Garrido 1968b). *Guantánamo:* Tortuguilla; Cajobabo (L.R.S.). *Isla de la Juventud:* La Fe; Siguanea; Sierra de Casas; Caleta de Carapachibey (Schwartz 1959a).

CONTENT

Polytypic endemic species; twelve subspecies are recognized:
  Leiocephalus stictigaster stictigaster
Leiocephalus stictigaster stictigaster Schwartz, 1959, *Bull. Florida State Mus.*

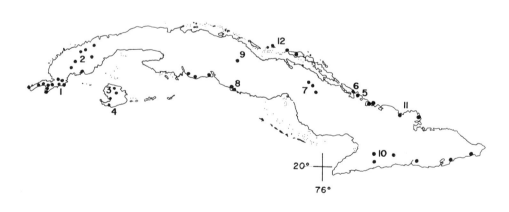

Map 9. Geographic distribution of *Leiocephalus stictigaster. 1, L. s. stictigaster; 2, L. s. sierrae; 3, L. s. exotheotus; 4, L. s. astictus; 5, L. s. lucianus; 6, L. s. parasphex; 7, L. s. ophiplacodes; 8, L. s. naranjoi; 9, L. s. lipomator; 10, L. s. celeustes; 11, L. s. gibarensis; 12, L. s. septentrionalis.*

4(4):123. Type locality: beach on Cabo Corrientes, Pinar del Río province, Cuba. Geographic range: Península de Guanahacabibes.

Leiocephalus stictigaster sierrae

Leiocephalus stictigaster sierrae Schwartz, 1959, *Bull. Florida State Mus.* 4(4): 126; holotype AMNH 77813. Type locality: San Vicente, Pinar del Río province, Cuba. Geographic range: Pinar del Río province, in Sierra de los Órganos, Sierra del Rosario, and lowlands.

Leiocephalus stictigaster exotheotus

Leiocephalus stictigaster exotheotus Schwartz, 1959, *Bull. Florida State Mus.* 4(4):130; holotype AMNH 81088. Type locality: 1.5 mi W Santa Fe, Isla de Pinos, La Habana province. Geographic range: Isla de la Juventud, north to the Ciénaga de Lanier.

Leiocephalus stictigaster astictus

Leiocephalus stictigaster astictus Schwartz, 1959, *Bull. Florida State Mus.* 4(4): 134; holotype AMNH 81095. Type locality: Caleta de Carapachibey, Isla de Pinos, La Habana province. Geographic range: Isla de la Juventud, south to the Ciénaga de Lanier.

Leiocephalus stictigaster lucianus

Leiocephalus stictigaster lucianus Schwartz, 1960, *Proc. Biol. Soc. Washington* 73:104; holotype AMNH 83583. Type locality: Playa Santa Lucía, Camagüey province, Cuba. Geographic range: known only from the type locality.

Leiocephalus stictigaster parasphex

Leiocephalus stictigaster parasphex Schwartz, 1964, *Quart. J. Florida Acad. Sci.* 17(3):212; holotype AMNH 92153. Type locality: Playa Bonita, eastern end of Cayo Sabinal, Camagüey province, Cuba. Geographic range: known only from the type locality.

Leiocephalus stictigaster ophiplacodes

Leiocephalus stictigaster ophiplacodes Schwartz, 1964, *Quart. J. Florida Acad. Sci.* 17(3):217; holotype AMNH 92771. Type locality: 2.7 mi SE Banao, Camagüey province, Cuba. Geographic range: serpentine savannas of the Camagüey province, to the area south of Sierra de Cubitas.

Leiocephalus stictigaster naranjoi

Leiocephalus stictigaster naranjoi Schwartz and Garrido, 1968, *Natl. Mus. Canada Nat. Hist. Papers* 37:3; holotype IZ 200. Type locality: Los Biasmones, Casilda, Las Villas province, Cuba. Geographic range: south of Sancti Spíritus and Camagüey provinces; vicinity of Juraguá; Casilda.

Leiocephalus stictigaster lipomator

Leiocephalus stictigaster lipomator Schwartz and Garrido, 1968, *Natl. Mus. Canada Nat. Hist. Papers* 37:11; holotype IZ 1230. Type locality: 3 km W of Santa Clara, Las Villas province, Cuba. Geographic range: known only from the type locality.

Leiocephalus stictigaster celeustes

Leiocephalus stictigaster celeustes Schwartz and Garrido, 1968, *Natl. Mus. Canada Nat. Hist. Papers* 37:14; holotype IZ 1182. Type locality: Contramaestre, Oriente province, Cuba. Geographic range: northern slope of Sierra Maestra, from the vicinities of Bueycitos to Contramaestre.

Leiocephalus stictigaster gibarensis

Leiocephalus stictigaster gibarensis Schwartz and Garrido, 1968, *Natl. Mus. Canada Nat. Hist. Papers* 37:18; holotype IZ 1236. Type locality: Gibara, Oriente province, Cuba. Geographic range: from the vicinities of Puerto Manatí to Punta de Mulas.

Leiocephalus stictigaster septentrionalis

Leiocephalus stictigaster septentrionalis Garrido, 1975, *Poeyana* 141:28; holotype IZ 3425. Type locality: Cayo Santa María, Archipiélago de Sabana-Camagüey, Las Villas province, Cuba. Geographic range: Cayo Santa María, Cayo Francés, Cayo Guillermo, in Archipiélago de Sabana-Camagüey.

DESCRIPTION

Medium size. Mean snout-vent length from 48.1 to 75.7 mm in males and from 46.0 to 64.1 mm in females.

Brown color in the dorsal region and limbs, with black isolated scales. Pale brown zones 1, 2, and 3. White or very pale brown zones 4 and 6. Brown postorbital stripe, generally with black borders; it continues (zone 5) brown with black, white, green, and orange dots. Lateral region (zone 7) pink, brown, or orange, with white, black, and red dots. Pale yellow gular region, with black or brown paramedian and diagonal continuous or discontinuous streaks. Pale yellow ventral region with black dots. Brown iris.

With longitudinal lateral skin fold. Vertically oval ear opening. Keeled, imbricate, lanceolate, dorsal scales; smooth, imbricate, circular ventral scales, a little larger than dorsals. Three or four large, triangular, preauricular scales. Enlarged postcloacal scales in males.

BIBLIOGRAPHIC SOURCES

Schwartz (1959*a*): original description; drawing of the hemipenis (fig. 6); four photographs each for one specimen in dorsal view (figs. 7–10); habitat data; evolution. Schwartz (1960*b*): detailed description; habitat data. Schwartz (1964*b*): drawings of the gular region (figs. 1–3); detailed description; phylogenetic relationships. Garrido and Schwartz (1968): habitat data; behavior; abundance. Schwartz and Garrido (1968*b*): drawings of the gular region (pl. 1. A–K); detailed description; habitat data. Garrido (1973*d*): habitat data. Garrido (1973*e*): habitat data; abundance. Garrido (1975*a*): photograph of the gular region (fig. 6). Coy Otero and Baruš (1979a): nematodes. Garrido (1980b): habitat data. Coy Otero and Lorenzo (1982): helminths. Rodríguez Schettino and Martínez Reyes (1985): habitat data; behavior; thermic relations; abundance; feeding; reproduction. Schwartz and Henderson (1985): photograph of a specimen (fig. 74). Silva Lee (1985): two photographs each for one specimen. Martínez Reyes *et al.* (1990): habitat data; behavior; thermic relations; feeding; reproduction in free life and captivity. Chamizo Lara *et al.* (1991): electrophoretic patterns of eleven proteic systems of muscle and liver; photograph of a specimen. Martínez Reyes (1994a): habitat data; feeding. Martínez Reyes (1994c): abundance. Chamizo Lara and Rodríguez Schettino (1996): habitat data.

Morphological Variation

Without metachromatism. Coloration is individually and geographically variable, as are the longitudinal lines and the gular patterns; scalation is not too variable.

Geographic Variation

The twelve subspecies are characterized as follows:

Leiocephalus stictigaster stictigaster (Península de Guanahacabibes): brown or pale brown weakly defined longitudinal zones, with isolate black dots. Dark brown postorbital stripe to the ear opening; it continues (zone 5) brown with many black dots. Pink lateral region. Pale gray gular region with black dots. Mean snout-vent length of 54.2 mm in males and of 50.4 mm in females.

L. s. sierrae (Pinar del Río): Pale brown, brown, and dark brown longitudinal lines, with black dots. Brown postorbital stripe, weakly defined; it continues (zone 5) dark brown with orange dots. Dark orange lateral region with black dots. Yellow gular region with black dots and dashes. Yellow ventral region with black dots. Mean snout-vent length of 70.9 mm in males and of 57.5 mm in females.

L. s. naranjoi (Casilda): dark brown, brown, and grayish brown longitudinal lines. Black postorbital stripe to the ear opening; it continues (zone 5) terra-cotta colored with red and green dots. Reddish brown lateral region, with green and black dots. Greenish gray gular region with dark gray discontinuous paramedian and diagonal lines. Yellow ventral region. Mean snout-vent length of 68.3 mm in males and of 54.3 mm in females.

L. s. lipomator (Santa Clara): grayish brown and dark brown longitudinal lines with black spots. Dark brown postorbital stripe (only the borders) to the ear opening; it continues (zone 5) reddish brown with green dots. Dark

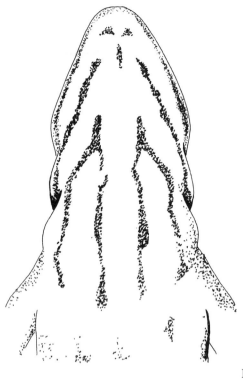

Fig. 19. Gular region of *Leiocephalus stictigaster* showing pattern of paramedian and diagonal gular lines.

0.3 cm

orange lateral region with green dots. Grayish white gular region with black continuous paramedian and diagonal lines. White ventral region with brown dots. Mean snout-vent length of 48.1 mm in (only two) males and of 53.9 mm in females.

L. s. ophiplacodes (Camagüey): pale brown and dark brown longitudinal lines. Brown postorbital stripe to the ear opening; it continues (zone 5) brown with orange flecks. Reddish orange lateral region. Ivory-colored gular region with black discontinuous paramedian and diagonal lines. Pale yellow ventral region with brown or orange dots. Mean snout-vent length of 67.7 mm in males and of 63.3 mm in females.

L. s. lucianus (Playa Santa Lucía): brown, pale brown, and white longitudinal lines. Dark brown postorbital stripe to the ear opening; it continues (zone 5) dark brown. Brown lateral region with black spots. Pale yellow gular region, with black paramedian and diagonal lines. White or pale yellow ventral region with brown or orange dots. Mean snout-vent length of 68.8 mm in males and of 46.0 mm in females.

L. s. gibarensis (Gibara): dark brown and pale brown longitudinal lines, with black longitudinal dashes. Dark brown postorbital stripe (only the borders); it continues (zone 5) dark brown with black and green dots. Pale yellow gular region, with black paramedian and diagonal lines. Pale yellow ventral region, with brown and reddish dots. Mean snout-vent length of 75.3 mm in males and of 52.6 mm in females.

L. s. celeustes (Contramestre): pale brown and dark brown longitudinal lines, with black spots. Dark brown postorbital stripe (only the borders); it continues (zone 5) brown with black longitudinal spots. Dark brown lateral region. Pale brown gular region, with black, paramedian lines shaped like inverted U. Pale brown ventral region, with thoracic brown or black dots. Mean snout-vent length of 75.5 mm in males and of 64.1 mm in females.

L. s. exotheotus (Isla de la Juventud, northern region): brown and yellow longitudinal lines. Without postorbital stripe; from the ear opening (zone 5), a red stripe with green dots. Red lateral region with yellow dots and a yellow inferior border. Pale yellow gular region, with black paramedian lines and isolated spots. Yellow ventral region, with or without black dots. Mean snout-vent length of 66.0 mm in males and of 56.1 mm in females.

L. s. astictus (Isla de la Juventud, southern region): pale brown, dark brown, and white longitudinal lines. Brown postorbital stripe, with black borders, to the ear opening; it continues (zone 5) brown with green isolated dots. Reddish lateral region with green dots. Grayish white gular region, with dark gray or black discontinuous paramedian and diagonal lines. Pale yellow ventral region. Mean snout-vent length of 60.1 mm in males and of 51.6 mm in females.

L. s. septentrionalis (Cayo Santa María): pale brown, brown, and grayish brown longitudinal lines. Brown postorbital stripe to the ear opening; it continues (zone 5) reddish brown, with white dots. Reddish brown lateral region, with white and red dots. White gular region, with dark brown paramedian and diagonal lines. Yellow ventral region with reddish dots. Mean snout-vent length of 63.5 mm in males and of 53.2 mm in females.

L. s. parasphex (Cayo Sabinal): dark brown and pale brown longitudinal lines. Dark brown postorbital stripe to the ear opening; it continues (zone 5) dark brown with red and orange flecks. Reddish orange lateral region. Grayish yellow gular region, with gray, discontinuous, and irregularly arranged paramedian and diagonal lines. Pale yellow ventral region. Mean snout-vent length of 63.7 mm in males and of 55.1 mm in females.

Populations living at Península de Zapata, Sierra de la Gran Piedra, and on the southern coast of Guantánamo province have not been assigned to any subspecies since they have not been taxonomically studied.

Sexual Variation

Coloration is similar in both sexes; generally, longitudinal lines in females are more defined than in males, which are distinguished by the enlarged postcloacal scales and their larger size (table 7.5).

Age Variation

Coloration of juveniles is similar to that of adult females. Males have enlarged postcloacal scales.

Natural History

This species lives in coastal zones in dry, sandy, and stony places where coastal vegetation or coastal shrubwood predominates; it also lives in inland semideciduous and coniferous forests. It is generally found in the lowlands, but it can inhabit uplands such as Sierra de los Órganos. The most commonly used substrate is always the ground, with either sand, stones, or earth, according to the locality. It takes refuge within hollows of ground rocks or buries in sand to hide or sleep. It is a fugitive species that quickly runs to its refuge or rapidly buries itself. The usual perches of males are a few exposed stones and fallen branches and trunks. Males compete for such places in terraria and once they are settled, they defend them.

It is a diurnal heliothermic and tigmothermic species that is accustomed to basking and resting on substrates previously heated by the sun. At Península de Guanahacabibes, 57.7% of observed individuals were found under direct sun. The mean body temperature of thirty-eight lizards, collected there during the wet season, was 34.5°C, while the means of air and

Table 7.5. Comparison of snout-vent length *(SVL)*, head length *(HL)*, and tail length *(TL)* between male and female adults of *Leiocephalus stictigaster* (symbols as in table 7.1).

Sex	N	X	CV	m	M	t	p
				SVL			
Males	51	66.4	15.69	41.0	88.1	7.03	<0.01
Females	48	54.1	12.05	34.4	76.4		
				HL			
Males	51	18.0	13.41	11.8	25.2	1.38	n.s.
Females	48	14.9	11.46	10.5	19.3		
				TL			
Males	41	112.8	16.53	48.1	152.0	4.39	<0.01
Females	31	96.4	13.57	47.3	121.4		

substrate temperatures were 30.8 and 32.2°C, respectively. The mean body temperature was statistically higher than the environmental ones and showed significant positive correlations with both, which indicates a direct dependence on the environment to achieve the needed body temperature. Once it is reached, the species looks for shaded places in the vegetation where it can feed and reproduce.

The greatest number of active adults was observed between 12:00 P.M. and 1:00 P.M.; most active juveniles were seen between 10:00 A.M. and 11:00 A.M. in the wet season. After 6:00 P.M. all individuals were in their nocturnal refuges.

The species takes refuge under stones or on the ground of the terrarium and goes out only when the air temperature increases, at about 10:00 A.M. If incandescent lamps are turned on in the terrarium to increase the supply of light and heat, it is observed to exhibit a marked preference for places near the lamps, even on hot days; if the lamps are not turned on, or on cool, cloudy days, it does not leave the refuge.

Differences between male and female diets were found by examining the stomach contents of eighty-one specimens of *L. s. stictigaster* from Península de Guanahacabibes, although in both sexes hymenopterans (formicids) and coleopterans were the most commonly eaten items. Altogether, eight orders of insects, insect larvae and pupae, spiders, acari, crustaceans, flowers, and fruits were found. In addition, saurophagy and cannibalism were evident, not only in the contents of the digestive tracts but also in the observed behavior of free-living animals. Formicids were found in 92.3% of the stomach contents of *L. s.* ssp. from Playa Girón, Matanzas province, while coleopterans composed 1.4% and crustaceans 4.8% of the ingested food. Most of the diet components were prey items easy to obtain while foraging over the ground, which is the typical strategy of species of the genus *Leiocephalus*. In captivity it has accepted larvae and adults of the rice moth (*Corcida cephalonica*) and other insects, besides wood lice, also known as pillbugs or sowbugs (Isopoda), and fruits such as guava (*Psidium guajava*) and "cundeamor" (*Momordica charantia*). When accustomed to taking food at a certain place in the terrarium, it also accepts small pieces of horseflesh.

The only known reproductive data for this species are those from Península de Guanahacabibes, where reproductive activity was observed only in April and July; during those months, 30.0 and 80.0% of the captured adult females, respectively, were gravid. Each had from one to four nonoviductal eggs measuring 3.0 to 8.5 mm (x = 5.7 mm) in diameter, and one or two oviductal eggs measuring 11.6 to 17.0 mm (x = 13.9 mm) in length. The majority of females (66.7%) had only one oviductal egg; the others had two of quite similar length.

Three females were captured alive in July and kept in separated terraria. All laid eggs in August and one of them laid two on the same day. These four eggs varied from 15.7 to 17.0 mm in length and the largest eclosioned within fifty-three days of being laid. Hatchling measurements were as follows: snout-vent length 21.4 mm; head length 8.6 mm; tail length 34.3 mm; and weight 0.5 g. Matings have not occurred in captivity; the only observations have been of displays by males in the presence of females, apparently a part of courtship behavior. In these displays the males made slight bobs and approached the females, which fled.

All populations are numerous in their geographic ranges; however, relative abundance seems to depend on habitat type, because on sandy coasts and in coastal shrubwoods there are more individuals than in semideciduous forests. At Península de Guanahacabibes, populations are much more numerous on cliffs and in coastal shrubwoods than in semideciduous forests on limestone rocks, and the relative abundance is higher in the wet season than in the dry. At Península de Zapata the relative abundance varied between 5.0 and 14.0 individuals/hour, depending on the locality (Playa Larga, Playa Girón, or Caleta Buena).

To date, five species of parasitic nematodes are known for *Leiocephalus stictigaster: Cyrtosomum sclepori, C. longicaudatum, Physaloptera squamatae, Oswaldocruzia lenteixeirai,* and *Porrocaecum* sp. larvae. Moreover, there is an ectoparasitic mite, the tick *Amblyomma albopictum.*

Leiocephalus onaneyi (plate 6)

Leiocephalus onaneyi Garrido, 1973, *Poeyana* 116:4; holotype IZ 2869. Type locality: top of Loma de Macambo, between San Antonio del Sur and Imías, Oriente province, Cuba.

GEOGRAPHIC RANGE

Isla de Cuba (known only from the type locality).

LOCALITIES (MAP 10)

Guantánamo: Loma de Macambo (Garrido 1973*a*).

CONTENT

Monotypic endemic species.

DESCRIPTION

Medium size, although it must be considered that adult males are not known.

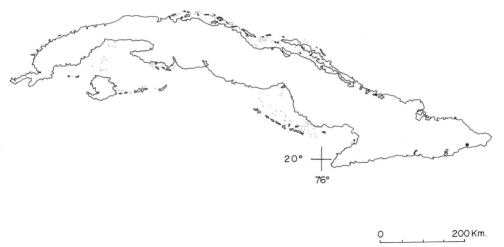

0 _____ 200 Km.

Map 10. Geographic distribution of *Leiocephalus onaneyi*.

All longitudinal lines present. Zones 1, 3, 5, and 7 dark brown; zones 2, 4, 6, and 8 very pale grayish brown. Dark brown postorbital stripe. Brown limbs. The longitudinal zones continue, narrower, to the first part of the tail. White gular and ventral region.

With longitudinal lateral skin fold. Vertically oval large ear opening. Keeled, lanceolate, imbricate, large dorsal scales; smooth, imbricate, circular ventral scales, near the same size as dorsals. Four or five large, triangular, preauricular scales. Enlarged postcloacal scales in the male.

Bibliographic Sources

Garrido (1973*a*): detailed description (original); coloration; photographs of the paratype in dorsal view (fig. 1), of the gular region scales (fig. 2), of the dorsal region of the head (fig. 3); habitat data; origin and evolution.

Morphological Variation

There are only three specimens: one subadult male, one female, and one juvenile; therefore, it is not possible to analyze variation. Coloration is similar in all of them; the subadult male has enlarged postcloacal scales.

Measurements of subadult male, female, and juvenile, in this order, are as follows: snout-vent length 68.0, 72.0, 37.9 mm; head length 15.5, 15.8, 9.7 mm; tail length broken; 110.0, 73.0 mm.

Natural History

This species lives only on the very erosive "diente de perro" limestone at the top of the Loma de Macambo, where the vegetation is xerophytic and the soil very scarce, existing only in the fissures of rocks with a predominance of spiny bushes or small leaves. This place is at 200 m above sea level on the southern coast of Guantánamo province.

This species has been collected on two occasions and has not been found again, despite recent searches; consequently, there are no data about its feeding, reproduction, behavior, or parasites. Apparently it is a diurnal heliothermic species, based on the arid place where it lives. Seemingly, its geographic range is limited.

Genus *Cyclura* (iguana)

Cyclura Harlan, 1824, *J. Acad. Nat. Sci. Philadelphia* 4:242, 250. Type species: *Cyclura carinata. Metapocerus* Wagler, 1830, *Natur. System. Amph.*:147. Type species: *Cyclura cornuta. Aloponotus* Duméril and Bibron, 1837, *Erp. Gén.* 4:190. Type species: *Cyclura ricordii.*

Geographic range: Greater Antilles (Cuba; Hispaniola; Jamaica); Cayman Islands; Anegada; Bahamas Islands; Turks and Caicos Islands.

CONTENT AND DESCRIPTION

Eight extant species are recognized. The genus is described as follows:

Very large size (snout-vent length to 745 mm in males and to 623 mm in females). In general, tail length is nearly the same as snout-vent length. Generally gray, brown, or black coloration; occasionally with transverse bands of contrasting shades. Red sclerotic, except in one species.

Dorsal crest of enlarged scales, like spines, which can be interrupted on scapular and sacral regions. The size of these spines varies according to their situation (nuchal, dorsal, or caudal); however, those over the tail are much smaller than the others. Verticilate tail, with verticils indicated by the spines of the caudal crest. Robust head, with protuberant enlarged scales. Large, vertically oval ear opening. From 26 to 80 femoral pores; generally aligned in a single row, interrupted at the midventral line.

Other features of the scalation can be found in the review of the genus by Schwartz and Carey (1977).

Cyclura nubila (iguana; plate 7)

Iguana (Cyclura) Nubila Gray, 1831, in Griffth's *Cuvier's Anim. Kingd.* 9:39. Holotype BMNH 1946.8.29.88. Type locality: "South America?" restricted to Cuba (Schwartz and Carey 1977). *Cyclura harlani* Duméril and Bibron, 1837 (*part.*), *Erp. Gén.* 4:218; syntypes MNHN A661, MNHN 2367. Type locality: Cuba. *Cyclura Macleayi* Gray, 1845, *Cat. Lizards Brit. Mus.*:190; holotype BMNH 1946.8.4.28. Type locality: Cuba. *Cyclura caymanensis* Barbour and Noble, 1916, *Bull. Mus. Comp. Zool.* 60:148; holotype MCZ 10534. Type locality: probably Cayman Brac, Cayman Islands. *Cyclura nubila:* Schwartz and Thomas, 1975, *Carnegie Mus. Nat. Hist. Special Public.* 1:113; holotype BMNH 1946.8.29. Type locality: "South America?"

GEOGRAPHIC RANGE

Greater Antilles (Isla de Cuba; Isla de la Juventud; Archipiélago de Sabana-Camagüey; Archipiélago de los Canarreos; Archipiélago de los Jardines de la Reina); Cayman Islands; introduced into Isla Magueyes, Puerto Rico.

LOCALITIES (MAP 11)

Pinar del Río: Cabo de San Antonio; Pan de Guajaibón; Rangel (Barbour 1914); Valle de Luis Lazo; Sumidero; Santa Cruz (Barbour and Ramsden 1919); Ensenada de Corrientes; El Veral (Garrido and Schwartz 1968); Cordillera de los Órganos (Baruš *et al.* 1969); Cayo Juan García (Varona and Garrido 1970); 45 km W of Cayuco (nowadays Manuel Lazo); 4.5 km W of San Vicente (Schwartz and Carey 1977). *La Habana:* Golfo de Batabanó (Barbour and Ramsden 1919); Jaimanitas (Cerny 1966); 3.2 km E of Boca de Jaruco (Schwartz and Carey 1977); Punta Jíjira (L.R.S.). *Matanzas:* key north of Cárdenas (Gundlach 1880); Punta Hicacos (Buide 1966); Cayos de las Cinco Leguas (Garrido and Jaume 1984); Cayo Cruz del Padre, Archipiélago de Sabana (Cubillas Hernández and Berovides Álvarez 1991). *Villa Clara:* keys north of Remedios (Barbour and Ramsden 1919); Cayo Francés, Cayo Santa María, Cayo Tío Pepe, Cayo Las Tocineras, Cayo Carenero, in Archipiélago de Sabana-Camagüey (Garrido 1973*d*); Cayo Bahía de Cádiz (Schwartz and Thomas 1975); Cayo Fragoso; Cayo Conuco (Martínez Reyes 1998). *Sancti Spíritus:* Trinidad; Cayo Macho de Tierra (Hardy 1956); Cueva de la Yagruma in Cayo Palma (Silva Taboada 1974); Punta Casilda (Schwartz and Carey 1977). *Ciego de Ávila:* San Juan de los Perros (Barbour and Ramsden 1919); Cayo Coco; Cayo Paredón Grande (Martínez Reyes 1998). *Camagüey:* Cayo Romano (Barbour 1914); Santa Cruz del Sur (Barbour and Ramsden 1919); Cayo Cabeza del Este; Cayo Cachiboca (Cochran 1934); key 3 km NW of Cachiboca (Schwartz and Carey 1977); Cayo Sabinal; Cayo Cruz (Martínez Reyes 1998); Cayo Juan Grín; Cayo Boca de Piedra Chica Este (L.R.S.). *Las Tunas:* Las Tunas (Díaz Castillo *et al.* 1991). *Granma:* Cabo Cruz; keys off Manzanillo (Barbour 1914); Belic (Barbour and Ramsden 1919). *Santiago de Cuba:* Santiago de Cuba (Barbour 1914); Ocujal (Cooper 1958); Aguadores; 27 km E of Siboney (Schwartz and Carey 1977); Las Cuevas, southern slope of Pico Turquino (L.R.S.). *Guantánamo:* Guantánamo; Cabo Maisí (Barbour 1914); near Baracoa (Barbour and Ramsden 1919); Kittery Beach, Naval Base of Guantánamo (Lando and Williams 1969); 13.8 km E of Imías (Schwartz and Carey 1977); Firing Point, Naval Base of Guantánamo (Phillips 1994); Boca de Boma (J. F. Milera, pers. comm.); Baitiquirí; Tortuguilla (L.R.S.). *Isla de la Juventud:* Isla de Pinos (Gundlach 1880); mountain ranges of the island (Barbour 1914); Cayo Cantiles; Cayo Matías (Cochran 1934); south of the island; Cayo la Piedra; Cayo Largo del Sur (Baruš *et al.* 1969); Cayo Ávalos; Cayo Hicacos; Cayo Majá (Schwartz and Thomas 1975); Punta del Este; Playa la Herradura; rocky beach between Bibijagua and Júcaro (Schwartz and Carey 1977); Cayo Rosario (Berovides Álvarez 1980); Cayo Campo (Estrada and Rodríguez 1985).

CONTENT

Polytypic species; three subspecies are recognized, one of them living in Cuba:

Cyclura nubila nubila (iguana; plate 7)

Cyclura nubila Gray, 1831, in Griffth's *Cuvier's Anim. Kingd.* 9:39. Holotype BMNH 1946.8.29.88. Type locality: "South America"?, restricted to Cuba (Schwartz and Carey 1977). *Cyclura harlani* Duméril and Bibron, 1837, *Erp. Gén.* 4:218; lectotype MNHN A661. Type locality: "Caroline." *Cyclura Macleayi* Gray, 1845, *Cat. Lizards Brit. Mus.*:190; holotype BMNH 1946.8.4.28. Type locality: Cuba. Geographic range: Isla de Cuba; Isla de la Juventud; Archipiélagos de Sabana-Camagüey, de los Canarreos, de los

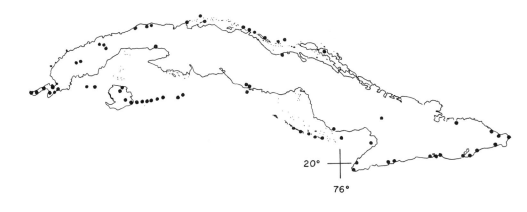

Map 11. Geographic distribution of *Cyclura nubila nubila.*

Jardines de la Reina; introduced on Isla Magueyes, Puerto Rico.

Description

The species is characterized in Cuba as follows:

Size very large. Maximal snout-vent length from 483 to 745 mm in males and from 405 to 623 mm in females.

Uniform gray or gray and pale brown color, with black diagonal transverse bands, spotted with pale brown. Pale brown limbs, black hands and feet; occasionally, with brown transverse bands. Pale brown head; gray labials; prefrontal shields very pale brown. Tail with ill-defined transverse bands, gray and pale brown. Nuchal and dorsal crests pale brown; color of caudal crest like that of transverse bands. Inferior lateral and ventral regions yellowish brown. Pale brown-and-gray dewlap. Dark brown iris. Red sclerotic. Juveniles with several oblique transverse bands, pale yellow and gray.

Thick limbs, with long, thin fingers and toes. One row of enlarged triangular scales on toes, the "comb." Stout body. Vertically oval, very large ear opening. Smooth, quadrangular, small dorsal scales; ventral scales smooth,

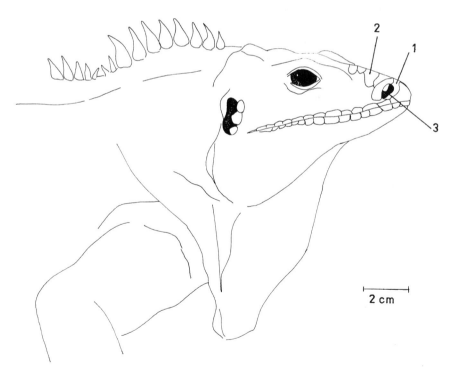

Fig. 20. Lateral view of *Cyclura nubila*. *1*, nasal scale; *2*, prefrontal shield; *3*, nostril.

quadrangular, a little imbricate, slightly larger than dorsals. Smooth, quadrangular, large limb scales, arranged as a mesh. Postcloacal scales not enlarged.

BIBLIOGRAPHIC SOURCES

Gray (1831): original description. Cocteau and Bibron (1838): detailed description; drawings of the body in lateral view, of the dorsal scales of the head, of two femoral pores and left hindlimb (pl. 8); feeding in captivity. Gundlach (1867, 1880): habitat data; behavior; feeding in captivity; human utilization. Barbour (1914): abundance. Barbour (1916): habitat data; abundance. Barbour and Ramsden (1919): coloration in life and scalation of adult and juvenile; photographs of the head in dorsal and lateral views (pl. 11, figs. 2 and 3); habitat data; behavior; human utilization. Pérez Vigueras (1934, 1956): ectoparasites. Pérez Vigueras (1935): parasitic nematode. Conant and Hudson (1949): longevity in captivity. Alayo Dalmau (1951, 1955): habitat data; coloration; anthropic damage. Aguayo (1951): paleontology. Buide (1951): behavior; feeding. Sutcliffe (1952): habitat data. Shaw (1954): description and photographs of juveniles (figs. 1 and 2); reproduction, feeding and behavior in captivity. Cooper (1958): habitat data. Buide (1966): abundance; anthropic damage. Cerny (1966, 1967): ectoparasites. Buide (1967): habitat data. Garrido and Schwartz (1968, 1969): coloration; habitat data; abundance. Baruš et al. (1969): helminths. Lando and Williams (1969): coloration in life; habitat data; behavior and feeding in captivity. Coy Otero (1970): nematodes. Varona and Garrido (1970): behavior; abundance. Garrido (1973b): habitat data. Garrido (1973d): anthropic damage. Buide González et al. (1974): anthropic damage. Schwartz and Carey (1977): coloration; scalation; drawing of the head in lateral view (fig. 7, up), in dorsal view (fig. 7, down); morphological variation; habitat data. Coy Otero and Baruš (1979a): nematodes. Berovides Álvarez (1980): feeding; predators. Coy Otero and Lorenzo (1982): helminths. Cruz (1984a, b): ectoparasites. Perera (1984): drawing of the left hindlimb (fig. 7); ecomorphology. Buide (1985): nine photographs of iguanas in Jardín Zoológico de La Habana; behavior; morphology; coloration; feeding; reproduction. Perera (1985a): behavior; abundance; habitat data. Perera (1985b): feeding. Rodríguez Schettino (1985): abundance. Rodríguez Schettino and Martínez Reyes (1985): habitat data; behavior; abundance; feeding; morphometric analysis. Silva Lee (1985): three photographs each of one iguana; habitat data; feeding; sexual variation; behavior; anthropic damage. Christian et al. (1986): thermoregulation; energetics. Guerra (1986): eight photographs of iguanas at Jardín Zoológico de La Habana; behavior. Rodríguez Schettino (1986b):

habitat data; abundance; feeding; anthropic damage. Rodríguez Schettino and Martínez Reyes (1986a): habitat data; abundance; feeding; anthropic damage. Christian and Torregrosa (1986): feeding in captivity. Christian (1987): feeding; reproduction. Martínez Reyes et al. (1989): bacteriosis in captivity. Rodríguez Schettino (1990): behavior and feeding in captivity. Christian and Lawrence (1991): microclimate of the nests. Christian et al. (1991): soil moisture in the nests. Cubillas Hernández and Berovides Álvarez (1991): refuges; density. Ehrig (1993): husbandry. Alberts (1994): behavior; conservation. Phillips (1994): behavior. Alberts (1995a): eggs. Alberts (1995b): husbandry; conservation. González et al. (1995): density. Berovides Álvarez (1995): conservation. Berovides Álvarez et al. (1996): conservation. Chamizo Lara and Rodríguez Schettino (1996): habitat data. Alberts and Grant (1997): body temperatures.

MORPHOLOGICAL VARIATION

Without metachromatism. Coloration varies a little individually and geographically. However, C. n. lewisi, from Grand Cayman, is typically a blue species. Scalation varies a little. Age is the main source of morphological variation in this species.

Geographic Variation

The Cuban subspecies is diagnosed the same as the species. Differs from the other two subspecies (C. n. caymanensis and C. n. lewisi) from the Cayman Islands in that it is the largest and adults have greenish or yellowish shades in the lateral regions. The male dorsal crest spines are very large; generally, the nuchal spines are the largest, although sometimes the caudal spines are longer and sharper.

In analyses of the scalation of eighteen specimens from Pinar del Río province, fourteen specimens from the remaining Isle of Cuba, and twenty-two specimens from Isla de la Juventud, it was determined that the mode of scales between frontal and interparietal scales was five for the Pinar del Río and Isla de la Juventud samples and six for the remaining Cuban ones. The first prefrontal and precanthal were in contact on one or both sides of the head in some of the Pinar del Río specimens, but not in the others. There are no other scalation or coloration characteristics that distinguish these samples.

Sexual Variation

In general, females are uniformly gray, with a reduced dewlap and dorsal spines less numerous and smaller than those of males, which have femoral pores. Differences in size are shown in table 7.6.

Table 7.6. Comparison of snout-vent length (*SVL*), head length (*HL*), and tail length (*TL*) between male and female adults of *Cyclura nubila* (symbols as in table 7.1). Measures after Perera (1984) and maxima after Schwartz and Carey (1977).

Sex	N	X	CV	m	M	t	p
			SVL				
Males	24	405.0	7.60	350.0	745.0	6.86	<0.01
Females	7	320.0	5.10	300.0	623.0		
			HL				
Males	24	76.7	8.90	75.0	110.9	7.44	<0.01
Females	7	56.1	6.60	50.0	86.1		
			TL				
Males	16	536.0	9.90	535.0	590.0	3.96	<0.01
Females	6	438.0	8.40	430.0	520.0		

Age Variation

Juveniles have from five to ten diagonal transverse bands on the body, which are pale brown with a black posterior border; these bands are broader at the middorsal line and continue, less defined, to the ventral region. The interspaces are black with brown and gray dots. The head is gray and pale brown. The limbs are black with pale brown oval spots and pale, dark brown transverse bands on fingers and toes. The spines of the dorsal crest are ill defined (plate 8).

The means of measurements of eight specimens from the herpetological collection of the Instituto de Ecología y Sistemática are as follows: snout-vent length 116.0 mm; head length 30.5 mm; tail length 194.8 mm. These means are about one-fourth of the adult mean size.

NATURAL HISTORY

This species lives preferentially on rocky coasts with "diente de perro," where there are many holes, fissures, and "casimbas" it uses as refuges. The xerophytic vegetation, composed of spiny grasses and bushes and succulent plants such as prickly pears and cacti, is used for shelter and food.

The species can be found in the majority of the keys of Cuban territory, in coastal xerophyllous vegetation, and on rocky or sandy ground. The iguana takes shelter in caves it digs with its limbs and on sandy coasts, where it can also be abundant. The predominant orientation of the refuge entrances at Cayo Cruz del Padre was southward. Such refuges contained from 0.6 to 1.1 iguanas/refuge.

The iguana can also be found in marshy zones such as mangrove forests, or in keys where this is the only available habitat. It is an excellent climber; it has been observed on trees or stones at great heights (up to 8 m above ground), where it usually perches to obtain food or to bask. It swims easily and can move from one key to another, if they are separated, by means of seawater channels.

In addition to occupying most of the coastal ecosystems, it lives in rocky uplands such as Sierra de Rangel, Pan de Guajaibón, Valle de Luis Lazo, Sumidero, and Santa Cruz in Pinar del Río province, although nowadays it is difficult to find in those places. At Isla de la Juventud, it lives on the rocky southern coasts and in the mountains, but it is more frequently found along the coasts.

It is a diurnal heliothermic species that spends several hours a day basking, remaining almost motionless. During the coldest months, in the dry season, it is not frequently seen on cloudy or rainy days; it leaves the refuge only when there is sun, at about noon, and individuals are rarely found in the morning or at sunset. However, during the wet season two activity peaks can be observed, one from 9:30 A.M. to 12:00 P.M. and the other from 3:30 P.M. to 6:00 P.M.; using this behavioral mechanism, the iguana is able to regulate its body temperature. It generally basks on the ground in open places such as coasts and keys, but in places where the vegetation is thick it uses high sites for basking, such as rocks and the crowns and branches of trees.

At Isla Magueyes, Puerto Rico, where this species has been successfully introduced, body temperature was constant between 9:15 A.M. and 6:00 P.M. (x = 38.6°C), with a high energy expenditure of 4,800 kJ/(ha x d). The skin surface temperatures of eight captive juveniles varied from 27.2 to 36.1°C, while internal body temperatures ranged from 24.5 to 32.3°C.

This is a fugitive species that rapidly escapes at the sign of any danger. It escapes to refuges by running short distances, stopping and running again until it reaches one; if does not find a nearby refuge, it faces the pursuer, opening the mouth threateningly and producing a harsh sound by expelling air. Such a confrontation can result in a fight in which it tries to bite the opponent. This behavior is usually observed in males, since females generally try to escape quickly. Juveniles flee toward refuges; when they are near the seashore, they often go to the water and hide among the seaweeds (*Thalassia testudinum*).

Four species of plants were found in the stomach contents of four individuals from Cayo Rosario, Archipiélago de los Canarreos, captured in July; three of them were the most common plants of the key. In the vicinity of Cayo Largo del Sur, Archipiélago de los Canarreos, it was found that feeding was variable, depending on the season of the year and the availability of food. During the wet season there was a notable preference for fleshy fruits of the coastal thistle (*Strumphia maritima*) and prickly pear (*Opuntia dilenii*); during the dry season, when these fruits were not available, the preference was for plants of the mangrove forests and gramineans from the keys.

At Península de Guanahacabibes the species feeds on at least nineteen types of plants, according to the stomach contents examined, and occasionally mollusks and coleopterans were found. In addition, 50% of the large intestine's volume consisted of nematodes, which apparently contribute to the digestion of cellulose. Iguanas have been observed in cultivated areas of Isla Magueyes eating sweet potato vines, fallen fruits of *Carica papaya,* and leaves and fruits of *Ipomaea pescaprae, Cordia caymanensis, Ernodea littoralis,* and *Mangifera indica.*

According to all of these observations, the Cuban iguana can be classified as a generalist phytophage with a certain degree of selectivity, and as an opportunistic omnivore. This wide trophic spectrum makes it easy to keep in captivity, where it has eaten fruits, cut meat, several kinds of vegetables, mice, cooked fish, sweetmeats, wildflowers, lettuce, and uncooked hens' eggs. At the Jardín Zoológico de La Habana, iguanas ate twenty-one of the twenty-five types of food that were offered during two years of observation. However, the most frequently eaten items were banana fruits, food used for mammals, lettuce, cabbage, uncooked fish, and chard. Some iguanas were fed in the confinement of the Instituto de Ecología y Sistemática with guava and banana fruits and food for mammals, in addition to wild plants such as *Bidens pilosa* and *Cordia sebestena,* which were offered as proof and sometimes accepted. Cuban iguanas given three different diets in a laboratory experiment in Puerto Rico excreted distinct amounts of nitrogenous wastes.

Little is known about reproduction in this species in free life; juveniles were observed only after May in the vicinity of Cayo Largo del Sur, which seems to indicate that layings occur in the first months of the year. However, according to observations at Península de Guanahacabibes and in the Jardín Zoológico de La Habana, courtship probably begins in April and layings occur between April and June, in sandy zones far away from the places where the population is normally found.

At Isla Magueyes, females dug nests from May to June, between 5:30 P.M. and 6:00 P.M., under bushes of *Tournefortia gnaphalodes* and *Suriana maritima.* The nests had depths of 26 to 70 cm and the females laid from two to fourteen eggs. Of the seventy-seven eggs found in fifteen nests, nine did not eclosion, two were decayed, and seven had been destroyed by ants. The temperatures of these nests, measured every three hours over a twenty-four hour period in August, varied between 29.5 and 34.2°C. Soil moisture had no effect on hatchling size.

According to data in the herpetological collection of the Instituto de Ecología y Sistemática, eggs laid in August by a captive female were of a cream color with a pinkish white sheen; their mean measurements were 52.1 x 32.6 mm. The eggs of another captive female, laid in May, had a thick shell and mean measurements of 67.7 x 40.7 mm.

At the San Diego Zoo in the United States, a female built a nest in July 1951. Digging a tunnel with a chamber at the end, situated approximately at 6 inches (13.2 cm) under the surface, she laid seventeen eggs, of which fourteen were on the floor of the chamber and the other three on top of the first. She did not cover the eggs with earth, but she filled the entrance of the tunnel. This female defended the nest by standing near it, opening her mouth and vigorously bobbing her head to draw away any intruder, although without direct aggression.

The seventeen eggs had flexible, leathery shells and were of an immaculate white color, most of them with a turgid appearance but some with one or two depressions. The mean measurements were 65.6 x 44.5 mm. The eggs were incubated in pots with wet sand, and at eighty-four days it was observed that the egg measurements had increased; the means were 65.9 x 45.9 mm. All eggs eclosioned after 119 to 123 days of incubation. Generally, the hatchlings stuck out their heads and emerged completely the next day. They had egg teeth with a mean of 1.3 mm. Their snout-vent lengths varied between 95 and 110 mm (x = 102.4 mm) and their weight between 48 and 59 g (x = 54.3 g); a case of twins was excluded from the sample to calculate the means of measurements and weights.

Although displays of reproductive behavior such as courtship, apparent matings, and nest digging were observed from March to May in the Jardín Zoológico de La Habana, neither egg laying nor hatchling emergences could be verified in that enclosure. A female laid eggs in the confinement of the Instituto de Ecología y Sistemática on two occasions in 1985; on the first, seven eggs were found and on the second, four. Their lengths varied from 69.0 to 77.8 mm with a mean of 73.6 mm. These eggs were gathered and incubated in pots with wet sand at environmental temperature, but in any case they completed their development. Another female, recently introduced into the confinement, deposited nine eggs in July 1986; their measurements varied between 67.8 and 75.5 mm (x = 72.9 mm) and they weighed between 72.5 and 82.5 g (x = 79.4 g). All were decayed after eight days of incubation.

A program is under way at the Center for Reproduction of Endangered Species in San Diego, California, for breeding Cuban ground iguanas captured at the U.S. naval base in Guantánamo Bay and reintroducing the offspring at their place of origin. Results have been successful, leading to the conservation of the species. The number of eggs per female was determined by means of radiographic measurement: 8.6 ± 0.84 in females with snout-vent lengths ranging from 300 to 465 mm. A positive correlation was found between snout-vent length and number of eggs, a negative correlation between snout-vent length and time of oviposition.

In captivity the Cuban iguana retains behavioral patterns observed in free life, not only daily but also seasonally. Nevertheless, the daily cycle of males is disturbed when more than two share an enclosure smaller than 200 m². During the adaptation period, some individuals refuse to eat; they are aggressive, extend their dewlaps, and open their mouths, trying to bite. Others do not leave their refuges, or, when they do, immediately retreat in the presence of people. After a while, they easily become accustomed to captivity and eat the food given them, not only vegetable but also animal in origin. Once they become tamed, they are docile and can be fed by hand. Juveniles spend most of the day in refuges and take more time in becoming accustomed to eat. The longest time in captivity reported for this species is three years and five months.

Some populations are numerous because of the gregarious habits of the species, but nowadays in most localities they are becoming scarce, diminished by anthropic activities that transform the landscape and, occasionally, by hunting. Nevertheless, in some keys and coastal zones with available food and refuges, populations are still numerous. At Cayo Rosario, a mean density of 9.6 iguanas/ha was found.

To date, six species of parasitic nematodes are known for *Cyclura nubila*: *Ozolaimus monhistera, Travassozolaimus travassosi, Paralaluris cyclurae, Atractis opeatura, Oswaldofilaria brevicaudata,* and *Oswaldocruzia lenteixeirai.*

The tick *Amblyomma torrei* has very frequently been found on this species in natural populations. In addition, there are two species of parasitic mites: one, *Ornithodoros cyclurae,* is found in the nasal ducts; the other, *Cyclurobia javieri,* is of the family Pterygosomidae.

Three iguanas, two females and one male kept in the confinement of the Instituto de Ecología y Sistemática, were diagnosed with ulcerous conjunctivitis; these iguanas tested positive for the bacteria *Klebsiella pneumoniae* and *Pseudomona aeruginosa.* This infection caused death by starvation, which indicates that it should be considered of great importance when keeping iguanas in captivity.

Genus *Chamaeleolis* (chipojos grises)

Chamaeleolis Cocteau 1838, *Hist. Fís. Pol. Nat. Isla de Cuba* 4:90. Type species: *Chamaeleolis fernandina. Chamaeolis (sic)* Cocteau, 1838, *Hist. Fís. Pol. Nat. Isla de Cuba* 4:90. *Pseudochamaeleon* Fitzinger, 1843, *Syst. Rept.*:63. Type species: *Pseudochamaeleon cocteaui.*

Geographic range: Isla de Cuba; Isla de la Juventud.

Content and Description

Endemic genus; four species are recognized, which are diagnosed as follows:

Intermediate or large size. Mean snout-vent length between 138.7 and 146.9 mm in males and between 114.3 and 144.7 mm in females.

Large head, with rugose cephalic casque ending in a posterior expanded arch. With a fleshy, small protuberance in the upper border of the ear opening. Large dewlap in both sexes. With digital pads and subdigital lamellae. Tail length nearly the same size as the snout-vent length; without caudal autotomy. Gray color, with brown or black spots.

Flat, circular, large, dorsal scales, with very small, granular, interstitial scales; smooth, circular, granular, and very small ventral scales. Gular edge scales conical, filamentous, or barbell-like. Enlarged postcloacal scales in males.

Chamaeleolis chamaeleonides (chipojo gris; plate 9)

Anolis chamaeleonides Duméril and Bibron, 1837, *Erp. Gén.* 4:168. Holotype MNHN 1004. Type locality: Cuba, restricted to the vicinity of La Habana, La Habana province, Cuba (Garrido and Schwartz 1967). *Chamaeolis (sic) fernandina* Cocteau, 1838, *Hist. Fís. Pol. Nat. Isla de Cuba* 4:90; holotype MNHN 1004. Type locality: Cuba. *Pseudochamaeleon cocteaui* Fitzinger, 1843, *Syst. Rept.*:63; holotype MNHN 1004. Type locality: Cuba. *Chamaeleolis chamaeleontides* Boulenger, 1885, *Cat. Lizards Brit. Mus.* 2:7. *Chamaeleolis chamaeleonides:* Barbour, 1935, *Zoologica* 19(3):105.

Geographic Range

Isla de Cuba (from Cabo de San Antonio to Sierra Maestra); Isla de la Juventud.

Localities (map 12)

Pinar del Río: Cabo de San Antonio; San Diego de los Baños (Stejneger 1917); Potrerito; Pica Pica; Sumidero; Baños de San Vicente; La Mulata (Garrido and Schwartz 1967); Sierra de Güira (Garrido *et al.* 1991). *La Habana and Ciudad de La Habana:* Santiago de las Vegas (Barbour 1914); Sierra de Anafe; Caimito del Guayabal; Arroyo Bermejo (Garrido and Schwartz 1967); Loma la Colorada, Madruga (Garrido 1973*f*); Caobí; Nueva Paz (Garrido 1980*b*); El Narigón, Puerto Escondido (L.R.S.). *Matanzas:* Río Hanábana (Barbour 1914); Punta de la Maya; Santo Tomás; Ciénaga de Zapata (Garrido and Schwartz 1967); San Miguel de los Baños (Garrido 1980*b*). *Cienfuegos:* Aguada de Pasajeros (Barbour and Ramsden 1919); Juraguá (Garrido and Schwartz 1967). *Camagüey:* 15 km *SW* of Camagüey; Sierra de Cubitas (Garrido and Schwartz 1967). *Holguín:* El Yayal (Agüero Cobiellas *et al.* 1993). *Granma:* Manzanillo (Barbour and Ramsden 1919); Buey Arriba, *SW* Bayamo (Garrido and Schwartz 1967). *Santiago de Cuba:* La Punta, Sierra Maestra (Garrido 1980*a*); Pico Turquino (Rodríguez Schettino 1985). *Isla de la Juventud:* Isla de Pinos (Schwartz and Thomas 1975); Cerro de Santa Isabel (Garrido 1980*b*).

CONTENT

Monotypic endemic species.

DESCRIPTION

Large size. Mean snout-vent length of 146.9 mm in males and of 139.8 mm in females.

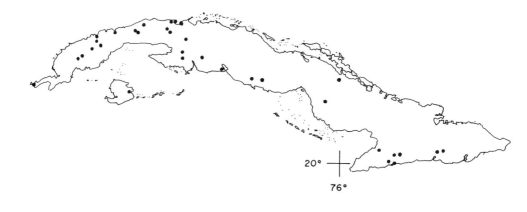

Map 12. Geographic distribution of *Chamaeleolis chamaeleonides*.

Coloration variable from gray to grayish white, with yellow, brown, reddish, violet, and black dots and streaks. Black suprascapular stripe in some individuals. Dewlap in both sexes, very pale pink, with four dark pink semicircular bars, and black and ochraceous dots. Pale yellowish gray tongue. Yellow iris.

Body and tail laterally compressed. Dorsal crest from the upper edge of the cephalic casque. Vertically oval, large ear opening, with a small fleshy protuberance in its superior border. Tail somewhat longer than body in adults, with a blunt tip. Flat, circular, large dorsal scales, separated by small, granular interstitial scales; granular ventral scales, much smaller than dorsals. Large and protuberant scales in the angular region of the jaw. Two rows of small, conical gular edge scales. Enlarged postcloacal scales in males.

BIBLIOGRAPHIC SOURCES

Duméril and Bibron (1837): original description. Cocteau and Bibron (1838): detailed description; measures; drawings of a specimen in dorsolateral view (pl. 15), of the dorsal region of the head (pl. 15, no. 1), of the lateral region of the head (pl. 15, no. 2), of the cloacal region (pl. 15, no. 3), of the ventral surface of the anterior limb (pl. 15, no. 4), of the ventral surface of the posterior limb (pl. 15, no. 5). Gundlach (1867): metachromatism; behavior; abundance. Gundlach (1880): coloration in life of the two phases of color; habitat data; behavior; abundance. Barbour (1914): taxonomy. Barbour and Ramsden (1916b): habitat data; behavior; abundance. Stejneger (1917): drawing of the lateral region of the head (fig. 35). Barbour and Ramsden (1919): coloration in life; drawings of the lateral region of the head (pl. 14, fig. 2), of the scalation of the dorsal region (pl. 14, fig. 3); habitat data; description of the juvenile. Alayo Dalmau (1955): habitat data; abundance; coloration; drawings of the lateral region of the head (pl. 6, fig. 7, no. 35), of the dorsal region of the head (pl. 6, fig. 7, no. 36), of the lateral region of the tail in fifth verticil (pl. 6, fig. 7, no. 37), of the dorsolateral scalation of the body (pl. 6, fig. 7, no. 38). Etheridge (1960): osteology; drawing of the jaw (fig. 1d). Ruibal (1964): drawing of the anterolateral region of the head (fig. 1); habitat data; behavior. Buide (1967): habitat data; abundance. Garrido and Schwartz (1967): drawing of the scales of the dewlap edge (fig. 1 left); morphological variation; habitat data; description of juveniles; origin; taxonomy. Williams (1969): origin. Coy Otero and Baruš (1979a): nematodes. Coy Otero and Lorenzo (1982): helminths. Garrido (1982a): aggressive display between males. De Queiroz (1982): drawing of the scleral right ring (fig. 1A). Buide (1985): two photographs of each of two specimens; habitat data. Schwartz and Henderson (1985): drawing of the lateral region of the head and of the anterior limb (pl. 3, no. 6); photograph of one specimen (fig. 69). Silva Lee (1985): photograph of one specimen; habitat data; behavior; anthropic damage. Williams (1989a): osteology; drawings of the right jaw in inner and ventral views (fig. 1C), of the interclavicule (fig. 3E). Agüero Cobiellas *et al.* (1993): habitat data; morphometry. Rodríguez Schettino (1996): habitat data.

Morphological Variation

Metachromatism from pale gray to dark gray, with the different-colored dots and streaks darkened; dewlap from pale pink to purplish pink with dark violet spots.

Coloration varies individually and geographically, as does the position of the spots.

Geographic Variation

Individuals from the area west of Cienfuegos city (Juraguá, Cinco Tiras) are grayish brown in color with black spots, vermiculations, and dots; they have a black suprascapular stripe, and the casque and the lateral region of the head have many black dots; the ventral region is white with yellowish brown spots that continue to the lateral region, limbs, gular region, and dewlap, which is purplish brown at the base and rosaceous at the edge.

The specimen from El Veral, Península de Guanahacabibes, has neither ventral spots nor any pattern of spots and dots. The specimens from Isla de la Juventud are brown in color, with pale brown spots.

Sexual Variation

Coloration is similar in both sexes; the dewlap is equal in males and females. Males are distinguished by their enlarged postcloacal scales. Differences in size are shown in table 7.7.

Age Variation

Coloration of juveniles is similar to that of adults. Males have enlarged postcloacal scales. Individuals with snout-vent lengths of less than 90 mm do not have protuberant scales in the angular region of the jaw, as is typical for adults, or well-developed cephalic casques.

Natural History

This species lives in semideciduous and mesophyllous evergreen forests at elevations of up to 1,000 m above sea level, although it also occupies fruit-

Table 7.7. Comparison of snout-vent length (SVL), head length (HL), and tail length (TL) between male and female adults of Chamaeleolis chamaeleonides (symbols as in table 7.1).

Sex	N	X	CV	m	M	t	p
				SVL			
Males	11	146.9	6.31	133.7	158.2	1.94	<0.10
Females	9	139.8	4.55	132.5	150.5		
				HL			
Males	11	53.9	7.52	47.2	59.2	2.35	<0.05
Females	9	50.0	6.09	43.5	55.4		
				TL			
Males	11	150.7	8.31	126.3	173.4	1.02	n.s.
Females	9	145.0	8.63	130.4	171.2		

bearing trees and coffee plantations. It has occasionally been observed in completely open areas: in a yard at San Vicente, Pinar del Río province; on the roadbed that leads to Santo Tomás, Península de Zapata; and in the area surrounding the swimming pool at the camping base El Narigón, on the northeastern coast of La Habana province (two individuals). In all of these localities, there was a contiguous forest. It uses tree and bush branches and trunks as perches, climbing to the top occasionally. To sleep, it lies on branches or vines of approximately the same diameter as its body. At El Yayal, Holguín province, it was found to be active at 4.2 m above ground as a mean, on tree trunks under the shade of vegetation.

The species exhibits slightly fugitive tendencies; at any sign of danger, individuals move slowly to the opposite side of the trunk where they are perched. They are gentle and tame in captivity, although when handled they open the mouth and extend the dewlap, but without trying to bite or escape. Apparently, males exhibit some territoriality: a resident male in a terrarium, facing a newly arrived male, slowly raised its body and carried its head back, extending the dewlap completely; then it rapidly bobbed two or three times and, following that, bobbed more quickly. These movements were repeated in two sequences.

This is a diurnal species, apparently nonheliothermic, since it has not been observed basking, but in shaded places within forests. However, the discovery of some individuals outside their vegetation shelters indicates that at least sometimes during its daily activity cycle, this species requires places to acquire the needed body temperature, either basking or in contact with a previously warmed substrate.

It has accepted slugs of the genus *Veronicella*, larvae of tree frogs (*Osteopilus septentrionalis*), coleopterans, lepidopteran larvae, and other insects, in the terraria of the Instituto de Ecología y Sistemática.

Reproduction in free-living members of this species is little known; only L. F. de Armas (pers. comm.) has observed a pair copulating, in April, at Caobí, Nueva Paz, La Habana province. The species has not reproduced in captivity; nor has any individual shown either courtship or mating behavior. A single female, recently introduced into the terrarium, laid one egg in May 1978 with a thick, hard shell measuring 21.0 x 14.7 mm.

In all cases in which this species has been collected, only lone individuals have been found, or as many as two or three near individuals. This seems to indicate that the species does not build numerous populations; however, its habits and cryptic coloration may contribute to underestimations of its abundance, making the findings difficult.

To date, two species of parasitic nematodes are known for *Chamaeleolis chameleonides: Travassozolaimus travassosi* and *Cyrtosomum longicaudatum*.

Chamaeleolis porcus (chipojo blanco; plate 10)

Chamaeleolis porcus Cope, 1864, *Proc. Acad. Nat. Sci. Philadelphia* 16:168; holotype ANSP 8133. Type locality: Cuba, restricted to the vicinity of Guantánamo city, Oriente province, Cuba (Garrido and Schwartz 1967).

GEOGRAPHIC RANGE

Isla de Cuba (Camagüey; Holguín; Granma; Santiago de Cuba; Guantánamo provinces).

LOCALITIES (MAP 13)

Camagüey: Florida (Schwartz and Henderson 1985). *Holguín:* Guardalavaca; Cayo Saetía (Garrido and Schwartz 1967); Gibara (Garrido 1973*a*); Banes (Garrido 1982*a*). *Granma:* Jiguaní (Barbour 1914). *Santiago de Cuba:* La Maya (Barbour 1914); Felicidad; Santa María del Loreto; El Cobre (Garrido and Schwartz 1967); Alto Songo; Río Frío (Garrido 1982*a*). *Guantánamo:* vicinities of Guantánamo (Barbour and Ramsden 1919); La Casimba, Maisí (Garrido 1967); Baracoa; Punta Caleta; mountains north of Imías; Campanario, Guaso; Los Hondones; Bayate; San Carlos (Garrido and Schwartz 1967); Monte Verde; Asunción (Garrido 1982*a*).

CONTENT

Monotypic endemic species.

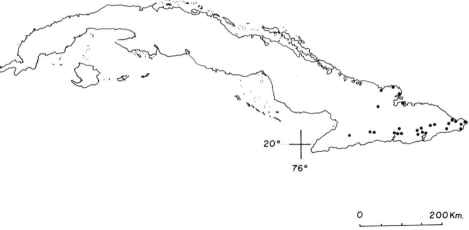

Map 13. Geographic distribution of *Chamaeleolis porcus*.

DESCRIPTION

Large size. Mean snout-vent length of 143.3 mm in males and of 144.7 mm in females.

Coloration variable: from grayish brown, to olivaceous brown, to grayish white, with black dots in the lateral regions of the head, the cephalic casque, and the neck in some specimens. Grayish white suprascapular stripe. Dewlap in both sexes, yellow or orange, pale at the base, with four semicircular brown bands. Black tongue with gray lateral regions. Brown iris.

Laterally compressed body and tail. Dorsal crest from the nuchal region. Vertically oval, large ear opening, with a very small or absent upper fleshy protuberance. Tail somewhat longer than body in adults, with a thin tip. Flat, circular, large, dorsal scales, separated by small, granular, interstitial scales; flat, small, ventral scales. Two rows of filamentous, thin gular edge scales, without reaching the jaw angular bone level and with a mean length of 3.1 mm. Enlarged postcloacal scales in males.

Six pairs of macrochromosomes and more than ten pairs of microchromosomes.

BIBLIOGRAPHIC SOURCES

Cope (1864): original description. Gundlach (1880): characteristics that differentiate this species from *C. chamaeleonides*. Garrido and Schwartz (1967): drawing of the scales of the longitudinal gular line (fig. 1, right); habitat data; origin. Baruš and Coy Otero (1969*b*): parasites. Gorman *et al.* (1969):

karyotype; photograph of a mitotic metaphase (fig. 2); feeding and behavior in captivity. Coy Otero (1970): helminths. Garrido (1982a): photographs of the holotype head, showing the casque and the scales of the longitudinal gular line (figs. 2 and 4); morphology; behavior in captivity. Buide (1985): photographs of an adult and a juvenile. Schwartz and Henderson (1985): photograph of a specimen (fig. 70). Silva Lee (1985): photograph of a specimen; coloration; metachromatism; behavior and feeding in captivity. Ruiz Urquiola and Gutiérrez (1997): reproductive behavior.

MORPHOLOGICAL VARIATION

Metachromatism from white or grayish brown to very dark gray; the brown bands of the dewlap darken almost to black. A high degree of individual and geographic variation in the coloration and pattern of the head dots.

Geographic Variation

Little variation, except for the cephalic dots. Individuals from Holguín, Guardalavaca, and Santa María del Loreto have black dots on the sides of the head, the cephalic casque, and the neck. Those from other localities have no dots.

Sexual Variation

Coloration is similar in both sexes; dewlap is equal in males and females. Males are recognized by their enlarged postcloacal scales. Differences in size are shown in table 7.8.

Table 7.8. Comparison of snout-vent length (SVL), head length (HL), and tail length (TL) between male and female adults of Chamaeleolis porcus (symbols as in table 7.1).

Sex	N	X	CV	m	M	t	p
				SVL			
Males	9	143.3	13.17	100.0	160.0	0.16	n.s.
Females	10	144.7	28.13	117.2	171.2		
				HL			
Males	9	45.0	38.40	32.8	54.8	0.97	n.s.
Females	10	50.7	12.76	40.0	59.3		
				TL			
Males	9	160.3	16.04	141.8	184.8	0.19	n.s.
Females	10	162.3	12.61	129.1	183.6		

Age Variation

Coloration of juveniles is similar to that of adults. Males have enlarged postcloacal scales. Mean snout-vent length of six juveniles deposited in the herpetological collection of the Instituto de Ecología y Sistemática was 73.1 mm; mean head length, 28.2 mm; and mean tail length, 74.1 mm. These means are about half of the adult size.

Natural History

This species lives in moist, shaded forests as well as in lowlands and mountains, where it is more frequent preferentially near rivers and streams. Individuals are generally found on trunks of the rose apple (*Syzygium jambos*) in moist gallery forests, perching also on trunks and branches of trees and bushes. A fugitive species in free life, it climbs trunks quickly up to the crown and jumps to the ground or to another tree when pursued farther. It is a diurnal species, apparently nonheliothermic, since it has been found only in shaded places.

There are no data about how this species feeds in free life, but some individuals have been kept in captivity and fed with flies, crickets, grasshoppers, cockroaches, moths, beetles, and larvae of lepidopterans and *Tenebrio molitor*. In some cases, this species catches insects by extruding the tongue; in others, by using both the mouth and the tongue. Furthermore, it has shown a taste preference for gastropods of the genus *Veronicella*. A female was fed for three months with beef and chicken meat and liver, slugs, and oranges. In general, the species stays motionless in captivity and only moves slowly toward food.

No information exists regarding reproduction in free-living members of this species. One female, captured in September 1965, had one oviductal egg 23 mm in length, with a shell, ready for oviposition. According to data from the herpetological collection of C. T. Ramsden of the Instituto de Ecología y Sistemática, a captive female, 170 mm in snout-vent length, laid one egg in September. Another semicaptive female laid two eggs, the second thirty days after the first one, white with a length of 22 mm; however, they did not eclosion. Copulation was observed in captivity between a female *C. porcus* from Holguín and a male *C. barbatus*. The latter exhibited the courtship pattern of its species (see *C. barbatus*) and the female answered in a similar way, but dropped the head more slowly with a different duration for each movement. Some captive females have laid eggs under the ground, masked with earth, so their eggs appeared to have brown spots on a white surface.

Males have not shown aggressive displays among themselves in captivity, either facing a *Chamaeleolis* decoy or a male of the species *A. equestris*. However, when they are handled, they try to bite and extend the dewlap.

Apparently this is not an abundant species; however, its cryptic coloration and the dense forests in which it lives allow it to go unnoticed.

To date, seven species of parasitic nematodes are known for *Chamaeleolis porcus: Cyrtosomum scelopori, C. longicaudatum, Piratuba digiticauda, Trichospirura teixeirai, Oswaldocruzia lenteixeirai, Parapharyngodon* sp., and *Porrocaecum* sp.; in addition, there is one trematode, *Mesocoelium monas*.

Chamaeleolis barbatus (chipojo blanco; plate 11)

Chamaeleolis porcus Cope, 1864 (*part.*), *Proc. Acad. Nat. Sci. Philadelphia* 16:183. *Chamaeleolis barbatus* Garrido, 1982, *Poeyana* 236:3; holotype IZ 5368. Type locality: limestone uplands at Ojo de Agua, Cinco Pesos, about 9 km NW of San Cristóbal, Pinar del Río province.

GEOGRAPHIC RANGE

Isla de Cuba (Pinar del Río, from Rangel to Cayajabos).

LOCALITIES (MAP 14)

Pinar del Río: Rangel (Garrido and Schwartz 1967); Ojo de Agua, San Cristóbal; Loma El Salón, Sierra del Rosario (Garrido 1982a); El Mulo, Sierra del Rosario (Martínez Reyes 1995); Soroa (Rodríguez Schettino *et al.* 1997); Mil Cumbres; Seboruco (L.R.S.).

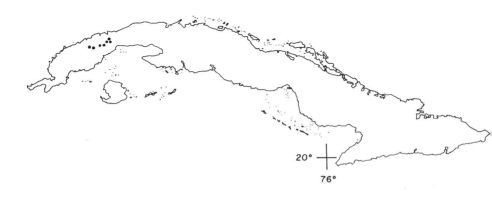

Map 14. Geographic distribution of *Chamaeleolis barbatus*.

CONTENT

Monotypic endemic species.

DESCRIPTION

Intermediate size. Mean snout-vent length of 138.8 mm in males and of 114.3 mm in females.

Gray color, with some pale yellow spots. Without suprascapular stripe. Dewlap in both sexes, white with four semicircular gray bands. Black tongue with yellow dots on tip and sides.

Laterally compressed body and tail. Dorsal crest from the upper edge of the cephalic casque. Vertically oval, narrow ear opening, with a large, fleshy protuberance in the superior part. Tail somewhat smaller than body in adults, with a blunt tip. Flat, circular, large dorsal scales, separated by small, granular interstitial scales; small, conical ventral scales. Two rows of filamentous, long and robust, barbel-like gular edge scales, reaching the jaw angular bone level, with a mean length of 5.3 mm. Enlarged postcloacal scales in males.

BIBLIOGRAPHIC SOURCES

Garrido (1982a): original description; photographs of the head showing the casque and the gular edge scales (figs. 1 and 3); morphology; behavior and courtship in captivity. Silva Lee (1985): photograph of a specimen. Ruiz Urquiola et al. (1997a): conservation. Rodríguez Schettino et al. (1997): habitat data. Ruiz Urquiola and Gutiérrez (1997): reproductive behavior.

MORPHOLOGICAL VARIATION

Metachromatism from pale gray to gray. Somewhat individually and geographically variable.

Geographic Variation

Little variation throughout its geographic range.

Sexual Variation

Coloration is similar in both sexes; dewlap is equal in males and females. Males are distinguished by their enlarged postcloacal scales. Measurements of both sexes are shown in table 7.9.

Age Variation

Coloration of juveniles is similar to that of adults. The snout-vent length of a juvenile was 79.0 mm.

Table 7.9. Comparison of snout-vent length *(SVL)*, head length *(HL)*, and tail length *(TL)* between male and female adults of *Chamaeleolis barbatus* (symbols as in table 7.1).

Sex	N	X	CV	m	M	t	p
				SVL			
Males	3	138.8	10.78	122.2	151.3	1.16	n.s.
Females	3	114.3	28.86	82.4	148.3		
				HL			
Males	3	49.3	2.43	48.1	50.5	1.13	n.s.
Females	3	40.6	32.93	27.5	54.2		
				TL			
Males	3	134.9	6.36	127.1	144.1	1.58	n.s.
Females	3	120.8	10.65	79.4	148.5		

NATURAL HISTORY

The species lives in moist, shaded forests on limestone outcroppings of Sierra del Rosario, Pinar del Río province, perching on tree trunks and branches. To sleep, it rests on horizontal branches and vines or on tree trunks. It is a diurnal species; apparently nonheliothermic, since it has been only observed in the shaded zones of moist forests. It moves slowly and does not try to escape when capture is attempted. Once in captivity, it stays almost motionless on perches, and when caught does not try to run away or bite, although it opens the mouth and extends the dewlap.

M. Martínez Reyes (pers. comm.) found one insect, one diplopod, some mollusks (*Zachrysia auricoma*), and abundant vegetable material, mainly fruits, in the stomach contents of five specimens collected at Sierra del Rosario. In general, the mean number of eaten articles and the mean volume of contents were larger for vegetable material than for animal. Some individuals kept in captivity in the Instituto de Ecología y Sistemática were fed with large insects, mainly hymenopterans, coleopterans, and lepidopterans, as well as with gastropods of the genus *Veronicella*, which were evidently the preferred food. At first, the food was given with forceps; after a while, it was placed in the terraria. Mollusks were caught with the mouth and scrubbed against trunks and stones, seemingly in an attempt to eliminate the slugs' mucous secretions. Where larvae of tree frogs (*Osteopilus septentrionalis*) were offered with forceps, lizards accepted them, but they did not eat the larvae if they were left alive in water in a glass dish. In addition, larvae and adults of *Tenebrio molitor* were avidly eaten.

Although reproduction is not known for free-living members of this species, courtship and mating behavior were observed in captivity. When two individuals of different sexes met, they looked at each other; the female raised her head, letting it fall abruptly, and repeated this behavior twice. The male responded in the same way, but alternating bobs with lateral head movements. Afterward, the male slowly approached the female until he was positioned on her, with his snout at the posterior edge of her casque, but she did not accept him; this forced the male to repeat courtship several times. During copulation, the pair stayed motionless. The whole process lasted approximately four hours. Some captive females have laid eggs under the ground, masked with earth, so their eggs appeared to have brown spots on a white surface.

The antagonistic behavior between two males was similar to courtship; however, they raised their heads more and extended their dewlaps and dorsal crests. The same reaction was observed in front of one male *C. chamaeleonides*. The aggressive display of a male of this species was observed when a snake (*Antillophis andreai*) was introduced into its terrarium. It was placed laterally to the snake and compressed its body, at the same time raising the nuchal and dorsal crests and extending the dewlap, opening the mouth in a threatening attitude.

Seemingly, it is not an abundant species; however, its cryptic coloration makes it difficult to detect, so some individuals may go unnoticed. Because of its limited geographical and ecological range, it is considered Endangered according to the IUCN Red List Categories.

Chamaeleolis guamuhaya (chipojo gris)

Chamaeleolis porcus Cope, 1864 (*part.*), *Proc. Acad. Nat. Sci. Philadelphia* 16:183. *Chamaeleolis guamuhaya* Garrido, Pérez Beato, and Moreno, 1991, *Carib. J. Sci.* 27(3–4):163; holotype MNHN-500. Type locality: mountain ranges of the Sierra de Trinidad (Macizos del Escambray), in the surroundings of Topes de Collantes.

Geographic Range

Isla de Cuba (Alturas de Trinidad, Macizo de Guamuhaya).

Localities (map 15)

Cienfuegos: 6 km *SW* of Aguacate, Sierra de Trinidad (Garrido 1982*a*). *Villa Clara:* Jibacoa, Sierra del Escambray (Garrido *et al.* 1991). *Sancti Spíritus:* Mina Carlota, Sierra de Trinidad (Wilson 1957); Topes de Collantes (Garrido and Schwartz 1967); La Chispa (Garrido *et al.* 1991).

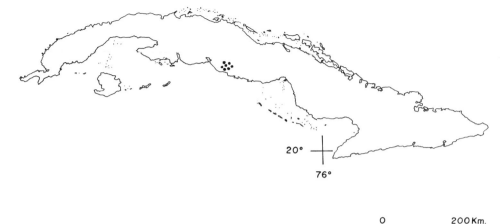

Map 15. Geographic distribution of *Chamaeleolis guamuhaya*.

CONTENT

Monotypic endemic species.

DESCRIPTION

Large size in the known adult female, adult males are still unknown.

Ash-gray color, with a greenish shade on the body and sides of the head. Tail with ill-defined brown transverse bands. Black tongue with some brown dots in its inferior region.

Laterally compressed body and tail. Dorsal crest from the upper edge of the cephalic casque. Vertically oval ear opening, with a large superior fleshy protuberance. Tail longer than body in adults, semiprehensile, with a blunt tip. Flat, circular, large dorsal scales; small, conical ventral scales. Two rows of filamentous, short, thin gular edge scales, reaching the posterior edge of the ocular orbit; the longest, of 2.4 mm.

BIBLIOGRAPHIC SOURCES

Wilson (1957): feeding and behavior in captivity. Garrido (1982a): description of a specimen. Garrido *et al.* (1991): detailed description (original); behavior; habitat data. Ruiz Urquiola *et al.* (1997b): conservation. Ruiz Urquiola and Gutiérrez (1997): reproductive behavior.

MORPHOLOGICAL VARIATION

Metachromatism from pale gray to dark brown. Few specimens are known, all found in proximate localities; consequently, it is not possible to value morphological variations due to geography, sex, or age.

Measurements of an adult female, a young male, and a young female, in this order, were as follows: snout-vent length 162, 117, 85 mm; head length 51, 44.1, 27.8 mm; tail length 173, 116, 86 mm. Coloration was similar in these specimens.

NATURAL HISTORY

This species lives in moist forests, where trees of 10 to 20 cm in diameter predominate, at an altitude of 200 to 300 m above sea level. Is a species of slow movements; if capture is attempted, it escapes toward the crown of the tree and hides on branches. In captivity it stays motionless for many hours without trying to escape. It has demonstrated activity only in the presence of food, approaching to a distance of 5 cm, from which it catches small insects by suddenly extending the tongue.

Although many attempts have been made to elicit aggressive displays in captive individuals—either by placing the individual in front of a decoy of *Chamaeleolis*, or by introducing live females of *C. chamaeleonides* and *C. barbatus* or a live male of *Anolis equestris*—they have not yielded results. Some captive females have laid eggs under the ground, masked with earth, so their eggs appeared to have brown spots on a white surface.

Because of its very limited geographical and ecological range, this species is considered Endangered according to the IUCN Red List Categories.

Genus *Anolis* (chipojos, camaleones, lagartijas; giant anoles, lizards, chamaeleons)

Anolis Daudin, 1802, *Hist. Nat. Rept.*, 4:50. Type species: *Anolis bullaris* Latreille. *Anolius:* Cuvier, 1817, *Le Régne Animal*, 2:41. *Xiphosurus* Fitzinger, 1826, *Neue Class Rept.*:17. Type species: *Anolis cuvieri* Merrem. *Dactyloa* Wagler, 1830, *Nat. Syst. Amph.*:148. Type species: *Anolis punctatus* Daudin. *Draconura* Wagler, 1830, *Nat. Syst. Amph.*:149. Type species: *Draconura nitens* Wagler. *Norops* Wagler, 1830, *Nat. Syst. Amph.*:149. Type species: *Norops auratus* Daudin. *Acantholis* Cocteau, 1836, *Compt. Rend. Acad. Sci. Paris*, 3:226. Type species: not given, posteriorly proposed *Anolis loysiana* Cocteau. *Chamaeleolis* Cocteau, 1838, *Hist. Fís. Pol. Nat. Isla de Cuba* 4:90. Type species: *Chamaeleolis fernandina. Pseudochamaeleon* Fitzinger, 1843, *Syst. Rept.*:63. Type species: *Pseudochamaeleon cocteaui. Eupristis* Fitzinger, 1843, *Syst. Rept.*:64. Type species: *Anolis equestris* Merrem. *Deiroptyx* Fitzinger, 1843, *Syst. Rept.*:66. Type species: *Anolis vermiculatus* Duméril and Bibron. *Trachypilus* Fitzinger, 1843, *Syst. Rept.*:67. Type species: *Anolis sagrei* Duméril and Bibron. *Ctenocercus* Fitzinger, 1843, *Syst. Rept.*:68. Type species: *Iguana bullaris* Latreille. *Gastrotropis* Fitzinger, 1843, *Syst. Rept.*:68. Type species: *Dactyloa nebulosa* Wiegmann. *Heteroderma* Fitzinger, 1843, *Syst. Rept.*:68. Type species: *Acantholis loysiana* Cocteau.

Geographic range: North America (southeastern United States); Central America (from Sonora and Tamaulipas, México, to the Isthmus of Panamá); South America (from the northern Caribbean coast to 20° S); Bermuda Islands; Bahama Islands; Greater Antilles; Lesser Antilles; Cayman Islands; Bay Islands; Pacific islands (Gorgona, Malpelo, Cocos, and Tres Marías).

CONTENT AND DESCRIPTION

More than 350 species are recognized; Cuban members of the genus (fifty-one species, forty-nine subspecies) are described as follows: size quite variable. Mean snout-vent length of males from 30.6 to 138.8 mm and of females from 28.5 to 147.9 mm. High morphological diversity. Tail length shorter than, equal to, or greater than that of the body, according to the species. Digital pads and subdigital lamellae. Dewlap in males; present in females of some species. Caudal autotomy. Caudal autotomic vertebrae with long transverse processes directed forward (beta condition), or without them (alpha condition).

Coloration variable, with shades of brown and green. Metachromatism in the majority of species. Quite variable scalation. Enlarged postcloacal scales in males of most species.

The Cuban species and subspecies are grouped here according to the classifications of Etheridge (1960) and Williams (1976), using their informal categories of sections, subsections, series, and species groups, which are described following the authors previously cited.

Section alpha (caudal autotomic vertebrae without transverse processes directed forward).

Subsection *carolinensis* (T-shaped interclavicle).

Series *carolinensis* (lateral processes of the interclavicle in close contact with the expanded proximal parts of the clavicles; three lumbar vertebrae as mode).

Species group *equestris* (splenial present; the superficial skull bones tend to become rugose; 9/10 anterior aseptate caudal vertebrae).

Anolis equestris (chipojo; giant anole; plate 12)

Anolis equestris Merrem, 1820, *Tentamen Syst, Amph.* 9:45; holotype unlocated. Type locality: unknown, restricted to the vicinity of La Habana, La Habana province, Cuba (Schwartz and Garrido 1972). *Xiphosurus equestris* Fitzinger, 1826, *Neue Class Rept.*:48. *Anolius rhodolaemus* Bell, 1827, *Zool. J.*:235; holotype unlocated. Type locality: Cuba.

GEOGRAPHIC RANGE

Isla de Cuba (from the vicinities of San Diego de los Baños, Pinar del Río, to Banes in the north and Cabo Cruz in the south); Cayos de las Cinco

Leguas; Cayo Santa María and others in the Archipiélago de Sabana-Camagüey.

LOCALITIES (MAP 16)

Pinar del Río: San Diego de los Baños; Herradura (Barbour 1914). *La Habana and Ciudad de La Habana:* Madruga (Barbour 1914); Habana (Barbour and Shreve 1935); Campo Florido; San José de las Lajas; Jaruco (Schwartz 1958); Bosque de La Habana (Collette 1961); Rancho Boyeros (Peters 1970); Marianao; Loma de Tierra; Nazareno; Güines (Schwartz and Garrido 1972); Arcos de Canasí; Nueva Paz (Garrido 1973*f*); Parque Lenin (Silva Lee 1985); Jardín Botánico Nacional (García Rodríguez 1989); Escaleras de Jaruco (González González 1989); Parque Zoológico Nacional (L.R.S.). *Matanzas:* Matanzas (Barbour and Shreve 1935); Península de Hicacos (Buide 1966); Cárdenas; Agramonte; Perico (Schwartz and Garrido 1972); Cayos Cinco Leguas (Garrido 1981); Bacunayagua (Soto Ramírez 1997). *Villa Clara:* Santa Clara (Barbour and Shreve 1935); Santo Domingo; Encrucijada; Caibarién; Sagua la Grande (Schwartz and Garrido 1972); Cayo Santa María; Cayo Las Brujas (Garrido 1973*d*); Cueva del Agua, Sagua la Grande (Silva Taboada 1974). *Cienfuegos:* Soledad (Smith 1929); Guajimico (Schwartz and Ogren 1956); Juraguá (Zajicek and Mauri Méndez 1969); Ciego Montero (Schwartz and Garrido 1972). *Sancti Spíritus:* Trinidad; San José del Lago; Yaguajay; Cabaiguán (Schwartz 1958); Punta Caguanes; Topes de Collantes; Mapos (Schwartz and Garrido 1972); Cueva Grande de Caguanes (Garrido 1981). *Ciego de Ávila:* Loma de Cunagua (Schwartz 1958); Morón (Schwartz and Garrido 1972); Cayo Coco (Garrido 1976*a*). *Camagüey:* Cueva 2, Sierra de Cubitas (Koopman and Ruibal 1955); Banao; Sierra de Cubitas; Martí;

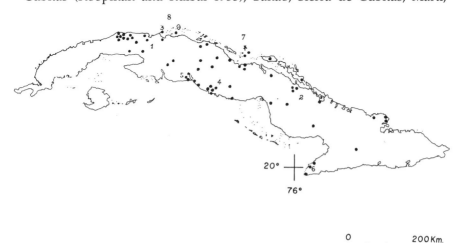

Map 16. Geographic distribution of *Anolis equestris*. 1, *A. e. equestris*; 2, *A. e. thomasi*; 3, *A. e. buidei*; 4, *A. e. persparsus*; 5, *A. e. juraguensis*; 6, *A. e. verreonensis*; 7, *A. e. potior*; 8, *A. e. cincoleguas*.

Camagüey (Schwartz 1958); Florida; Nuevitas (Schwartz and Garrido 1972); Sierra de Najasa (Garrido 1981); Cayo Guajaba (Garrido *et al.* 1986); Cayo Sabinal (Martínez Reyes 1998). *Las Tunas:* Las Tunas (Díaz Castillo *et al.* 1991). *Holguín:* Gibara; Banes (Schwartz 1958); El Jobo (El Jobal), between Holguín and Bayamo (Schwartz and Garrido 1972). *Granma:* Cabo Cruz (Schwartz 1964*a*); Niquero (Schwartz and Garrido 1972); Bayamo (Buide 1985). *Santiago de Cuba:* Santiago de Cuba (Garrido and Jaume 1984).

CONTENT

Polytypic endemic species; eight subspecies are recognized:

Anolis equestris equestris

Anolis equestris Merrem, 1820, *Tentamen Syst. Amph.* 9:45. Type locality: unknown, restricted to the vicinity of La Habana, La Habana province, Cuba (Schwartz and Garrido 1972). *Anolis equestris equestris:* Barbour and Shreve 1935, *Occ. Papers Boston Soc. Nat. Hist.* 8:249. Geographic range: from San Diego de los Baños to the area north of Villa Clara; introduced in Miami, Florida, United States of America.

Anolis equestris thomasi

Anolis equestris thomasi Schwartz, 1958, *Herpetologica* 14(1):3; holotype AMNH 78148. Type locality: 2 km *SE* of Banao, Camagüey province, Cuba. Geographic range: from Ciego de Ávila province to Banes and Bayamo.

Anolis equestris buidei

Anolis equestris buidei Schwartz and Garrido, 1972, *Studies Fauna Curaçao Carib. Isl.* 39(134):34; holotype IZ 1294. Type locality: about 0.5 km of Rincón Francés, Península de Hicacos, Matanzas province, Cuba. Geographic range: Península de Hicacos.

Anolis equestris persparsus

Anolis equestris persparsus Schwartz and Garrido, 1972, *Studies Fauna Curaçao Carib. Isl.* 39(134):36; holotype AMNH 78116. Type locality: 4 km *E* of Trinidad, Las Villas province, Cuba. Geographic range: eastern half of Cienfuegos province; Villa Clara province; Sancti Spíritus province.

Anolis equestris juraguensis

Anolis equestris juraguensis Schwartz and Garrido, 1972, *Studies Fauna Curaçao Carib. Isl.* 39(134):39; holotype IZ 1152. Type locality: 3 km *SW* of Juraguá, Las Villas province, Cuba. Geographic range: vicinities of the type locality.

Anolis equestris verreonensis

Anolis equestris verreonensis Schwartz and Garrido, 1972, *Studies Fauna Curaçao Carib. Isl.* 39(134):44; holotype IZ 488. Type locality: Verréon, Cabo Cruz, Oriente province, Cuba. Geographic range: vicinities of Cabo Cruz.

Anolis equestris potior

Anolis equestris santamariae Garrido, 1975, *Poeyana* 141:14. Holotype IZ 3098.

Type locality: Cayo Santa María, Archipiélago de Sabana-Camagüey, Las Villas province, Cuba. *Anolis equestris potior:* Schwartz and Thomas 1975, *Carnegie Mus. Nat. Hist. Spec. Publ.* 1:81. Geographic range: Cayo Santa María and Cayo Las Brujas, in Archipiélago de Sabana-Camagüey.

Anolis equestris cincoleguas

Anolis equestris cincoleguas Garrido, 1981, *Poeyana* 232:3; holotype IZ 5398. Type locality: Cayos Cinco Leguas, Matanzas province, Cuba. Geographic range: Cayos de las Cinco Leguas, to the area north of Cárdenas.

DESCRIPTION

Large size. Mean snout-vent length from 131.7 to 149.3 mm in males and from 124.6 to 136.5 mm in females.

Green color, with some very pale green isolated scales. Brown and dark green upper and lateral regions of the head. Greenish yellow labials, with green spots; yellow postlabial stripe to the ear opening. Yellow suprascapular stripe, dark green bordered, with a greenish yellow inner "tongue." Dewlap in both sexes, pink with the edge scales yellow and those of the gular region dark green. Without transverse bands in tail. Brown iris.

Rugose cephalic casque. Small, diagonal, oval ear opening. Quadrangular, smooth, large dorsal scales, separated by very small, granular interstitial scales; smooth, circular, small ventral scales. Postcloacal scales not enlarged.

Six pairs of macrochromosomes and twelve pairs of microchromosomes (2n = 36).

BIBLIOGRAPHIC SOURCES

Merrem (1820): original description. Cocteau and Bibron (1838): detailed description; drawing of a specimen in lateral view (pl. 9), of the head in dorsal view (pl. 9, no. 2); of the cloacal region (pl. 9, no. 3), of the ventral surface of the hindlimb (pl. 9, no. 4). Gundlach (1867): habitat data; abundance; metachromatism; feeding. Gundlach (1880): detailed description; five phases of coloration of an alive individual; description of the juvenile; habitat data; abundance; behavior. Barbour (1914): abundance. Barbour and Ramsden (1916*b*): abundance; habitat data. Stejneger (1917): drawings of the dorsal region of the head (fig. 42), of the lateral region of the head (fig. 43), of the lateral region of the tail in the fifth verticil (fig. 44), of the scales of the dorsal crest (fig. 45). Barbour and Ramsden (1919): detailed description; coloration in life; description of the juvenile; drawing of the head in dorsal view (pl. 14, fig. 5); habitat data; behavior; feeding; predators. Conant and Hudson (1949): longevity. Alayo Dalmau (1955): drawings of the head in dorsal view (pl. 7, fig. 9), in lateral view (pl. 7, fig. 9, no. 43), of the tail in lateral view in the fifth verticil (pl. 7, fig. 9, no. 44), of the dorsal crest (pl. 7, fig. 9, no. 45); habitat data; coloration of the adult and juvenile. Koopman and Ruibal (1955): fossil remains. Schwartz (1958): coloration in life; measures of adults and juveniles. Etheridge (1960): osteology. Collette (1961); morphology of the subdigital lamellae; coloration; description of the pigments of the peritoneal cavity; drawing of the subdigital lamellae of the third toe of the left hindlimb (fig. 2f); habitat data; behavior; morphometry. Ruibal (1961): habitat data. Lynch and Smith (1964): tooth replacement; lateral view of the lower jaw (fig. 1), of the upper jaw (fig. 2). Ruibal (1964): morphological variation; description of the juvenile; habitat data; behavior; feeding; predators. Schwartz (1964a): description of the morphology and the coloration of several populations. Ruibal and Ernst (1965): description of the ultrastructure of the subdigital lamellae. Buide (1966): habitat data; abundance. Gorman et al. (1967): karyotype. Zajicek and Mauri Méndez (1969): hemoparasites; drawing of Hemogregarina sp. (fig. 2D). Williams (1969): behavior. Coy Otero (1970): helminths. Schoener (1970a): interspecific relations in size. Dubois and Macko (1972): trematode. Schwartz and Garrido (1972): morphology and coloration of several populations; drawing of the head and anterior region of the body in dorsolateral view (fig. 6 A-F). Webster et al. (1972): karyotype. Cihar (1973): photograph of a specimen. Garrido (1973d): coloration; habitat data. Garrido (1975a): morphology and coloration; photographs of the head and anterior region of the body in lateral view (figs. 3 and 4). Brach (1976): structure of the ocular papillary conus; photographs of the ocular papillary conus (figs. 2–6). Garrido (1980a): description of an egg laid in captivity. De Smet (1981): karyotype; photograph of the idiogram (fig. 5). Rundquist (1981): longevity. Berghe (1982): photograph of two individuals, in lateral and dorsal views (fig. 1). Coy Otero and Lorenzo (1982): helminths. Buide (1985): four photographs of three individuals; morphology; behavior; reproduction; feeding. Schwartz and Henderson (1985): drawings of the lateral region of the head and the forelimb (pl. 3, nos. 7 and 8). Silva Lee (1985): two photographs of a specimen; habitat data; behavior. Espinosa López, Posada García et al. (1990): electrophoretic pattern of muscle bands. Espinosa López, Sosa Espinosa, and Berovides Álvarez (1990): electrophoretic patterns of eighth proteic and enzymatic systems. Schwartz and Henderson (1991): behavior; feeding. Fernández Méndez et al. (1997): conservation.

MORPHOLOGICAL VARIATION

Metachromatism from green to dark green to dark brown. Coloration is quite variable individually and geographically. However, age is the greatest source of variation in each population.

Geographic Variation

The eight subspecies are characterized as follows:

Anolis equestris equestris (western region): green, without isolated scales of another color. Pale brown cephalic casque, with yellow spots in central and occipital regions. Green labials. Postlabial band yellow. Pink dewlap. Mean snout-vent length of 143.1 mm in males and of 137.1 mm in females.

A. e. buidei (Península de Hicacos): green, with isolated pale green scales. Brown cephalic casque, with yellow central and occipital regions. Yellow labials. Cervical and loreal regions blue. Ochraceous yellow suprascapular stripe, black bordered, with a white inner "tongue." Pink dewlap. Mean snout-vent length of 144.4 mm in males and of 129.1 mm in females.

A. e. cincoleguas (Cayos de las Cinco Leguas): blue in the anterior region of the body and green posteriorly, with white isolated scales. Greenish cephalic casque, with yellow canthal, occipital, and supraocular regions. Brown with yellow spots labials. Blue temporal region. Bluish white suprascapular stripe, black bordered and with a yellow inner "tongue." Pink dewlap. Yellow fingers and toes. Mean snout-vent length of 131.7 mm in males and of 130.3 mm in females.

A. e. persparsus (central region): green with white, pale green or blue isolated scales. Brown cephalic casque. Yellow labials. Very long and narrow yellow suprascapular stripe. Pink dewlap. Mean snout-vent length of 140.5 mm in males and of 128.5 mm in females.

A. e. juraguensis (Juraguá, Cienfuegos): green with white isolated scales in the posterior part of the body and tail. Brown cephalic casque. White postlabial band. White labials with green spots. White suprascapular stripe, with a yellow inner "tongue." One grayish green stripe in the mid-dorsal longitudinal line, and another white lateral stripe. Pink dewlap. Mean snout-vent length of 149.3 mm in males and of 124.6 mm in females.

A. e. potior (Cayo Santa María): blue and green, with yellow transverse bands on body, limbs, and tail. Dark blue cephalic casque. Yellow labials with green and black spots. Yellow postlabial band. Yellow suprascapular stripe. Pale pink dewlap. Mean snout-vent length of 136.5 mm in females; males are not known.

A. e. thomasi (central eastern region): green with white vermiculations. Green cephalic casque, with yellow in the occipital and canthal regions. Yellow labials and postlabial band. White suprascapular stripe. Very pale pink dewlap. Mean snout-vent length of 146.3 mm in males and of 135.5 mm in females.

A. e. verreonensis (Cabo Cruz): green with white vermiculations. Dark brown or black cephalic casque, with white dots to the neck and scapular region. Yellow labials with green spots. Yellow postlabial band. Yellow suprascapular stripe. Pink dewlap. Maximum snout-vent length of 168.0 mm in males and of 152.0 mm in females.

The specimens collected at Finca La Celia, Bayamo, and in Santiago de Cuba have not been assigned to any subspecies since they have not been taxonomically studied.

Sexual Variation

Coloration pattern is similar in both sexes; the dewlap is equal in males and females. Differences in size are shown in table 7.10.

Age Variation

Juveniles are green with four white transverse bands on body and tail, extended to a white longitudinal lateral line. Some of them have white spots resembling diamonds on the dorsal surface of the tail. Head has a bluish shade (plate 13). The cephalic casque is ill developed, with smooth head scales.

NATURAL HISTORY

This species lives in all kinds of forests, except in cloud and montane rain forests, at elevations from sea level to nearly 1,000 m above sea level. It is also found in cultivated areas, savannas, and groves or gardens in urban zones. It occupies mangroves on the southern coast of Península de Hicacos and sometimes lives among the sea grapes of the northern coast. It is found on the largest plants of the keys, such as the mastic tree (*Bursera simaruba*) and fan palms (*Coccothrynax* spp.). An arboricolous species, it perches on

Table 7.10. Comparison of snout-vent length *(SVL)*, head length *(HL)*, and tail length *(TL)* between male and female adults of *Anolis equestris* (symbols as in table 7.1).

Sex	N	X	CV	m	M	t	p
				SVL			
Males	23	143.3	9.50	116.3	171.4	3.57	<0.01
Females	29	131.9	7.14	114.4	157.2		
				HL			
Males	23	49.1	7.84	42.4	55.2	2.98	<0.01
Females	29	45.7	9.44	38.1	57.5		
				TL			
Males	13	273.2	11.70	145.5	316.0	2.43	<0.05
Females	20	248.3	10.69	186.0	317.2		

trunks and high branches of trees and spends the greater part of the time on the crowns. It has also been observed in caves sleeping in fissures, near the entrance, and on an interior stalactite.

It is a diurnal heliothermic species that basks in tree crowns during the morning. At noon it descends the trunks, seemingly as a thermoregulatory behavior, to avoid sunlight. It is most frequently found basking in a great amount of sunlight; the black coloration of the peritoneum seems to serve as protection for the internal organs.

The species is very agile and flees on tree crowns, where it spends many hours. Its behavioral displays mainly occur on branches. When surprised on a tree trunk, it slowly places itself on the opposite side of the trunk, trying to make itself inconspicuous; afterward, however, it quickly escapes upward. It can jump and run among branches and to nearby tree crowns.

It is an aggressive species that attempts to bite its pursuers, opening the mouth and extending the dewlap as a defensive behavior. When biting, it does not wound but strongly presses the skin and may cause pain. It bites and mutilates other lizards in collecting bags. Its appearance, size, and biting habits when disturbed or frightened are the basis of the false belief that its bite causes fever in humans. However, this species does not have sharp teeth, but conic, blunt, and very small ones, which do not secrete toxins or poisons. Consequently, the bite is completely inoffensive; it is only disturbing because of the pressure, which can cause a small wound.

Males settle territories in the canopy, at great heights above ground. Fights between two males are frequent when they meet on the same tree. Both extend their dewlaps, open their mouths, and laterally face each other, trying to bite. They also bob and push up their bodies by limb flexion and extension, as well as lifting the tail and body. Both males and females remain in a pale green phase while performing their behavioral displays.

The species feeds on fruits, large insects, and small vertebrates such as frogs, lizards, and caged birds. It is an opportunist omnivorous species, with a wide spectrum of consumption possibilities. It usually sits and waits for its prey and jumps to catch it. In addition, it can eat uncultivated fruits, biting them slightly. Small vertebrates are captured and firmly bitten with the mouth in an attempt to strangle them. The species makes a series of cyclic movements with the hyoid apparatus during feeding that seem to disturb swallowing. It accepts several types of food in captivity, such as butterfly larvae, grasshoppers, crickets, cockroaches, adults and larvae of *Tenebrio molitor*, tree frogs (*Osteopilus septentrionalis*), guava, "cundeamor" seeds, and minced horseflesh. It is fed with lizards (*Anolis lucius* and *A. porcatus*) in the Zoological Station of Rakovnik, Czechoslovakia.

The reproductive cycle of this species is not known, only that courtship and mating take place in the high branches of trees and oviposition generally occurs on the ground, in burrows that females make near the bases of trees, or in tree holes. An egg laid by a captive female from Cayo Santa María measured 23.5 x 13.3 mm and was cream colored. Seemingly, eclosions mainly occur toward the end of the wet season, since a greater number of juveniles have been observed from the last days of July to the first days of August.

Apparently it is a common species throughout its geographic range. In general, it does not establish large familiar groups; it is more frequent to find one or two individuals in each tree; however, there are groves that support numerous individuals. The population from Cayos de las Cinco Leguas, to the north of Cárdenas, is exceptionally numerous. The status of populations of Cayo las Brujas and *A. e. potior* is considered critical because of loss of habitat and low relative abundance, according to the IUCN Red List Categories.

It is a long-lived species; one individual survived in captivity for three years and five months, and another still living in November 1979 was twelve years, ten months, and sixteen days old.

To date, four species of parasitic nematodes are known for *Anolis equestris: Cyrtosomum sclepori, C. longicaudatum, Physaloptera squamatae,* and *Oswaldocruzia lenteixeirai,* as well as one trematode, *Neodiplostomum centuri,* and one species of cestode, *Sparganum* sp. A species of the genus *Hemogregarina* has also been found in its red cells.

Seemingly, the smallest individuals, juveniles and subadults, are subject to predation by various birds. Those living in tree crowns are more easily captured by the Sparrow Hawk (*Falco sparverius*), the usual predator of lizards, than are other terrestrial lizard species of the same size. Furthermore, an attack by a Red-Legged Thrush (*Turdus plumbeus*) on a subadult was reported, and M.S. Buide (pers. comm.) found a Lizard Cuckoo (*Saurothera merlini*) that had suffocated trying to swallow a subadult giant anole.

Anolis luteogularis (chipojo; giant anole; plate 14)

Anolis luteogularis Noble and Hassler, 1935, *Copeia* 3:113; holotype AMNH 46502. Type locality: San Diego de los Baños, Pinar del Río province, Cuba. *Anolis equestris hassleri* Barbour and Shreve, 1935, *Occ. Papers Boston Soc. Nat. Hist.* 8:251. *Anolis luteogularis luteogularis:* Schwartz and Garrido, 1972, *Stud. Fauna Curaçao Carib. Isl.* 39(134):8.

Geographic Range

Isla de Cuba (from Cabo de San Antonio to Península de Zapata); Isla de la

Juventud; Cayo Juan García, Cayería de San Felipe; Cayo Cantiles, Archipiélago de los Canarreos.

LOCALITIES (MAP 17)

Pinar del Río: San Diego de los Baños (Noble and Hassler 1935); San Vicente (Barbour and Shreve 1935); Isabel Rubio; Cabezas; Monte Magota (Schwartz 1958); Sierra del Rosario (Buide 1967); Península de Guanahacabibes (Garrido and Schwartz 1968); Pica-Pica (Peters, 1970); Herradura; Soroa; Candelaria; Cabañas (Schwartz and Garrido 1972); Cayo Real (Garrido 1973*b*); Cueva 1, Viñales (Garrido 1981); Río San Juan, El Taburete (González Bermúdez and Rodríguez Schettino 1982); El Salón, Sierra del Rosario (Silva Rodríguez and Estrada 1982); Sierra de los Organos (Silva Lee 1985); El Rubí; Las Peladas (Martínez Reyes 1995); El Abra, Mil Cumbres (L.R.S.). *La Habana and Ciudad de La Habana:* Guanajay (Stejneger 1917); San Antonio de los Baños (Barbour and Shreve 1935); Artemisa (Schwartz 1958); Ceiba del Agua; Laguna de Ariguanabo; Puentes Grandes; Bosque de La Habana; south of Güines (Schwartz and Garrido 1972); Bauta; Arroyo Arenas; El Husillo; Atabey; Batabanó (Garrido and Jaume 1984). *Matanzas:* Santo Tomás, Ciénaga de Zapata; Playa Larga (Schwartz and Garrido 1972); Soplillar (Garrido 1980*b*). *Isla de la Juventud:* Sierra de Casas (Barbour 1916); Los Indios (Barbour and Shreve 1935); Cayo Largo (Zajicek and Mauri Méndez 1969); Cayo Cantiles (Garrido and Schwartz 1969); La Fe; La Vega; Santa Isabel; Pedernales; Carapachibey; Jacksonville; Cayo Piedra (Schwartz and Garrido 1972); La Daguilla (Garrido and Jaume 1984).

CONTENT

Polytypic endemic species; nine subspecies are recognized:

Anolis luteogularis luteogularis

Anolis luteogularis Noble and Hassler, 1935, *Copeia* 3:113. Type locality: San Diego de los Baños, Pinar del Río province, Cuba. *Anolis luteogularis luteogularis:* Schwartz and Garrido, 1972, *Stud. Fauna Curaçao Carib. Isl.* 39(134):8. Geographic range: from Isabel Rubio to the area south of Güines.

Anolis luteogularis nivevultus

Anolis equestris guanahacabibensis Peters, 1970, *Mitt. Zool. Mus. Berlin* 46(1):203 (*nomen nudum*); holotype IZ 339. Type locality: El Veral, Península de Guanahacabibes, Pinar del Río province, Cuba. *Anolis luteogularis nivevultus* Schwartz and Garrido, 1972, *Studies Fauna Curaçao Carib. Isl.* 39(134):11. Geographic range: Península de Guanahacabibes, east to La Jaula.

Anolis luteogularis hassleri

Anolis equestris hassleri Barbour and Shreve, 1935, *Occ. Papers Boston Soc. Nat. Hist.* 8:251. *Anolis luteogularis hassleri:* Schwartz and Garrido, 1972,

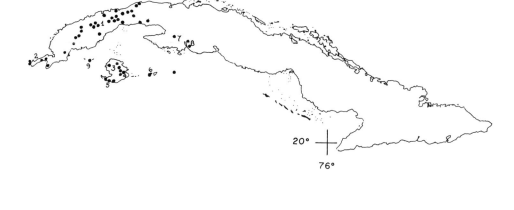

Map 17. Geographic distribution of *Anolis luteogularis. 1, A. l. luteogularis; 2, a. l. nivevultus; 3, A. l. hassleri; 4, A. l. delacruzi; 5, A. l. sectilis; 6, A. l. coctilis; 7, A. l. calceus; 8, A. l. jaumei; 9, A. l. sanfelipensis.*

Studies Fauna Curaçao Carib. Isl. 39(134):14; holotype MCZ 11178. Type locality: Los Indios, Isla de Pinos, La Habana province. Geographic range: Isla de la Juventud, north of the Ciénaga de Lanier.

Anolis luteogularis delacruzi

Anolis luteogularis delacruzi Schwartz and Garrido, 1972, *Studies Fauna Curaçao Carib. Isl.* 39(134):16; holotype IZ 1277. Type locality: Santa Isabel, southeastern portion of the northern two-thirds of Isla de Pinos, north of Ciénaga de Lanier, Isla de Pinos, La Habana province, Cuba. Geographic range: known only from the type locality.

Anolis luteogularis sectilis

Anolis luteogularis sectilis Schwartz and Garrido, 1972, *Studies Fauna Curaçao Carib. Isl.* 39(134);19; holotype IZ 388. Type locality: Pedernales, Isla de Pinos, La Habana province, Cuba. Geographic range: Isla de la Juventud, south of the Ciénaga de Lanier.

Anolis luteogularis coctilis

Anolis luteogularis coctilis Schwartz and Garrido, 1972, *Studies Fauna Curaçao Carib. Isl.* 39(134):22; holotype IZ402. Type locality: Punta del Inglés, Cayo Cantiles, Archipiélago de los Canarreos, La Habana province, Cuba. Geographic range: Cayo Cantiles.

Anolis luteogularis calceus

Anolis luteogularis calceus Schwartz and Garrido, 1972, *Studies Fauna Curaçao Carib. Isl.* 39(134):25; holotype IZ 1295. Type locality: Santo Tomás, Ciénaga de Zapata, Las Villas province, Cuba. Geographic range: Península de Zapata.

Anolis luteogularis jaumei

Anolis luteogularis jaumei Schwartz and Garrido, 1972, *Studies Fauna Curaçao Carib. Isl.* 39(134):27; holotype IZ 369. Type locality: Playa Larga, Ciénaga de Zapata, Las Villas province, Cuba. Geographic range: vicinity of Playa Larga.

Anolis luteogularis sanfelipensis

Anolis luteogularis sanfelipensis Garrido, 1975, *Poeyana* 141:23; holotype IZ 2972. Type locality: Cayo Real, Cayos de San Felipe, Pinar del Río province, Cuba. Geographic range: known only from the type locality.

DESCRIPTION

Large size. Mean snout-vent lengths from 120.5 to 158.5 mm in males and from 109.3 to 150.2 mm in females.

Yellowish green color, with some yellow scales; pale brown transverse bands on body, limbs, and tail. Upper and lateral regions of the head pale brown and rosaceous yellow. White or pale yellow labials and jaws; a brown blotch on the angular region of jaw. White or yellow suprascapular stripe, with an ochraceous inner "tongue." Dewlap in both sexes, pale pink, yellow, or orange, with white inner scales. Brown and yellow transverse bands on tail. Brown iris.

Rugose cephalic casque. Vertically oval, large ear opening. Smooth, circular, large dorsal scales; smooth, circular, small ventral scales. Postcloacal scales not enlarged.

BIBLIOGRAPHIC SOURCES

Noble and Hassler (1935): original description; coloration of the dewlap. Conant and Hudson (1949): longevity. Alayo Dalmau (1955): coloration of the dewlap; scalation. Schwartz (1958): coloration of the dewlap. Buide (1967): coloration of the dewlap. Garrido and Schwartz (1968, 1969): coloration; habitat data. Zajicek and Mauri (1969): hemoparasites. Schwartz and Garrido (1972): drawing of the head and anterior part of the body in dorsolateral view (fig. 5, A-H); description of juveniles; habitat data. Garrido (1973*b*): coloration; habitat data. Garrido (1975*a*): photograph of a specimen in lateral view (fig. 5). Coy Otero and Baruš (1979*a*): parasites. Schwartz and Henderson (1985): drawing of the lateral region of the head and the forelimb (pl. 3, no. 5). Silva Lee (1985): photograph of a specimen; coloration. Socarrás *et al.* (1988): feeding. González Grau *et al.* (1989): predators. Martínez Reyes (1994*c*): habitat data. Rodríguez Schettino (1996): habitat data. Rodríguez Schettino *et al.* (1997): habitat data.

Morphological Variation

Metachromatism from yellowish green, with pale brown transverse bands, to brown with dark brown transverse bands.

Coloration pattern is quite variable individually and geographically.

Geographic Variation

The nine subspecies are described as follows:

Anolis luteogularis nivevultus (Península de Guanahacabibes): yellowish gray or greenish yellow, with brown or black isolated scales. Ivory-colored labials. White nuchal and temporal regions. Brown angular blotch. White, ill-defined suprascapular stripe. Brown and greenish transverse bands on tail. Pink dewlap. Mean snout-vent length of 145.8 mm in males and of 144.3 mm in females.

A. l. luteogularis (Pinar del Río, La Habana, and Ciudad de La Habana provinces): yellowish green, with yellow isolated scales. Ivory-colored labials. Without angular blotch. Yellow suprascapular stripe. Brown and gray transverse bands on tail. Yellow, rosaceous yellow, or orange dewlap. Mean snout-vent length of 141.9 mm in males and of 142.4 mm in females.

A. l. calceus (Península de Zapata): uniformly green. Grayish yellow labials, with gray spots. Without angular blotch. Black postocular, loreal, and nuchal regions; a white occipital blotch. Yellow or greenish yellow suprascapular stripe. Green tail, without crossbands. Pink dewlap. Mean snout-vent length of 147.5 mm in males and of 144.4 mm in females.

A. l. jaumei (Playa Larga): uniformly green. Grayish yellow labials, with gray spots. Without angular blotch. Blue head. White suprascapular stripe, with a yellow inner "tongue." Green tail, without crossbands. Pink dewlap. Mean snout-vent length of 155.5 mm in males and of 141.1 mm in females.

A. l. sanfelipensis (Cayo Real): pale green with brown isolated scales. White labials, with brown spots. Brown angular blotch. Grayish green head. White suprascapular stripe, with a yellowish green inner "tongue." Brown and green transverse bands on tail. Pink dewlap. Mean snout-vent length of 122.2 mm in males and of 109.3 mm in females.

A. l. hassleri (Isla de la Juventud, to the north): brown or green, with four grayish blue crossbands on body. Ivory-colored labials. Without angular blotch. Brown head with white spots. Ill-defined greenish suprascapular stripe. Green and grayish blue transverse bands on tail. Pink dewlap, white edged. Mean snout-vent length of 158.5 mm in males and of 146.7 mm in females.

A. l. sectilis (Isla de la Juventud, to the south): grayish green. Ivory-colored labials. Without angular blotch. Dark brown head. White supras-capular stripe, with a yellow inner "tongue." Grayish green and gray or white crossbands on tail. Pink or pale yellow dewlap. Mean snout-vent length of 151.5 mm in males and of 150.2 mm in females.

A. l. delacruzi (Santa Isabel, Isla de la Juventud): brown with white transverse bands. White labials, with green or brown spots. Dark brown angular blotch. Dark brown head. Ill-defined white suprascapular stripe, with an anterior greenish spot. Brown and white crossbands on tail. Pink dewlap. Mean snout-vent length of 154.5 mm in males and of 140.5 mm in females.

A. l. coctilis (Cayo Cantiles): pale green with blue isolated scales. Brown head, with pale brown between canthal ridges. White labials. Without angular blotch. Black nuchal, postorbital, and loreal regions. A pale brown occipital spot. White suprascapular stripe, with an ochraceous inner "tongue." Green and brown transverse bands on tail. Pale pink dewlap. Mean snout-vent length of 120.5 mm in males and of 118.5 mm in females.

Sexual Variation

Coloration pattern is similar in both sexes; dewlap is equal in males and females. Size differences are shown in table 7.11.

Age Variation

Juveniles are yellowish green, with four yellow transverse bands or without them. Some individuals with snout-vent length from 67.0 to 84.0 mm have crossbands and others do not, irrespective of size. Differently colored isolated scales on bodies of some subspecies are present in lizards with snout-vent lengths larger than 112.0 mm.

Natural History

This species lives in all types of forests except rain and cloud forests, at altitudes ranging from sea level to about 700 m. It also dwells in savannas, cultivated areas, and groves and gardens of urban zones. It is an arboricolous species that perches on tree crowns and high branches, where it spends most of its time. In addition, individuals have been found in the following places: in a hole in the dead bark of a tree, at 1 m above ground; in mangroves of the northern coast of the Península de Guanahacabibes; in a "mogote" complex of vegetation; on the hill of Santa Isabel, Isla de la Juventud; in mangroves (*Laguncularia racemosa*) from Cayo Real, Cayos de San Felipe; and on the inside of a cave 1 m deep.

It is an agile species on tree crowns, over which it rapidly runs and jumps

Table 7.11. Comparison of snout-vent length *(SVL)*, head length *(HL)*, and tail length *(TL)* between male and female adults of *Anolis luteogularis* (symbols as in table 7.1).

Sex	N	X	CV	m	M	*t*	p
			SVL				
Males	25	144.7	10.56	106.0	165.4	0.84	n.s.
Females	30	141.5	9.55	114.5	165.2		
			HL				
Males	25	49.8	20.01	37.1	56.5	4.37	<0.01
Females	30	46.9	8.01	37.1	52.5		
			TL				
Males	18	273.4	12.09	208.5	326.2	0.54	n.s.
Females	20	278.1	6.93	211.3	313.8		

when pursued. It escapes upward and can jump onto other nearby trees. If it is surprised on a trunk, it tends to move to the opposite side, trying to remain unnoticed and escaping quickly. It is aggressive when disturbed or captured; defensive behavior includes mouth opening, attempting to bite, and dewlap extension. The bite of this species is painful because of its pressure; however, it is neither toxic nor poisonous and does not cause fever in humans.

A diurnal species, it is apparently heliothermic. It receives a great amount of sunlight in the canopy of the forests, although during the warmest hours of the day it descends the trunks to the shadows.

It feeds on insects, fruits, and small vertebrates. M. Martínez Reyes (pers. comm.) found insects, diplopods, mollusks (*Zachrysia auricoma*), and abundant vegetable material, mainly fruits, in the stomach contents of ten specimens collected at Sierra del Rosario. In general, the mean number of eaten items and the mean volume of food contents were greater for vegetable material than for animal. In the Zoological Station of Rakovnik, Czechoslovakia, this species was fed with lizards of the species *Anolis lucius* and *A. porcatus*. It ate lepidopteran larvae, grasshoppers, crickets, and cockroaches, larvae and adults of *Tenebrio molitor*, tree frogs (*Osteopilus septentrionalis*), lizards (*A. sagrei*, *A. porcatus*), guava and "cundeamor" seeds, and minced horseflesh in the enclosure of the Instituto de Ecología y Sistemática.

This species is seemingly common throughout its geographic range. It does not generally build large familiar groups; more commonly, one or two individuals are found in each tree. At Playa Larga, Matanzas province, it was found in relative abundance of fewer than one individual/hour seen. It is a long-lived species: one individual lived two years and eleven months in captivity.

To date, nine species of parasitic nematodes are known for *Anolis luteogularis: Parapharyngodon cubensis, Cyrtosomum sclepori, C. longicaudatum, Piratuba* sp., *Porrocaecum* sp., larvae, *Skrjabinoptera (Didelphysoma) phrynosoma, Abbreviata* sp., larvae, *Cyrnea* sp., larvae, and *Oswaldocruzia lenteixeirai,* as well as one species of parasitic mite, the tick *Amblyomma torrei.* In addition, a species of the genus *Hemogregarina* has been found in its red cells.

The Lizard Cuckoo (*Saurothera merlini*) hunts *A. luteogularis* at the entrance of the Cueva de los Majáes, Sierra del Rosario, to which the lizard is attracted by the cockroaches (*Periplaneta americana*) abundant in the cave.

Anolis noblei (chipojo; giant anole; plate 15)

Anolis equestris noblei Barbour and Shreve, 1935, *Occ. Papers Boston Soc. Nat. Hist.* 8:250; holotype MCZ 26653. Type locality: Sierra de Nipe, Oriente province, Cuba. *Anolis equestris galeifer* Schwartz, 1964, *Bull. Mus. Comp. Zool.* 131(12):409. *Anolis noblei*: Schwartz and Garrido, 1972, *Studies Fauna Curaçao Carib. Isl.* 39(134):51.

GEOGRAPHIC RANGE

Isla de Cuba (Sierra de Nipe; Sierra del Cristal; Sierra Maestra; Sierra de la Gran Piedra).

LOCALITIES (MAP 18)

Holguín: Sierra de Nipe (Barbour and Shreve 1935); Bahía de Nipe (Barbour 1937); Cupeyal, Sagua de Tánamo (Schwartz and Garrido 1972). *Granma:* Buey Arriba; Las Mercedes; Guisa (Schwartz 1964a); Sierra Maestra (Buide 1967); Jiguaní (Schwartz and Garrido 1972). *Santiago de Cuba:* Guamá; Santiago de Cuba (Schwartz 1958); Loma del Gato, Hongolosongo (Schwartz 1964a); Pico Turquino; Alto Songo; Limonar; El Cristo; Sierra de La Gran Piedra (Schwartz and Garrido 1972).

CONTENT

Polytypic endemic species; two subspecies are recognized:
   Anolis noblei noblei
Anolis equestris noblei Barbour and Shreve, 1935, *Occ. Papers Boston Soc. Nat. Hist.* 8:250. Type locality: Sierra de Nipe, Oriente province, Cuba. *Anolis noblei noblei*: Schwartz and Garrido, 1972, *Studies Fauna Curaçao Carib. Isl.* 39(134):52. Geographic range: Sierra de Nipe; Sierra del Cristal.

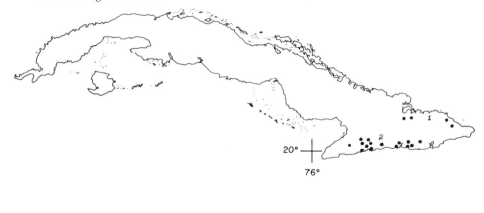

Map 18. Geographic distribution of *Anolis noblei. 1, A. n. noblei; 2, A. n. galeifer.*

Anolis noblei galeifer

Anolis equestris galeifer Schwartz, 1964, *Bull. Mus. Comp. Zool.* 131(12):409; holotype MCZ 59326. Type locality: near Buey Arriba, southwest of Bayamo, Oriente province, Cuba. *Anolis noblei galeifer:* Schwartz and Garrido, 1972, *Studies Fauna Curaçao Carib. Isl.* 39(134):53. Geographic range: Sierra Maestra; Sierra de La Gran Piedra.

DESCRIPTION

Large size. Mean snout-vent length from 154.4 mm to 162.2 mm in males and of 158.7 mm in females.

Green with white isolated scales on head, body, and limbs. Brown upper and lateral regions of the head. Pale green labials with green spots. Pale green postlabial band, to the nuchal region. Pale green suprascapular stripe, with a small blue inner "tongue." Dewlap in both sexes, pale pink, blue toward the gular region and yellow edged. Tail without crossbands. Brown iris.

Rugose cephalic casque. Small, round ear opening. Smooth, circular, small dorsal scales; circular ventral scales, smaller than dorsals. Postcloacal scales not enlarged.

BIBLIOGRAPHIC SOURCES

Barbour and Shreve (1935): original description; coloration. Alayo Dalmau (1955): coloration; scalation. Schwartz (1964a): drawings of the head in lateral view (figs. 2 and 3). Buide (1967): coloration of the dewlap. Schwartz and Garrido (1972): drawing of the head and the anterior part of the body in dorsolateral view (figs. 7A and B); description of the juveniles. Rodríguez Schettino (1985): abundance. Schwartz and Henderson (1985): drawing of the lateral region of the head and of the forelimb (pl. 3, no. 2). Williams (1989a): osteology; drawings of the right jaw in lingual view (fig. 1A), of the parietal bones in dorsal view (fig. 2C).

Morphological Variation

Metachromatism from green with white spots to dark green with pale yellow spots; postlabial band and suprascapular stripe from pale green to grayish green, with dark blue inner "tongue."

Coloration pattern varies individually and geographically.

Geographic Variation

The two subspecies are described as follows:

Anolis noblei noblei (north of the Holguín province): green with white spots. Brown cephalic casque with white dots. Green labials. Pale green postlabial band and suprascapular stripe. Pink dewlap, blue toward the gular region. Mean snout-vent length of 154.4 mm in males; one female of 156.0 mm (Schwartz and Garrido 1972).

A. n. galeifer (Sierra Maestra and La Gran Piedra): uniformly green. Brown cephalic casque with some white dots. Green labials and postlabial band. Pale green suprascapular stripe. Pink dewlap. Mean snout-vent length of 162.2 mm in males and of 158.7 mm in females.

Table 7.12. Comparison of snout-vent length *(SVL)*, head length *(HL)*, and tail length *(TL)* between male and female adults of *Anolis noblei* (symbols as in table 7.1).

Sex	N	X	CV	m	M	t	p
				SVL			
Males	9	149.4	10.35	121.0	170.4	0.80	n.s.
Females	8	145.4	14.09	115.6	172.1		
				HL			
Males	9	48.2	7.65	39.6	54.5	0.07	n.s.
Females	8	48.0	14.03	39.1	57.0		
				TL			
Males	7	262.1	17.04	200.0	311.0	0.81	n.s.
Females	7	242.9	18.05	168.2	280.0		

Sexual Variation

Coloration pattern is similar in both sexes; dewlap is equal in males and females. Measurements are shown in table 7.12.

Age Variation

Juveniles are uniformly green with pale green postlabial band and supras-capular stripe. Some have very pale crossbands on the body. Snout-vent length of one juvenile was 46.0 mm.

NATURAL HISTORY

This species lives in semideciduous and evergreen forests as well as in the rain forests of the eastern mountains, perching on tree trunks and crowns. It is a diurnal species, apparently heliothermic based on its preference for forest canopies, where it receives a great amount of sunlight.

Individuals flee in free life as well as in captivity, going to the opposite side of the trunk to remain unnoticed and then escaping upward. Defensive behavior includes mouth opening, dewlap extension, and attempting to bite. The bite is painful because of its pressure; however, it is neither toxic nor poisonous and does not cause fever in humans.

Members of this species have been fed in captivity with cockroaches and other arthropods, guava and "cundeamor" fruits, and minced horseflesh.

The species is seemingly scarce throughout its geographic range, mainly *A. n. noblei*. However, *A. n. guleifer* is common at Sierra del Turquino. Since it is not easily detected, perhaps it is more abundant than it appears to be in other places as well.

Anolis smallwoodi (chipojo; giant anole; plate 16)

Anolis equestris smallwoodi Schwartz, 1964, *Bull. Mus. Comp. Zool.* 131(12): 412; holotype AMNH 89526. Type locality: Laguna de Baconao, Oriente province, Cuba. *Anolis smallwoodi:* Schwartz and Garrido, 1972, *Studies Fauna Curaçao Carib. Isl.* 39(134):56.

GEOGRAPHIC RANGE

Cuba (from Hongolosongo to Imías; Sierra del Guaso; Cuchillas de Moa).

LOCALITIES (MAP 19)

Holguín: Moa (Schwartz 1964a); Cupeyal del Norte, Sagua de Tánamo (Schwartz and Garrido 1972); Farallones de Moa; Calentura Arriba; La Melba (Estrada *et al.* 1987). *Santiago de Cuba:* Laguna de Baconao; Hongolosongo; Playa Juraguá; Santiago de Cuba (Schwartz 1964a); Gran Piedra (Schwartz

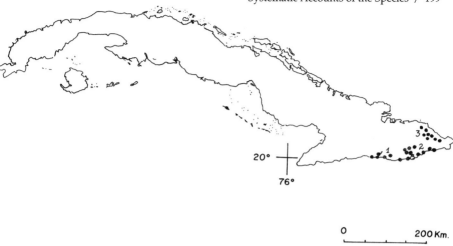

Map 19. Geographic distribution of *Anolis smallwoodi*. *1, A. s. smallwoodi; 2, A. s. palardis; 3, A. s. saxuliceps.*

and Garrido 1972). *Guantánamo:* San Carlos (Barbour and Ramsden 1919); Yateras; Baitiquirí; Base Naval (Schwartz 1964a); Bayate; Sierra del Guaso; Monte Líbano; Imías; Sierra del Purial (Schwartz and Garrido 1972); Cayo Probado; Cabezadas del Jaguaní; Ojito de Agua; Cabezadas del Yarey; Cayo Fortuna; Alto del Yarey (Estrada *et al.* 1987).

DESCRIPTION

Large size. Mean snout-vent length from 150.5 to 151.1 mm in males and from 128.6 to 142.0 mm in females.

Green with many yellow spots on body, limbs, and tail. Dark brown upper and lateral regions of the head. Yellow labials, with black spots. Pale green postlabial band to the nuchal region. Pale green suprascapular stripe, with a pale blue inner "tongue." Dewlap in both sexes, pale pink, blue toward the gular region and yellow edged. A pale green longitudinal lateral band, which separates lateral and ventral regions. Green and yellow ill-defined crossbands on tail. Brown iris.

Rugose cephalic casque. Small, round ear opening. Smooth, circular, small dorsal scales; smooth, circular ventral scales, smaller than dorsals. Postcloacal scales not enlarged.

BIBLIOGRAPHIC SOURCES

Schwartz (1964a): drawings of the head (figs. 4, 5, and 7). Buide (1967): coloration of the dewlap. Lando and Williams (1969): coloration in life; habitat data; behavior; abundance; feeding and behavior in captivity. Schwartz and Garrido (1972): drawing of the head and anterior region of the body in dorsolateral view (fig. 8 A-C); description of juveniles. Coy Otero and Baruš (1979a): parasites. Schwartz and Henderson (1985): drawings of the lateral region of the head and of the forelimb (pl. 3, nos. 3 and 4). Socarrás *et al.*, 1988): feeding in captivity.

Morphological Variation

Metachromatism from green with yellow spots to black with ochraceous spots; postlabial band and suprascapular stripe from green to grayish green, with blue inner "tongue."

Coloration pattern varies individually and geographically.

Geographic Variation

The three subspecies are described as follows:

Anolis smallwoodi smallwoodi (west of Guantánamo Bay): green with large pale green or greenish yellow spots. Gray, brown, or blue cephalic casque. Green labials. Pale green postlabial band with dark green border. Yellow suprascapular stripe. Pink or pale orange dewlap. Mean snout-vent length of 150.7 mm in males and of 128.6 mm in females.

A. s. palardis (east of Guantánamo province): green or brown, with large green or yellow spots. Brown or black cephalic casque, with small white spots. Pale blue labials, with green spots. Yellowish green postlabial band and suprascapular stripe. Pink dewlap. Mean snout-vent length of 151.1 mm in males and of 142.0 mm in females.

A. s. saxuliceps (Cuchillas de Moa): green with yellow spots. Brown cephalic casque with yellow spots. Yellow labials with black spots. Green postlabial band and suprascapular stripe. Pale pink dewlap, blue toward the gular region. Mean snout-vent length of 150.5 mm in males and a maximum of 162.0 mm in females.

Sexual Variation

Coloration pattern is similar in both sexes; dewlap is equal in males and females. Size differences are shown in table 7.13.

Age Variation

Juveniles are uniformly green, with a pale green postlabial band and suprascapular stripe. Four paler crossbands on body, the last one U-shaped, in individuals with snout-vent lengths of 44.0 to 100.0 mm. These bands disappear in subadults with snout-vent lengths of 100.0 to 116.0 mm, on which spots have already appeared.

Table 7.13. Comparison of snout-vent length (SVL), head length (HL), and tail length (TL) between male and female adults of *Anolis smallwoodi* (symbols as in table 7.1).

Sex	N	X	CV	m	M	t	p
			SVL				
Males	12	149.1	12.50	104.0	181.6	2.09	n.s.
Females	8	133.6	8.32	93.5	152.8		
			HL				
Males	12	51.0	11.12	36.1	59.6	1.26	n.s.
Females	8	47.8	11.33	32.6	54.9		
			TL				
Males	10	292.3	16.72	150.2	360.3	0.72	n.s.
Females	7	276.3	5.69	184.2	311.4		

NATURAL HISTORY

The species lives in semideciduous and evergreen forests as well as in the rain forests of the eastern provinces, where it perches on crowns, trunks, and high branches. It is found in coastal zones, lowlands, and uplands and also can live in mangroves, gallery forests, and coffee plantations. A diurnal species, it is apparently heliothermic based on its preference for tree crowns, where it spends most of its time; however, it descends the trunks during the warmest hours of noon.

Individuals flee in free life, trying to remain unnoticed by placing themselves on the opposite side of the trunk, and afterward running away upward. As an aggressive display, they extend the dewlap and open the mouth, attempting to bite. The bite is painful because of its pressure; however, it is neither toxic nor poisonous and does not cause fever in humans. This is a species of slow movements, even when it eats. When pursued, however, individuals quickly run and jump among the tree branches and can jump to another tree. They show some territoriality, since they always perch on the same tree. Males defend territories with aggressive displays such as limb extension and flexion, dewlap extension, and direct attack. For instance, two males were observed in their respective trees, separated by a distance of more than 10 m. After several limb flexions and dewlap extensions, they descended and attacked each other on the ground, attempting to bite. However, four adult lizards kept in a small terrarium at the Instituto de Ecología y Sistemática always occupied the same places, from which they almost never moved, without showing any aggressive behavior; occasionally, they even positioned themselves on top of each other.

This species has been fed in captivity with flowers (*Hibiscus* sp.), lettuce, apple slices, pieces of meat, chicken liver, *Anolis* lizards, and small mice (*Mus musculus*); it has rejected mango, guava, papaya, and prickly pear fruits. It was fed in the Zoological Station of Rakovnik, Czechoslovakia, with lizards of the species *A. porcatus* and *A. lucius*. The lizards in the enclosure of the Instituto de Ecología y Sistemática were fed with "cundeamor" (*Momordica charantia*) and *"galán de día"* (*Cestrum diurnum*) fruits, small *Anolis* lizards, and pieces of horseflesh. They slowly and carefully examined the live prey before catching it, using their tongues.

Seemingly, this species is common throughout its geographic range. It does not build large familiar groups; more commonly, one or two individuals are observed in each tree.

To date, one species of parasitic nematode is known for *Anolis smallwoodi: Porrocaecum* sp.

Anolis baracoae (chipojo, saltacocote; giant anole; plate 17)

Anolis equestris baracoae Schwartz, 1964, *Bull. Mus. Comp. Zool.* 131(12): 419; holotype MCZ 57404. Type locality: Baracoa, Oriente province, Cuba. *Anolis baracoae:* Schwartz and Garrido, 1972, *Studies Fauna Curaçao Carib. Isl.* 39(134):66.

GEOGRAPHIC RANGE

Isla de Cuba (eastern end of the island, east of an imaginary line between Cayo Güín in the north and Imías in the south).

LOCALITIES (MAP 20)

Guantánamo: Baracoa (Schwartz 1964a); Cayo Güín; Jauco; Maisí; Imías (Schwartz and Garrido 1972).

CONTENT

Monotypic endemic species.

DESCRIPTION

Large size. Mean snout-vent length of 132.5 mm in males and of 116.2 mm in females.

Green with orange isolated scales on body, limbs, and tail. Pale brown upper head and greenish sides of head. Yellow labials with green spots. Yellow postlabial band, with some green dots, to the posterior border of the ear opening. Black, ill-defined suprascapular stripe, with many inner dark orange scales. Dewlap in both sexes, pink; blue toward the gular region and green-and-yellow edge scales. Tail without crossbands. Brown iris.

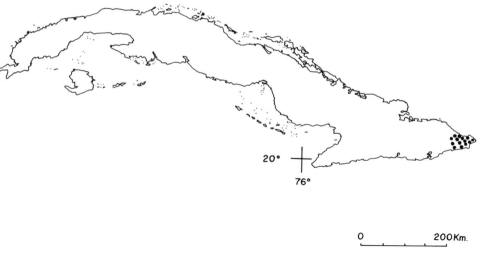

Map 20. Geographic distribution of *Anolis baracoae*.

Cephalic casque not very rugose. Vertically oval, small ear opening. Smooth, circular, small dorsal scales; smooth, circular ventral scales, somewhat smaller than dorsals. Postcloacal scales not enlarged.

BIBLIOGRAPHIC SOURCES

Schwartz (1964*a*): original description; drawing of the head in lateral view (fig. 6). Buide (1967): coloration of the dewlap. Baruš and Coy Otero (1968): parasites. Sonin and Baruš (1968): parasites. Schwartz and Garrido (1972): drawing of the head and the anterior region of the body in dorsolateral view (fig. 8D); coloration; description of juveniles; morphological variation. Coy Otero (1979): helminths. Coy Otero and Baruš (1979*b*): parasites. Silva Lee (1984): photograph of a specimen. Silva Lee (1985): three photographs; feeding. Valdés Zamora *et al.* (1986): feeding. Socarrás *et al.* (1988): feeding in captivity.

MORPHOLOGICAL VARIATION

Metachromatism from green with orange spots to dark brown with pale yellowish orange spots. Some individuals change to green with white spots or to brown with red spots.

Coloration is individually variable, as is the distribution of spots. Dewlap color also varies from bluish green, pale blue, brown with green spots, or pink. These variations may be caused by metachromatic changes.

Geographic Variation

Individual variations previously mentioned occur at different localities; therefore, they do not create geographic races.

Sexual Variation

Coloration pattern is similar in both sexes; the dewlap is equal in males and females. Variations in the dewlap colors do not seem to be related to sex. Differences in size are shown in table 7.14.

Age Variation

Juveniles are green, without crossbands; instead, they have pale spots arranged in some suffused bands across the dorsum and sides of the body. The snout-vent lengths of two juveniles were 45.0 and 48.0 mm, respectively.

NATURAL HISTORY

This species lives in forests of the easternmost region of Cuba, as well as in lowlands and uplands, on tree trunks and high branches. It is a diurnal species, apparently heliothermic based on its preference for tree crowns, where it receives a great amount of sunlight; however, it descends the trunks during the warmest hours of the day.

It is a slow but fugitive species in free life and in captivity; individuals try to hide by positioning themselves on the opposite side of the trunk, afterward running away upward. As an aggressive display, they open the

Table 7.14. Comparison of snout-vent length *(SVL)*, head length *(HL)*, and tail length *(TL)* between male and female adults of *Anolis baracoae* (symbols as in table 7.1).

Sex	N	X	CV	m	M	t	p
			SVL				
Males	10	132.5	9.94	112.4	150.4	2.78	<0.05
Females	10	116.2	11.23	98.1	135.5		
			HL				
Males	10	46.3	9.27	37.5	51.2	2.07	n.s.
Females	10	41.7	13.40	34.1	48.9		
			TL				
Males	6	258.6	9.23	227.5	286.5	1.65	n.s.
Females	6	230.7	14.78	105.6	288.6		

mouth, attempting to bite, and extend the dewlap. The bite of this species is not dangerous, only painful because of the pressure; it is not toxic and does not cause fever in humans.

Feeding in free-living members of this species is unknown; they have been observed to prey on gastropods (*Polymita picta*) and hummingbirds (*Chlorostilbon ricordii*). The species has been fed with lizards of species of *A. lucius* and *A. porcatus* in the Zoological Station of Rakovnik, Czechoslovakia, and with "cundeamor" fruits, pieces of horseflesh, and large insects in the enclosure of the Instituto de Ecología y Sistemática.

This species is seemingly abundant throughout its limited geographic range. It does not build great familiar groups; one or two individuals live on each tree. In some groves and forests there are numerous populations.

To date, one species of parasitic cestode (*Sparganum* sp.) and nine species of parasitic nematodes are known for *Anolis baracoae*: *Cyrtosomum scelopori, C. longicaudatum, Piratuba digiticauda, Oswaldofilaria brevicaudata, Physaloptera squamatae, Trichospirura teixeirai, Oswaldocruzia lenteixeirai, Porrocaecum* sp. larvae, and *Abbreviata* sp. larvae.

Anolis pigmaequestris (chipojo enano; dwarf giant anole; plate 18)

Anolis pigmaequestris Garrido, 1975, *Poeyana* 141:4; holotype IZ 2884. Type locality: Cayo Francés, Archipiélago de Sabana-Camagüey, Caibarién, Las Villas province, Cuba.

Geographic Range

Archipiélago de Sabana-Camagüey (Cayo Francés and Cayo Santa María).

Localities (Map 21)

Villa Clara: Cayo Francés; Cayo Santa María (Garrido 1973*d*).

Content

Monotypic endemic species.

Description

Intermediate size. Mean snout-vent length of 131.5 mm in males and of 115.4 mm in females.

Yellowish green with some green and yellow isolated scales. Grayish brown upper head and loreals. Yellow labials. Yellow postlabial band, to the ear opening. Black postorbital band. Yellow suprascapular stripe. Dewlap in both sexes, pink with pale green edge. Green and brown crossbands on tail. Brown iris.

Cephalic casque not very rugose. Oblique oval ear opening. Quadrangular, smooth, large dorsal scales; small ventral scales. Postcloacal scales not enlarged.

Bibliographic Sources

Garrido (1973*d*): coloration; measurements; habitat data. Garrido (1975*a*): photograph of the head and the anterior part of the body in lateral view (fig. 2); habitat data; behavior; data of an egg; morphological variation. Fernández Méndez (1997): conservation.

Morphological Variation

Metachromatism from green to dark green to brown. Coloration pattern varies individually.

Geographic Variation

One female, captured at Cayo Santa María, had a blue head and a green body, limbs, and tail; yellow ochre suprascapular stripe and pinkish orange dewlap. These features differ from those of individuals from Cayo Francés.

Sexual Variation

Coloration pattern is similar in both sexes; the dewlap is equal in males and females. Differences in size are shown in table 7.15.

Age Variation

Juveniles are not known.

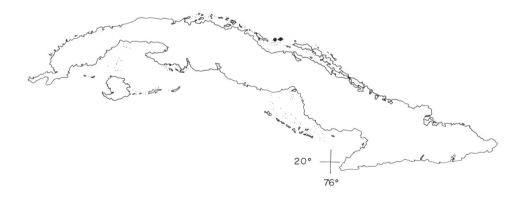

20°

76°

0   200 Km.

Map 21. Geographic distribution of *Anolis pigmaequestris*.

Table 7.15. Comparison of snout-vent length *(SVL)*, head length *(HL)*, and tail length *(TL)* between male and female adults of *Anolis pigmaequestris* (symbols as in table 7.1).

Sex	N	X	CV	m	M	*t*	p
				SVL			
Males	2	131.5	1.94	133.3	140.0	3.48	<0.05
Females	3	115.4	5.18	109.1	121.0		
				HL			
Males	2	41.5	1.87	40.9	42.0	2.92	<0.05
Females	3	36.1	5.03	34.8	37.0		
				TL			
Males	2	159.1	13.55	143.8	174.3	2.60	n.s.
Females	2	215.4	10.08	200.1	230.8		

NATURAL HISTORY

This species lives on trees, bushes, and palms of Cayo Francés and Santa María, in the Archipiélago de Sabana-Camagüey. Males are mainly on trunks of palmettos (*Pseudophoenix* sp.) and females in the shrubby vegetation adjacent to the palmetto area. It is a diurnal species, apparently heliothermic based on its preference for sunny places. It moves rapidly among the vegetation of the keys. Females stay motionless among bushes and afterward escape by jumping; males escape upward.

Feeding in free-living members of the species is not known. There are no data on its reproduction; only that one egg laid by a female captured in October was white and measured 23.5 x 13.3 mm.

It was a common species at Cayo Francés and scarce at Cayo Santa María. However, it has not been captured again, despite many careful searches. Fernández Méndez (1997*a*) reported some sightings at Cayo Santa María. Because of the extreme situation of the species, its status is considered critical according to the IUCN Red List Categories.

Species group *carolinensis* (splenial absent; 6–8 anterior aseptate caudal vertebrae; long head).

Anolis porcatus (camaleón, caguayo; Cuban green anole; plate 19)

Anolis carolinensis Cocteau and Bibron, 1837, *Hist. Fís. Pol. Nat. Isla Cuba* 4:79 (*non* Cuvier, 1829:50). *Anolis porcatus* Gray, 1840, *Ann. Mag. Nat.* ser. 1, 5:112; syntypes BMNH 1946.8.12.66–70. Type locality: Cuba. *Anolis carolinensis porcatus* Gray, 1845, *Cat. Lizards:* 201. *Anolis principalis porcatus* Cope 1887, *Proc. U.S. Natl. Mus.* 10:437.

GEOGRAPHIC RANGE

Isla de Cuba; Isla de la Juventud; Cayos de San Felipe; Archipiélago de los Canarreos; Archipiélago de los Colorados; Archipiélago de Sabana-Camagüey; Archipiélago de las Doce Leguas; introduced into Santo Domingo, República Dominicana (Powell *et al.* 1990; Powell 1992) and into Florida, United States of America (Meshaka *et al.* 1997).

LOCALITIES (MAP 22)

Pinar del Río: Bahía Honda (Garman 1887); San Diego de los Baños; Pinar del Río; Guane; Cabañas (Stejneger 1917); La Güira (Cochran 1934); Viñales; Isabel Rubio; Cayo La Reina, Archipiélago de los Colorados; Herradura; Dimas (Ruibal and Williams 1961*a*); Península de Guanahacabibes (Garrido and Schwartz 1968); Cayo Juan García (Garrido *in* Varona and Garrido 1970); Soroa (Peters 1970); Cayo Real (Garrido 1973*b*): Quiebrahacha (Pérez-Beato and Berovides Álvarez 1979); Pica Pica; La Fe; San Waldo; La Coloma; Los Palacios; Candelaria; Sierra del Rosario, Cayo Juan García (Rodríguez González 1982); Cayo Inés de Soto, Archipiélago de los Colorados (Estrada and Novo Rodríguez 1984*a*); Santa Damiana; Santa Cruz de los Pinos (Pérez-Beato and Berovides Álvarez 1986*b*); El Rubí; Las Peladas; Las Terrazas; El Salón; El Taburete (Martínez Reyes 1995). *La Habana and Ciudad de La Habana:* Habana (Garman 1887): Caimito (Stejneger 1917); Santiago de Las Vegas (Otero 1950); Bosque de La Habana (Collette 1961); San Antonio de los Baños; San José de las Lajas (Ruibal and Williams 1961*a*); Sierra de Camarones; Arcos de Canasí; Jaruco; Madruga; Nueva Paz (Garrido 1973*f*); Jibacoa (Llanes Echevarría 1978); Jardín Zoológico de La Habana (Sánchez Oria 1980); Sierra de Anafe; Guanajay; Playa Caimito; Playa del Rosario; Guanímar; Surgidero de Batabanó (Rodríguez González 1982); Laboratorio Biológico Docente, Boyeros (Granda Martínez 1987); Cueva de Sandoval, Vereda Nueva (Silva Taboada 1988); Jardín Botánico Nacional (García Rodríguez 1989); Escaleras de Jaruco (González González 1989); El Narigón, Puerto Escondido (Alarcón Chávez *et al.* 1990); Managua; Quivicán; Ceiba del Agua (L.R.S.). *Matanzas:* Matanzas (Garman 1887); Alacranes (Ruibal and Williams 1961*a*); Playa Larga; Santo Tomás (Garrido 1980*b*); Punta Sabanilla; Varadero; Versalles; Bellamar; Valle del Yumurí (Rodríguez González 1982); Bacunayagua (Soto Ramírez 1994). *Villa Clara:* Caibarién (Garman 1887); Vega Alta (Cochran 1934); Cayo Santa María; Cayo Francés (Garrido 1973*d*); Cayo Conuco; Cayo Las Brujas (Rodríguez González 1982); Cayo Las Tocineras (Martínez Reyes 1998). *Cienfuegos:* Soledad (Schwartz and Ogren 1956); Rodas; Ciego Montero; Cienfuegos

(Ruibal and Williams 1961*a*). *Sancti Spíritus:* Topes de Collantes; Pico de Poterillo (Rodríguez Schettino *et al.* 1979); Cabaiguán (Palau Rodríguez and Pérez Silva 1995). *Ciego de Ávila:* Loma de Cunagua (Ruibal and Williams 1961*a*); Cayo Coco (Martínez Reyes 1998). *Camagüey:* Camagüey (Barbour and Ramsden 1919); Martí; Playa Santa Lucía; Sierra de Cubitas; Sierra de Najasa (Ruibal and Williams 1961*a*); Vertientes; Archipiélago de las Doce Leguas (Rodríguez González 1982); Cayo Guajaba (Garrido *et al.* 1986); Cayo Sabinal; Cayo Romano (Martínez Reyes 1998). *Holguín:* Banes; Sagua de Tánamo; Moa (Ruibal and Williams 1961*a*); Monte Iberia; Nuevo Mundo; El Guayabo (Rodríguez González 1982); La Zoilita; El Culebro (Abreu *et al.* 1989). *Granma:* Buey Arriba; Yara; Jiguaní (Ruibal and Williams 1961*a*); Cabo Cruz (Rodríguez González 1982). *Santiago de Cuba:* Santiago de Cuba; Guamá (Stejneger 1917); Bueycito (Schwartz and Ogren 1956); Pico Turquino (Ruibal and Williams 1961*a*); Palma Soriano (Peters 1970); Contramaestre El Caney; Loma del Gato; El Cobre; La Mula (Rodríguez González 1982); Siboney (Fong and del Castillo 1997); Baconao (L.R.S.). *Guantánamo:* Río Ovando; Imías; Baracoa; Guantánamo; Río Yumurí (Ruibal and Williams 1961*a*); Base Naval de Guantánamo (Lando and Williams 1969); Yacabo (Peters 1970); Mandinga; Cupeyal; Confluente; Monte Verde; Jaguaní; Punta de Maisí; Río Duaba (Rodríguez González 1982); Cajobabo (L.R.S.). *Isla de la Juventud:* Isla de la Juventud (Stejneger 1917); Cayo Cantiles (Garrido and Schwartz 1969); Sierra de Caballos; Siguanea (Rodríguez González 1982); Sierra de Casas; Nueva Gerona (L.R.S.).

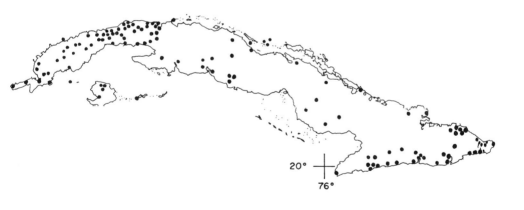

Map 22. Geographic distribution of *Anolis porcatus*.

CONTENT

Monotypic endemic species.

DESCRIPTION

Small or medium size. Mean snout-vent length of 64.0 mm in males and of 53.0 mm in females.

Green with some white isolated scales, or brown with black vermiculations. Pale green or pale blue sides of head. Green labials with dark green spots. Dark green or black suprascapular patch, white bordered, in some individuals. Yellow or white longitudinal middorsal stripe in females and some males. White ventral region. Dewlap in males, mauve with greenish white inner scales. Brown iris.

Canthal and frontal ridges in males; frontals higher than canthals in lateral view. Circular or horizontally oval ear opening. Smooth, circular, very small dorsal scales; smooth, circular ventral scales, somewhat larger than dorsals. Keeled and rugose head scales. Enlarged postcloacal scales in males.

Six pairs of macrochromosomes and twelve pairs of microchromosomes (2n = 36).

BIBLIOGRAPHIC SOURCES

Cocteau and Bibron (1838): (as *carolinensis*) external morphology; drawings of a specimen in lateral view (pl. 11), of dorsal scalation of the head (pl. 11, no. 1), of gular scalation (pl. 11, no. 2), of femoral and cloacal regions (pl. 11, no. 3), of ventral surface of the right hindlimb (pl. 11, no. 5), of the body scalation (pl. 11, no. 6); coloration; metachromatism. Gray (1840): original description. Gundlach (1867): coloration; habitat data; abundance. Gundlach (1880): coloration of a male; habitat data; abundance. Garman (1887): morphology. Barbour (1914): habitat data; abundance. Barbour and Ramsden (1916*b*): habitat data; abundance. Stejneger (1917): coloration of adult and juvenile. Barbour and Ramsden (1919): coloration; scalation; drawing of the dorsal scalation of the head (pl. 4, fig. 9); habitat data; abundance. Dunn (1926): reproduction. Hadley (1929): metachromatism. Barbour (1930*a*): abundance. Barbour (1930*b*): abundance; marketing. Otero (1950): photograph of a specimen in lateral view; feeding. Alayo Dalmau (1951): habitat data; abundance. Sutcliffe (1952): habitat data. Alayo Dalmau (1955): coloration; drawings of the dorsal scalation of the head (pl. 7, fig. 10, no. 55), of the head in lateral view (pl. 7, fig. 10, no. 56), of the tail in lateral view, in the fifth verticil (pl. 7, fig. 10, no. 57); measurements; habitat data; abundance. Schwartz and Ogren (1956): measurements. Etheridge (1960): osteology. Collette (1961): morphology of the subdigital lamellae; drawing of subdigital lamellae of the third toe of the left hindlimb (fig. 2e); behavior; habitat data; abundance; ecomorphology; metachromatism. Ruibal and Williams (1961a): coloration; drawings of the dorsal scalation of the anterior part of the head (fig. 1), of the head in lateral view (fig. 3), of the ear opening (fig. 6), of specimens in dorsolateral view (figs. 7–9); morphology; geographic variation. Ruibal (1964): coloration; drawing of the dorsal scalation of the anterior part of the head (fig. 2), of the head in lateral view (fig. 3); habitat; abundance. Buide (1967): abundance. Rand (1967): reproduction. Ruibal (1967): coloration; metachromatism; intra- and interspecific aggressive behavior. Garrido and Schwartz (1968): abundance. Gorman and Atkins (1968): karyotype. Garrido and Schwartz (1969): coloration; habitat data; abundance. Lando and Williams (1969): coloration in free life; habitat data; abundance; behavior; daily activity. Williams (1969): external morphology; drawing of a specimen in lateral view (fig. 3); habitat data; behavior. Coy Otero (1970): helminths. Peters (1970): habitat data; behavior; abundance. Ramos Vecín (1970): morphology of the nemasperm; photographs of the nemasperm (figs. 3 and 5). Schoener (1970a): interspecific relationships in size. Garrido (1973d): coloration; habitat data; abundance. Williams and Rand (1977): coloration; ecomorphology. Llanes Echevarría (1978): coloration in free life; measurements; habitat data; thermic relations; daily activity. Ortiz Díaz (1978): coloration in free life; measurements; feeding. Pérez-Beato and Berovides Álvarez (1979): scalation; cephalic index; pattern of mental scalation. Espinosa López et al. (1979): electrophoretic patterns of general proteins and esterases; drawings of the pattern of general proteins (fig. 1), of the pattern of esterases (fig. 2). Cepero Chaviano (1980): coloration; scalation; electrophoretic pattern of general proteins and esterases; photographs of males and females in dorsal view (figs. 1–5), of the pattern of general proteins (fig. 9, nos. 1, 4–8; fig. 10, nos. 1, 3, 4, 7, 8; fig. 11, nos. 2–4, 7, 8), of the pattern of esterases (fig. 13, nos. 1–4, 6; fig. 14, nos. 2, 5, 7, 8). Garrido (1980a): reproductive data. González Suárez (1980): electrophoretic pattern of hemoglobin. Sánchez Oria (1980): coloration in free life; scalation; drawings of a specimen (fig. 1), of five patterns of the dorsal scalation of the head (fig. 6, nos. 1–5); photograph of a male, a female, and a juvenile (fig. 2), of the head and anterior part of the body of a male in lateral view (fig.3), of the ventral surface of the right hindlimb (fig. 5); measurements; habitat data; abundance; population dynamics; experimental thermoregulation. Silva Rodríguez (1980): electrophoretic pattern of general proteins. De Smet (1981): ideogram (fig. 5). González Martínez (1981): histostructure of the liver. Mugica Valdés (1981): electrophoretic patterns of esterases. Rodríguez González (1981): habitat data; daily activity; feeding. González Martínez (1982): histostructure of the liver. Mugica Valdés *et al.* (1982): electrophoretic pattern of esterases (fig. 1a). Pérez-Beato (1982*b*): measurements of adults and juveniles; development indexes. Rodríguez González (1982): morphometry; photographs of the anterior part of the body in lateral view (figs. 14–16); scalation; geographic variation; reproduction. Estrada and Silva Rodríguez (1984): morphometry; general index of locomotion. Pérez-Beato and Berovides Álvarez (1984): morphometry; cephalic index; pattern of mental scalation. Silva Lee (1984): photograph of a specimen. Buide (1985): photographs of two males and a female; coloration; habitat data; abundance. Espinosa López, González Suárez, and Berovides Álvarez (1985): electrophoretic pattern of albumin and esterases of muscle; photographs of both patterns (fig. 1). Regalado Morera (1985): tonic immobility. Rodríguez Schettino (1985): habitat data; abundance; altitudinal distribution. Rodríguez Schettino and Martínez Reyes (1985): habitat data; abundance; thermic relations. Sánchez Oria and Berovides Álvarez (1985): behavior; experimental thermoregulation. Schwartz and Henderson (1985): drawing of the lateral region of the head and of the forelimb (pl. 2, no. 2). Silva Lee (1985): three photographs of three specimens; metachromatism; aggressive behavior; reproduction; territoriality; predators; feeding. Armas and Alayón (1986): feeding. Espinosa López *et al.* (1986): electrophoretic patterns of five protein systems. Estrada and Novo Rodríguez (1986*a*): habitat data. Pérez-Beato and Berovides Álvarez (1986*a*): morphology. Estrada *et al.* (1987): habitat data. Granda Martínez (1987): morphometry; morphometric indexes. Quintana Ferrer (1987): morphometry. Sánchez Oria and Berovides Álvarez (1987): drawing of five patterns of dorsal scalation of the head (fig. 1). Abreu and Cruz (1988): predators. García Rodríguez (1989): habitat data; abundance; feeding; reproduction; thermic relations; photograph of a male (fig. 4). Manójina *et al.* (1989): predators. Porter *et al.* (1989): karyotype; photographs of the ideogram (fig. 1D), of the secondary spermatocyte (fig. 2D). Alarcón Chávez *et al.* (1990): abundance. Espinosa López, Posada García *et al.* (1990): electrophoretic patterns of general proteins of muscle. Espinosa López, Sosa Espinosa, and Berovides Álvarez (1990): electrophoretic patterns of eight protein and enzymatic systems. Pérez-Beato and Arencibia (1991): morphological relationships among populations. Quesada Jacob *et al.* (1991): habitat data; abundance. Schwartz and Henderson (1991): habitat data. Powell (1992): definition; distribution. Sampedro Marín *et al.* (1993): habitat data; feeding; electrophoretic patterns of several proteins. Martínez Reyes (1994*c*): habitat data. Palau Rodríguez and Pérez Silva (1995): morphometry. Rodríguez

Schettino (1996): habitat data. Fong and del Castillo (1997): habitat data; feeding. Rodríguez Schettino *et al.* (1997): habitat data. Sanz *et al.* (1997): histology.

MORPHOLOGICAL VARIATION

Metachromatism from green with white dots to brown with black vermiculations. Coloration is quite variable individually and geographically, as are the intensity and distribution of the dots and spots.

Geographic Variation

Population variations exist regarding coloration pattern and shape of the ear opening. In the western region of the isle of Cuba, lizards have green bodies with white scales on the nuchal region, black vermiculations, and a black suprascapular patch; their ear openings are circular or horizontally oval. In the central region of Cuba, the body is green with many white spots and a circular ear opening. In the eastern region of the island, the body is green with a black suprascapular patch, and males have a yellow middorsal longitudinal line; the ear opening is circular.

In addition, there is a clinal variation in the cephalic index (head length/head width) and in the morphological pattern of the mental scales in western populations.

The electrophoretic pattern of general proteins has six bands in lizards from La Habana and eight bands in those from Pinar del Río province.

Lizards from the eastern zone of Pinar del Río province (from Pinar del Río and La Coloma to Candelaria) have horizontally oval ear openings; the others, including those from Isla de la Juventud and the archipelagoes, have circular ear openings.

Among the lizards from the Jardín Zoológico de La Habana, there are five scalation patterns for scales between supraorbitals: complete contact; partial contact; one row of scales; one and two rows of scales; and two rows of scales. The most frequent pattern is one row of scales, in juveniles as well as in adults. In addition, surrounding the interparietal scale are one, two, or three rows of scales; however, the most frequent pattern is two rows, in juveniles as well as in adults.

Sexual Variation

Females are green with yellow labials, eyelids, and middorsal longitudinal lines; their canthal and frontal ridges are not as high as those in males (plate 20). Males have a greater number of subdigital lamellae than females, and they are also discerned by their dewlaps, their larger size, and their enlarged postcloacal scales. Differences in size are shown in table 7.16.

Table 7.16. Comparison of snout-vent length *(SVL)*, head length *(HL)*, and tail length *(TL)* between male and female adults of *Anolis porcatus* (symbols as in table 7.1).

Sex	N	X	CV	m	M	t	p
			SVL				
Males	37	64.0	9.76	41.5	74.3	5.13	<0.01
Females	11	53.0	11.55	40.5	61.4		
			HL				
Males	37	21.2	11.77	12.5	25.5	6.07	<0.01
Females	11	16.4	8.19	13.4	18.3		
			TL				
Males	35	120.1	8.59		150.1	4.23	<0.01
Females	11	105.5	8.31		118.2		

Age Variation

Juveniles have the female coloration pattern; they do not have high frontal and canthal ridges. Males are recognized by their enlarged postcloacal scales. Lizards smaller than 45.0 mm are juveniles; their mean snout-vent length is 34.8 mm. The mean snout-vent length of five newly born lizards was 22.7 mm and that of nine adult males was 72.5 mm. This shows a calculated development index (mean s-v length of adults/mean s-v length of juveniles) of 3.19.

The liver's histological structure varies according to age; juveniles have greater nuclear volume and less diversity of cell types.

Natural History

This species lives in gardens and groves of urban zones, in open areas and savannas, and in secondary and semideciduous forests, mainly at roadsides and on roads that cross them. It perches on tree and bush trunks and branches, on wooden and cement fence posts, and on the outside walls and windows of houses, although occasionally it is also found inside. In addition, it is found on the long leaves of species of the genera *Agave* and *Fourcraea* and on trunks of different species of palms, such as *Roystonea regia, Cocos nucifera,* and *Coccothrynax* sp., of algarrobo *Samanea saman* and *Ficus* spp.

Adults position themselves on tree trunks and fence posts at greater heights than do juveniles, which are more associated with bushes and tall grasses. The species perches between 1 and 5 m above ground, although in places where this variable was quantified, the means were relatively low:

0.95 m at Jibacoa; 1.14 m for males at Guanajay; and 1.85 m at the cliffs of Península de Guanahacabibes, where individuals occupy the trunks of palmettos, the tallest plants. Where the species is found syntopically with *A. sagrei*, it occupies higher tree trunks and branches than those used by *A. sagrei*, but with smaller diameters.

Usually it escapes upward, quickly climbing the trunks toward the crown or climbing to the tops of fence posts, then jumping to another adjacent tree or post. In some instances, it tries to hide on the opposite side of the trunk; it is also able to jump to and from branches separated by a distance of 1 m (L. Moreno, pers. comm.).

According to the mean of the general index of locomotion calculated for one population (0.94), the species is a crawler that shows a great capacity for climbing to great heights on trees. However, the mean index calculated for another population (0.79) was much smaller, as was the index of expected arboreality (1.32), which is very small in comparison with those of other arboricolous species. Seemingly, these differences appeared because the second sample was collected on human buildings, which impose limits on the types and heights of available perches.

Males defend their territories, which are not very large. The aggressive display between two males begins with head bobbing, which lasts 2.5 sec.; this is followed by complete extension of the dewlap and raising of the nuchal and dorsal crests; individuals sometimes flex and drop the head during the last bob.

It is a diurnal heliothermic species that basks frequently on trunks, branches, and leaves of trees and bushes. The mean air temperatures at which this species was found at different localities in La Habana and Ciudad de La Habana provinces were 31.3°C for males, 31.6°C for females, and 34.0°C for juveniles. In Guantánamo, most lizards were observed under direct or filtered sunlight. The mean body temperature of eight individuals captured on the second cliff of Península de Guanahacabibes was 32.7°C, higher than the means of the air and substrate, which were 30.4 and 31.0°C, respectively. On the other hand, in the coastal shrubwood of the Jardín Botánico Nacional, the air temperature in which these lizards were observed varied between 27.4 and 38.4°C; the relative humidity between 54 and 80%; and the light intensity between 5,000 and 30,258 lux.

In experiments conducted on adult males in a porcelain pot, movements of individuals began 45.59 sec. after a 200-watt lamp was turned on, with air temperature increased to 35.1°C. Similarly, panting began after 1.57 min., at an air temperature of 41.4°C, and thermal shock at 2.65 min., with an air temperature of 47.0°C. The maximum temperature tolerated was 50.0°C.

All of these data indicate that *A. porcatus* is a species that efficiently thermoregulates, increasing its body temperature by obtaining radiant energy from the sun. This is a less costly strategy because in the places where it lives it is easy to find adequate sites for basking.

This species is more active during the morning and afternoon; at noon it diminishes its activity, sheltering in shaded places, under fallen palm fronds, and among big leaves of plantain (*Musa paradisiaca*). At La Habana and Ciudad de La Habana provinces two activity peaks were found, one at 10:30 A.M. and the other at 1:00 P.M., followed by peaks in the mean weight of stomach contents at 11:30 A.M. and 3:00 P.M. Activity was moved to around noon during February at Guanajay, with only one peak at 12:30 P.M. This indicates that the species' level of activity is regulated by environmental temperature and the need for food, because in February the majority of lizards were found to be active during the warmest hours of the day.

It sleeps during the night, with the body extended over long leaves, small branches, or thin trunks. Most individuals observed during the first hours of the morning are found under tree bark and crevices, which indicates that these are also possible sites for nocturnal resting.

It is a facultative omnivorous species, although mainly insectivorous. It approaches houses searching for insects, and some individuals have been seen biting guava fruits. Based on the stomach contents of specimens collected at several localities of La Habana and Ciudad de La Habana provinces, hymenopterans (mainly formicids) are the most frequently consumed insects, as well as dipterans (domestic flies and mosquitoes), coleopterans, lepidopterans, orthopterans, larvae of dipterans, insect eggs, and spiders. In addition, it is possible that it occasionally eats small lizards: there is one report of a male swallowing an *A. homolechis* female.

During a year of collection in the coastal shrubwood of the Jardín Botánico Nacional, it was found that formicids were the most frequent and important prey, followed by coleopterans, dipterans, lepidopterans, orthopterans, blatopterans, and arachnids. Nevertheless, the frequency and importance of these trophic components varied: during the dry season hymenopterans were more often eaten, while during the wet season the importance of coleopterans increased. Furthermore, formicids were the most abundant prey items found in the stomach contents of lizards captured at Siboney, Santiago de Cuba province.

Based on observations of free-living individuals, the species bites fruits of custard apple (*Annona squamosa*) and guava (ripe and open); takes prepared foods such as meal for captive mammals and birds; and forages around waste deposits, which attract numerous insects. L. Moreno (pers. comm.) observed an *A. porcatus* hunting tree frog larvae (*Osteopilus septentrionalis*) from the edge of a receptacle in which rainwater had accumulated. It has been fed in captivity with moths, domestic flies, fruit flies, crickets, larvae of beetles, cockroaches, and land isopods.

Based on the reproductive condition of 251 adult specimens captured at different localities, egg production is accomplished mainly between February and August. The mean length of oviductal eggs varied between 6.5 and 10.8 mm, and the testes length was greater than 3.0 mm in all months. The population in the coastal shrubwood of the Jardín Botánico Nacional reproduces between April and September, the period during which males have a mean testes length of 5.0 mm and females have oviductal eggs with a mean length of 8.4 mm; both sexes have small fat bodies.

The longest period of copulation observed in captivity was 59 min., with a mean of 20.0 min., while the mean of the duration of precopulatory behavior was 26.0 sec. The members of the pair were gently separated. Eggs are white and elliptical and measure from 11.8 x 5.2 to 15.5 x 9.0 mm.

Females of this species lay eggs in the interstices of leaves in trees and bushes. Eggs have been found among the leaves of the coconut tree (*Cocos nucifera*) and royal palm (*Roystonea regia*) in groups of twenty-nine. The number of small juveniles, with snout-vent lengths between 21.0 and 31.0 mm, increases from July to September, and it is at this time that the greatest number of eclosions occurs.

This species is abundant in gardens, urban zones, and plantations, while in forests it is scarce. Fluctuations in the number of individuals were found for the population from Bosque de La Habana, in relation to the growth of gramineans near the road that crosses this place. During the time when grasses were cut, there were many lizards on bush trunks and fence posts, but when the gramineans grew tall, lizards disappeared from those places.

For the population group that inhabits the palmettos on the second cliff of Península de Guanahacabibes, the relative abundance was 12.0 individuals/hour during the dry season and 4.0 individuals/hour during the wet season. On the other hand, the relative abundance calculated for the coastal shrubwood of El Narigón, on the northeastern coast of La Habana, was 1.9 individuals/hour in June 1990. In the coastal shrubwood of the Jardín Botánico Nacional, the mean abundance was 23.6 individuals/hour during a nine-month period of observation. At Playa Larga, Matanzas province, the relative abundance was 4.0 individuals/hour.

In general, the abundance of this species depends on habitat conditions:

populations are larger in zones where the vegetation is not too dense. Therefore, urbanization has not only disturbed the species but allowed it to use a greater diversity of environmental resources.

To date, two species of parasitic cestodes are known for *Anolis porcatus: Urotrema wardi* and *U. scabridum;* in addition, there are two species of parasitic nematodes, *Parapharyngodon cubensis* and *Cyrtosomum scelopori.*

Snakes seem to be important predators of *A. porcatus.* Several individuals of *Tropidophis maculatus, T. pardalis, Antillophis andreai,* and *Alsophis cantherigerus* were fed in captivity with *A. porcatus.* In addition, the Sparrow Hawk (*Falco sparverius*) feeds on this species, both in free life and in captivity. Other birds, for instance, some species of Blackbirds (Icterinae), attack these lizards in urban zones. Domestic cats are customary predators of small lizards, and the behavior known as tonic immobility in males of *A. porcatus* has been observed in front of cats, but not facing *Antillophis andreai.* Hutias (*Capromys pilorides*), in the enclosure of the Instituto de Ecología y Sistemática, accepted these lizards when they were offered as part of a feeding experiment, an indication that this mammal, mainly phytophagous, could be a predator of lizards in natural conditions. The same is true of Almiquí (*Solenodon cubanus*), which eats *A. porcatus* in captivity.

Anolis isolepis (plate 21)

Ctenocercus isolepis Cope, 1861, *Proc. Acad. Sci. Philadelphia* 13:214; holotype formerly in USNM, now lost. Type locality: Monte Verde, Oriente province, Cuba. *Anolis isolepis:* Gundlach, 1867, *Repertorio Fís. Nat. Isla de Cuba* 2:110.

GEOGRAPHIC RANGE

Isle of Cuba (isolated populations in central and eastern regions).

LOCALITIES (MAP 23)

Cienfuegos: Buenos Aires, Sierra de Trinidad (Garrido and Jaume 1984); San Blas (Garrido 1985). *Sancti Spíritus:* Sierra de Jatibonico (Barbour 1914). *Camagüey:* Camagüey (Ruibal 1964). *Holguín:* Moa (Alayo Dalmau 1955); Gibara; Monte Iberia; Piloto (Garrido 1985); Cupeyal del Norte (Estrada *et al.* 1987). *Santiago de Cuba:* Hongolosongo; El Cobre (Alayo Dalmau 1955); Pico Turquino (Garrido 1980*a*). *Guantánamo:* Monte Verde (Cope 1861); Monte Líbano (Barbour and Ramsden 1919); Cupeyal; Bayate (Garrido 1985); springs of the Yarey River (Estrada *et al.* 1987).

CONTENT

Polytypic endemic species; two subspecies are recognized:

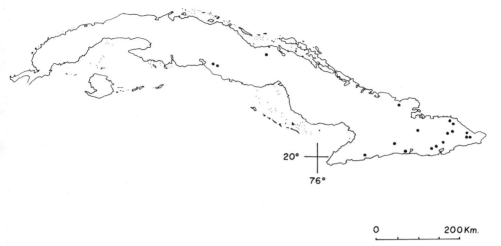

Map 23. Geographic distribution of *Anolis isolepis*.

Anolis isolepis isolepis
Anolis isolepis (Cope), 1861, *Proc. Acad. Nat. Sci. Philadelphia* 13:214. Type locality: Monte Verde, Oriente province, Cuba. Geographic range: Isle of Cuba (isolated populations in central and eastern regions, except for the Pico Turquino).
Anolis isolepis altitudinalis
Anolis isolepis altitudinalis Garrido, 1985, Doñana, *Acta Vertebrata* 12:42; holotype CZACC 4.7028. Type locality: Alto del Cardero (1,050 m), Pico Turquino, Sierra Maestra, Santiago de Cuba province (formerly Oriente), Cuba. Geographic range: Alto del Cardero; Pico Cuba.

DESCRIPTION

Very small size. Mean snout-vent length from 37.6 to 46.5 mm in males and from 35.0 to 48.0 mm in females.

Green with black dots and spots. White supralabial stripe to the suprascapular region. Pale green ventral region. Dewlap in both sexes, rosaceous yellow. Brown iris.

Very small, circular ear opening. Short limbs and tail. Smooth, circular, small dorsal scales; smooth, circular, small ventral scales, the same size as dorsals. Polygonal, flat head scales, arranged as a mosaic. Enlarged postcloacal scales in males.

BIBLIOGRAPHIC SOURCES

Cope (1861): original description. Gundlach (1880): coloration. Barbour (1914): coloration; habitat data; abundance. Barbour and Ramsden (1916):

habitat data; abundance. Barbour and Ramsden (1919): coloration in alive adult; drawing of the dorsal scalation of the head (pl. 7, fig. 2); scalation; habitat data; abundance. Barbour (1930a, b): habitat data; abundance. Alayo Dalmau (1955): coloration; scalation; drawing of the dorsal scalation of the head (pl. 3, fig. 1, no. 2); habitat data; abundance. Etheridge (1960): osteology. Ruibal (1964): coloration; scalation; metachromatism; habitat data; behavior. Williams (1969): habitat data; abundance. Schoener (1970a): intraspecific relationships in size. Williams and Rand (1977): coloration; ecomorphology. Garrido (1980a): data of an egg laid in captivity. Garrido (1985): morphometry; habitat data; behavior; morphological variation. Rodríguez Schettino (1985): habitat data; abundance; altitudinal distribution. Estrada *et al.* (1987): habitat data.

MORPHOLOGICAL VARIATION

Metachromatism from green to olivaceous brown to violet with black dots and spots; the supralabial stripe ranging from white to yellow. Little individual variation in coloration patterns.

Geographic Variation

The two subspecies are described as follows:

Anolis isolepis isolepis (Cuba, except for the Pico Turquino): coloration as in the species. Mean snout-vent length of 37.6 mm in males and of 35.0 mm in females.

A. i. altitudinalis (Pico Turquino): green with black vermiculations; white supralabial stripe to the suprascapular region. Dewlap in both sexes, yellowish white. Brown iris. Rugose head scales; keeled dorsal and ventral scales. Mean snout-vent length of 46.5 mm in males and of 48.2 mm in females.

Sexual Variation

Coloration pattern is similar in both sexes; the dewlap is the same color in males and females, slightly larger in males.

Males are recognized by their enlarged postcloacal scales. Size is similar in both sexes (table 7.17).

Age Variation

Juveniles have broad transverse bands on body and tail, dots on nape, and light and dark spots on head. Their snout-vent length is about 28 mm. Subadults are green, without black spots and with a pale brown middorsal longitudinal line.

Table 7.17. Comparison of snout-vent length *(SVL)*, head length *(HL)*, and tail length *(TL)* between male and female adults of *Anolis isolepis* (symbols as in table 7.1).

Sex	N	X	CV	m	M	t	p
			SVL				
Males	5	35.7	14.19	32.6	41.8	0.65	n.s.
Females	11	38.0	18.28	24.6	49.2		
			HL				
Males	5	13.0	14.62	10.3	15.5	0.19	n.s.
Females	11	13.2	16.33	9.6	16.8		
			TL				
Males	4	42.3	19.74	31.6	51.7	3.34	<0.05
Females	5	54.7	14.90	43.8	70.3		

NATURAL HISTORY

This species lives only in mountainous cloud forests and rain forests at more than 800 m above sea level, generally near the forest borders and at the roadsides of roads that cross them. Males perch on trunks and high branches of trees at heights up to 4 m above ground. Females dwell among dead leaves or plants of the herbaceous stratum.

It is a diurnal species, apparently nonheliothermic based on the shaded and moist places where it lives.

When capture is attempted, females escape by climbing trees or running on the ground. On being captured, they emit squeaks and try to bite.

There are no data about the feeding habits of this species or on its reproductive cycle. The only information known is that a female collected in July laid an egg the day following her capture, which was white and measured 10.0 x 5.7 mm.

Seemingly, this is a scarce species throughout its geographic range; however, it is difficult to detect because of the inaccessibility of its habitat.

Anolis allisoni (camaleón azul, caguayo azul; plate 22)

Anolis carolinensis porcatus: Gundlach, 1880 *(part.)*, *Contr. Erp. Cubana*:43. *Anolis allisoni* Barbour, 1928, *Proc. New England Zool. Club* 10:58; holotype MCZ 26725. Type locality: Coxen Hole, Roatan Island, Honduras.

GEOGRAPHIC RANGE

Bay Islands, Honduras; Half Moon Cay and Turneffe Islands, Belice; Isle of Cuba (from south of La Habana province to Banes in the north and Cabo Cruz in the south).

LOCALITIES (MAP 24)

La Habana: Nueva Paz; Playa Tasajera (Peters 1970); Playa Caimito; Playa Mayabeque; Playa del Rosario; Guanímar (Rodríguez González 1982). *Matanzas:* Rincón Francés, Península de Hicacos (Buide 1967); Ciénaga de Zapata (Garrido 1980*b*); Australia sugarcane factory (Rodríguez González 1982); Cárdenas (Garrido and Jaume 1984). *Villa Clara:* Caibarién (Ruibal and Williams 1961*a*); Santo Domingo (Rodríguez González 1982); Los Caneyes, Santa Clara (L.R.S.). *Cienfuegos:* Santa Isabel de las Lajas; Ciego Montero; Soledad; Rodas; Cruces (Ruibal and Williams 1961*a*); Yumurí (Rodríguez González 1982). *Sancti Spíritus:* Trinidad; San José del Lago (Ruibal and Williams 1961*a*); Casilda (Peters 1970); Topes de Collantes; Cabaiguán (Palau Rodríguez and Pérez Silva 1995); Sancti Spíritus (Palau Rodríguez 1997); Hotel Zaza (L.R.S.). *Ciego de Ávila:* Morón (Ruibal and Williams 1961*a*); Isla de Turiguanó (Rodríguez González 1982); Velazco (L.R.S.). *Camagüey:* Camagüey; Sierra de Cubitas (Ruibal 1961); Martí; Cascorro; Playa Santa Lucía; Santa Cruz del Sur; Nuevitas (Ruibal and Williams 1961*a*); Florida (Peters 1970); Vertientes (Rodríguez González 1982). *Las Tunas:* Victoria de las Tunas (Peters 1970); Manatí (Coy Otero 1979); Amancio Rodríguez; Puerto Padre (Rodríguez González 1982); El Cornito (Pérez-Beato and Berovides Álvarez 1986*b*). *Holguín:* Gibara (Garrido 1973*d*); Holguín (Garrido 1980*b*); Bahía de Naranjo (Rodríguez González 1982); Velasco (Garrido and Jaume 1984); Guardalavaca (L.R.S.). *Granma:* Birama; San Ramón; Cabo Cruz; Manzanillo (Ruibal and Williams 1961*a*); Pilón; Belic (Rodríguez González, 1982); Nicaro (Garrido and Jaume 1984); Embalse Leonero (Montañez Huguez *et al.* 1985).

CONTENT

Monotypic species.

DESCRIPTION

Cuban populations are described as follows:

Small or medium size. Mean snout-vent length of 69.6 mm in males and of 51.5 mm in females.

Green with blue head, anterior region of the body, and forelimbs in males. White labials, with blue spots. White postlabial stripe to the ear opening. Black suprascapular patch in some individuals. Yellow middorsal longitudinal line in females. Greenish white ventral region. Dewlap in males, mauve with very pale blue inner scales. Brown iris.

High canthal and frontal ridges in males; canthals higher than frontals in lateral view. Circular ear opening, followed by an elongate, triangular skin

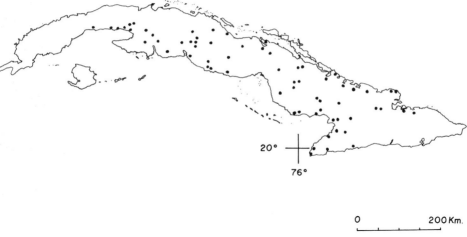

Map 24. Geographic distribution of *Anolis allisoni* in Cuba.

depression. Smooth, circular, small dorsal scales; smooth, circular ventral scales, somewhat larger than dorsals. Rugose and keeled head scales. Enlarged postcloacal scales in males.

Six pairs of macrochromosomes and twelve pairs of microchromosomes (2n = 36).

Bibliographic Sources

Gundlach (1880): coloration. Barbour (1928): original description. Ruibal (1961): habitat data; behavior; measurements; daily activity; thermic relations; reproduction. Ruibal and Williams (1961*a*): coloration; external morphology; drawings of the head in lateral view (figs. 2 and 4), of the ear opening (fig. 5), of a specimen in lateral view (fig. 10). Ruibal (1964): coloration; external morphology; habitat data; abundance. Gorman (1965): karyotype. Buide (1966): abundance. Ruibal (1967): drawing of males in aggressive display (fig. 3, up and down); coloration; metachromatism; aggressive behavior between males; courtship; interspecific behavior. Williams (1969): external morphology; habitat data. Peters (1970): habitat data. Schoener (1970*a*): interspecific relations in size. Coy Otero and Baruš (1973*b*): nematode. Williams and Rand (1977): coloration; ecomorphology. Coy Otero (1979): nematodes. Coy Otero and Baruš (1979*a*): nematodes. Espinosa López *et al.* (1979): electrophoretic pattern of general proteins and esterases; drawings of the pattern of general proteins (fig. 1), of the pattern of esterases (fig. 2). Garrido (1980*a*): data of an egg laid in captivity. Garrido (1980*b*): coloration; habitat data. Cepero Chaviano (1980): coloration; scalation; photographs of a female (fig. 3, left), of a male (fig. 4, right; fig. 5, left), of the pattern of general proteins (fig. 9, nos. 2 and 3; fig. 10, nos. 2, 5, and 6; fig. 11, nos. 1, 5, and 6), of the pattern of esterases (fig. 13, nos. 3, 7, and 8; fig. 14, nos. 1, 3, 4, and 6). González Suárez (1980): elctrophoretic pattern of hemoglobin. Mugica Valdés (1981): polymorphism in electrophoretic patterns of esterases of liver. Rodríguez González (1981): habitat data; daily activity; feeding. Mugica Valdés *et al.* (1982): electrophoretic pattern of liver esterases; drawing of the pattern of esterases (fig. 1b). Rodríguez González (1982): morphometry; photographs of the anterior part of the body in lateral view (figs. 17 and 18); morphological variation; reproduction. Estrada and Silva Rodríguez (1984): morphometry; general index of locomotion. Buide (1985): coloration; habitat data. Schwartz and Henderson (1985): drawing of the lateral region of the head and of the forelimb (pl. 2, no. 1). Silva Lee (1985): photograph of a specimen. Pérez-Beato and Berovides Álvarez (1986a): morphology. Pérez-Beato and Berovides Álvarez (1986b): drawing of five patterns of coloration of the head and body (fig. 1). Granda Martínez (1987): morphometry; morphometric indexes; coefficient of arboreality. Quintana Ferrer (1987): morphometry. Espinosa López, Posada García *et al.* (1990): electrophoretic patterns of general proteins of muscle. Espinosa López, Sosa Espinosa, and Berovides Álvarez (1990): electrophoretic patterns of eight proteic systems. Pérez-Beato and Arencibia (1991): morphologic relations between populations. Garrido and Pareta (1994): communal egg laying. Martínez Reyes (1994c): habitat data. Palau Rodríguez and Pérez Silva (1995): morphometry. Palau Rodríguez (1997): morphometry; feeding.

Morphological Variation

Metachromatism from green and blue to dark brown with black vermiculations. Coloration pattern is quite variable individually and geographically, as are the distribution and intensity of the blue color. Up to five coloration patterns were observed in one sample of twenty specimens collected at Las Tunas province, which suggests that this species shows great color polymorphism.

Geographic Variation

Lizards from Cabo Cruz, Granma province, have round ear openings like *A. porcatus,* but other features like *A. allisoni;* for this reason, they have been considered hybrids. Lizards from the southern coast of La Habana province have deep blue on the body to the hindlimbs, including thighs in some individuals. Lizards from the northern coast of Holguín province, between Gibara and Banes, have smaller skin depressions behind their ear openings than do others.

Nevertheless, these variations are not enough to consider it a subspecies, because the populations are polymorphic in color and scalation.

Sexual Variation

Females are green, without blue, and with a very pale brown middorsal longitudinal stripe. They do not have high canthal and frontal ridges and are discerned from *A. porcatus* females by the skin depressions behind their ear openings.

Males are recognized by their dewlaps, enlarged postcloacal scales, and larger size (table 7.18).

Age Variation

Juveniles have the coloration pattern of adult females; they do not have high canthal or frontal ridges. Males have enlarged postcloacal scales.

NATURAL HISTORY

This species lives in coastal and subcoastal shrubwoods, at the limits of forests and secondary shrubs, and in urban zones. It occupies tree trunks and branches, fence posts, and outside walls of houses, at heights of 1 to 5 m above ground with a mean of 1.7 m, in Playa Caimito on the southern coast of La Habana province.

It escapes upward, climbing tree trunks or posts. Sometimes it tries to hide on the opposite sides of trunks. When on house walls it runs away toward the ceiling, hiding in fringes and cornices.

The general index of locomotion calculated for one population (0.90) suggests that the species is a crawler capable of climbing to great heights on trees. However, the index calculated for another population was much

Table 7.18. Comparison of snout-vent length *(SVL)*, head length *(HL)*, and tail length *(TL)* between male and female adults of *Anolis allisoni* (symbols as in table 7.1).

Sex	N	X	CV	m	M	t	p
				SVL			
Males	27	66.1	10.72	59.4	80.1	6.99	<0.01
Females	16	52.1	9.25	44.4	60.2		
				HL			
Males	27	22.1	13.13	19.2	27.3	7.51	<0.01
Females	16	16.3	9.17	14.0	19.2		
				TL			
Males	27	121.2	10.01		146.5	3.36	<0.01
Females	15	108.9	8.81		128.3		

lower (0.65), as was the coefficient of expected arboreality (1.15). This are very low indices compared with those of other arboricolous species. Seemingly, these differences appeared because the second sample was collected on human buildings, which impose limits on the types and heights of available perches.

Generally, males occupy their own perches and defend them against other males. The threat display of captive males begins with head bobs that last about 3 sec., often accompanied by dewlap extension and elevation of the nuchal crest; the whole process is repeated several times. In front of other species, such as *A. porcatus* and *A. sagrei*, this species has not exhibited aggressive reactions.

It is a diurnal heliothermic species. During the months of July and August, the mean body temperature for individuals from two localities of Camagüey province was 33.06°C, higher than the mean air temperature (31.6°C). The highest number of individuals was observed between 12:00 P.M. and 2:00 P.M. in May 1978.

Based on the stomach contents of thirteen specimens collected at Playa Caimito in May, the eaten insects were hymenopterans (mainly formicids), dipterans (domestic flies and mosquitoes), coleopterans, orthopterans, and insect eggs, in addition to spiders. Furthermore, formicids were the most important items found in the stomach contents of *A. allisoni* from Sancti Spíritus province, followed by coleopterans and dipterans.

In the enclosure of the Instituto de Ecología y Sistemática, the species has been fed with moths, domestic flies, and fruit flies (*Drosophila* spp.); sometimes it has bitten guava fruits.

A. allisoni can reproduce during the entire year, although with an increment in egg production during the wet season, based on the increase in the percentage of gravid females observed in those months. The mean length of the oviductal eggs varies from 8.3 to 11.5 mm. Females lay eggs in decayed wood or in hollows of stone walls. Eggs are white with flexible, leathery shells. The weight of four eggs laid by two captive females varied from 0.3 to 0.5 g and their length, from 12.6 x 5.8 to 18.0 x 8.0 mm.

This species is abundant throughout its geographic range, but mainly in open places and urban zones; it is less numerous in forest zones. At Playa Larga, Matanzas province, the relative abundance was fewer than 1.0 individual/hour.

To date, five species of parasitic nematodes are known for *Anolis allisoni*: *Parapharyngodon cubensis, Cyrtosomum sclepori, Oswaldocruzia lenteixeirai, Physaloptera squamatae,* and *Porrocaecum* sp. larvae.

Anolis angusticeps (plate 23)

Anolis angusticeps Hallowell, 1856, *Proc. Acad. Nat. Sci. Philadelphia* 8:228; holotype ANSP 7789. Type locality: Cienfuegos, Las Villas province, Cuba.

Geographic Range

Isla de Cuba; Isla de la Juventud (southern region); Archipiélago de los Colorados; Archipiélago de Sabana-Camagüey (Cayo Largo, Cayo Francés, Cayo Las Brujas, Cayo Santa María, and Cayo Guillermo); Archipiélago de los Canarreos (Cayo Real, Cayo Juan García, and Cayo Cantiles); Cayos de San Felipe; Bahama Islands.

Localities (map 25)

Pinar del Río: Sierra de Guane (Barbour and Ramsden (1916*a*); San Vicente; Ensenada de Corrientes; Viñales; Las Martinas; Herradura (Hardy 1966); Península de Guanahacabibes (Garrido and Schwartz 1968); San Waldo; San Diego de los Baños; Rangel (Schwartz and Thomas 1968); Cayo Juan García (Garrido *in* Varona and Garrido 1970); Cayo Real (Garrido 1973*b*); Nortey (Garrido 1975*d*); Cayo Inés de Soto, Archipiélago de los Colorados (Estrada and Novo Rodríguez 1984*a*); El Mulo; Las Peladas; Las Terrazas; Soroa (Martínez Reyes 1991); Mil Cumbres (L.R.S.). *La Habana and Ciudad de La Habana:* Sitio Perdido (Cochran 1934); Bosque de La Habana (Collette 1961); San José de Las Lajas; Güines; Guanabo (Schwartz and Thomas 1968); Santiago de las Vegas; Cotorro; Cojímar; Tapaste; Jibacoa; Jaruco; Aguacate; Nueva Paz (Garrido 1975*d*); Jardín Botánico Nacional (García Rodríguez 1989); Escaleras de Jaruco (González González 1989); El Narigón, Puerto Escondido (Hechevarría *et al.* 1990). *Matanzas:* Matanzas; Cárdenas (Gundlach 1880); Península de Hicacos (Buide 1966); Carlos Rojas; Península de Zapata (Garrido 1975*d*); San Miguel de los Baños (Estrada and Hedges 1995); Bacunayagua (L.R.S.). *Villa Clara:* Cayo Lanzanillo (Schwartz and Thomas 1968); Cayo Francés; Cayo Las Brujas (Garrido 1973*d*); Cayo Santa María; Caibarién (Garrido 1975*d*); Cayo Fragoso (Martínez Reyes 1998). *Cienfuegos:* Cienfuegos (Hallowell 1856); Soledad (Barbour and Ramsden 1919); Pasa Caballos; Limones; Sierra de Trinidad; Guajimico (Schwartz and Thomas 1968); Río Calabazas, Placetas (Estrada and Hedges 1995). *Sancti Spíritus:* Sierra de Jatibonico (Barbour and Ramsden 1919); Trinidad; San José del Lago (Schwartz and Thomas 1968); Caguanes; Jatibonico (Garrido 1975*d*). *Ciego de Ávila:* Cayo Guillermo (Garrido 1973*d*); Cayo Coco (Martínez Reyes 1998). *Camagüey:* Camagüey;

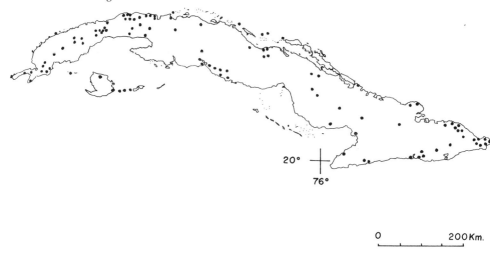

Map 25. Geographic distribution of *Anolis angusticeps angusticeps*.

Playa Santa Lucía; Martí; Río Jigüey (Hardy 1966); Sierra de Cubitas (Garrido 1975*d*); Cayo Guajaba (Garrido *et al.* 1986); Cayo Sabinal; Cayo Paredón Grande (Martínez Reyes 1998). *Las Tunas:* Las Tunas (Díaz Castillo *et al.* 1991); 32 km *SW* of Victoria de Las Tunas (Estrada and Hedges 1995). *Holguín:* Moa (Hardy 1966); Vita; Cananova; Potosí (Garrido 1975*d*); Farallones (Estrada *et al.* 1987); Mayabe, Holguín; Guardalavaca (L.R.S.). *Granma:* Birama; San Ramón (Schwartz and Thomas 1968); Bajada al Pesquero de la Alegría (Estrada and Hedges 1995); Cabo Cruz (L.R.S.). *Santiago de Cuba:* Playa Juraguá; La Socapa (Alayo Dalmau 1951); Pico Turquino (Hardy 1966); La Mula (Rodríguez Schettino 1985); Santiago de Cuba; Playa Siboney; Sardinero (Estrada and Hedges 1995). *Guantánamo:* Río Ovando; Cuchillas de Guajimero; Baracoa (Hardy 1966); Maisí (Garrido 1967); Sierra del Purial (Garrido and Schwartz 1967); Tacajó; Cupeyal (Peters 1970); Bayate; Yateras (Garrido 1975*d*); Ojito de Agua; Cayo Fortuna (Estrada *et al.* 1987); Majayara; Bahía de Boma (J. F. Milera, pers. comm.). *Isla de la Juventud:* Cayo Cantiles (Schwartz and Thomas 1968); Punta del Este (Garrido 1973*b*); Cayo Matías (Acosta Cruz *et al.* 1985); Cayo Campo (Estrada and Rodríguez 1985).

CONTENT

Polytypic species; two subspecies are recognized, one of them living in Cuba:

Anolis angusticeps angusticeps
Anolis angusticeps Hallowell, 1856, *Proc. Acad. Nat. Sci. Philadelphia* 8:228.

Type locality: Cienfuegos, Las Villas province, Cuba. *Anolis angusticeps angusticeps:* Barbour, 1937, *Bull. Mus. Comp. Zool.* 82(2):128. Geographic range: Isla de Cuba; Isla de la Juventud (southern region); Cayo Inés de Soto, in Archipiélago de los Colorados; Cayos Lanzanillo, Francés, Las Brujas, Santa María, and Guillermo, in the Archipiélago de Sabana-Camagüey; Cayos Real and Juan García, in Cayos de San Felipe; Cayos Cantiles, Campo, and Matías, in the Archipiélago de los Canarreos.

DESCRIPTION

The Cuban subspecies is described as follows:

Very small size. Mean snout-vent length of 42.5 mm in males and of 37.7 mm in females.

Greenish brown, with dark brown spots on body and dark brown transverse bands on tail. Gray labials, with some brown spots; pale gray postlabial stripe with brown spots, which continues laterally to the groin. A pale gray dorsolateral and longitudinal line joins the postlabial stripe and continues through the tail. Grayish white ventral region. Dewlap in males, yellowish pink, with white inner scales. Brown iris.

Circular, small ear opening. Short limbs. Smooth, circular, very small dorsal scales; smooth, circular ventral scales, somewhat larger than dorsals. Rugose and striate head scales in females; smooth in males. Enlarged postcloacal scales in males.

Six pairs of macrochromosomes and twelve pairs of microchromosomes (2n = 36).

BIBLIOGRAPHIC SOURCES

Hallowell (1856): original description. Gundlach (1867): habitat data. Gundlach (1880): coloration. Barbour and Ramsden (1919): coloration; habitat data. Alayo Dalmau (1955); coloration; drawing of the dorsal scalation of the head (pl. 3, fig. 1, no. 1); habitat data. Etheridge (1960): osteology. Collette (1961): coloration; morphology of the subdigital lamellae; drawing of the subdigital lamellae of the third left toe (fig. 2b); abundance; habitat data. Ruibal (1961): habitat data. Ruibal (1964): coloration; habitat data. Buide (1966): habitat data; behavior. Hardy (1966): coloration; drawing of the gular scalation (mental and postmental scales) (fig. 4B). Garrido and Schwartz (1967, 1968): habitat data. Gorman and Atkins (1968): karyotype. Schwartz and Thomas (1968): behavior of nocturnal resting; habitat data; behavior. Garrido and Schwartz (1969): behavior. Williams (1969): habitat data. Varona and Garrido (1970): coloration. Schoener (1970*a*): interspecific relationships in size. Garrido (1973*b*): habitat data.

Garrido (1975*d*): coloration; scalation; habitat data. Williams and Rand (1977): coloration; ecomorphology. Garrido (1980*a*): reproductive data in captivity. Estrada and Silva Rodríguez (1982): abundance; sexual ratio; habitat data; thermic relations; daily activity. Estrada and Silva Rodríguez (1984): morphometry; general index of locomotion. Novo Rodríguez 1985): reproductive data. Rodríguez Schettino (1985): habitat data; altitudinal distribution. Rodríguez Schettino and Martínez Reyes (1985): habitat data; thermic relations; abundance. Silva Lee (1985): photograph of a specimen. Estrada *et al.* (1987): habitat data. Alarcón Chávez *et al.* (1990): habitat data; abundance. Hechevarría *et al.* (1990): abundance. Martínez Reyes (1994*c*): habitat data. Rodríguez Schettino (1996): habitat data. Fong and del Castillo (1997): habitat data. Rodríguez Schettino *et al.* (1997): habitat data.

MORPHOLOGICAL VARIATION

Metachromatism from gray and pale brown to dark brown. Coloration is quite variable individually, as is the disposition of the brown spots.

Geographic Variation

A clinal variation is observed in the scalation of Cuban populations. The number of scales between the first canthals and the frequency of one row of scales separating the supraorbital semicircles increase from west to east, while the number of postmental scales decreases.

Coloration pattern, with individual variations, is similar throughout the geographic range.

Sexual Variation

Coloration pattern is similar in both sexes. Females have blackish transverse bands, while males have ill-defined ones. Females have rugose and striate head scales, and males, smooth. Males are recognized by their enlarged postcloacal scales and their dewlaps. Differences in size are shown in table 7.19.

Age Variation

Juvenile coloration is similar to that of adults. Mean snout-vent length of twenty-eight newly born lizards was 17.4 mm, the mean of the tail length 26.5 mm; the mean weight was 112.9 mg. These data show a development index of 2.4 and 2.8 for size and weight, respectively.

NATURAL HISTORY

This species lives in shrubwoods, evergreen and semideciduous forests, cloud forests, and rain forests, generally at their borders, where the vegeta-

Table 7.19. Comparison of snout-vent length *(SVL)*, head length *(HL)*, and tail length *(TL)* between male and female adults of *Anolis angusticeps* (symbols as in table 7.1).

Sex	N	X	CV	m	M	*t*	p
				SVL			
Males	10	42.5	9.64	35.6	49.0	3.22	<0.01
Females	10	37.7	6.21	34.2	41.3		
				HL			
Males	10	14.0	10.12	12.0	16.6	3.15	<0.01
Females	10	12.1	10.84	9.5	13.8		
				TL			
Males	6	53.4	16.82	41.6	69.3	1.59	n.s.
Females	5	46.4	9.00	41.6	50.5		

tion is not very dense. It is also found in rocky places with little vegetation, at roadsides in rural areas, in pine forests, and in semiarid montane serpentine shrubwoods.

It perches on trunks and branches of trees and bushes, at heights of more than 1 m above ground, as well as on fence posts, rocks, and walls. In the semideciduous forest of Península de Guanahacabibes it was observed at a mean height above ground of 2.28 m. Adults use more trunks and high branches, and juveniles use grasses and small bushes in open areas. According to the general index of locomotion (0.92), the species is a crawler, indicating that it has a great capacity to climb trees. When on tree trunks, it generally has the head upward and runs toward the crowns, after being motionless for a short time. It moves slowly, remaining very close to branches and twigs; this behavior, and its cryptic coloration, make detection difficult.

It is a diurnal species that is most frequently found in shaded places, or where the sunlight is filtered by vegetation. In addition, it can be observed in sunny, open sites.

The mean body temperature of eight specimens captured in the semideciduous forest of Península de Guanahacabibes during the wet season was 32.4°C, higher than the air and substrate temperatures (28.8 and 29.3°C, respectively). This suggests that the species is efficient in thermoregulation.

It sleeps during the night, lengthened longitudinally on thin branches; the tail is posteriorly extended with the end lying on the branch. It demonstrates no preference for type, size, or height above ground of branches used for sleeping.

Feeding in free life is unknown, but in captivity it has accepted fruit flies (*Drosophila* spp.) and ants. *A. a. oligaspis* from Bimini, Bahama Islands, consumes more dipterans than hymnopterans, homopterans, larvae of lepidopterans, spiders, and vegetable material, based on examined stomach contents.

There are few data about its reproductive cycle. The mean time spent during nineteen copulations observed in captivity was 9 min., and the longest one lasted 23 min. A female collected in October laid a yellowish white egg measuring 10.0 x 5.0 mm. A communal nest of this species was found in September in a hollow of a palmetto trunk at Cayo Francés, Archipiélago de Sabana-Camagüey. In it were 1,625 empty shells, open at one pole, and 65 white recently laid eggs with a mean length of 9.9 x 7.0 mm. The mean weight of fifty-one eggs was 209.25 mg. Eclosion occurred between six and twenty-two days after collection. The mean weight of the newly born lizards was 112.95 mg.

These data indicate that reproduction is maintained at least until the end of the wet season, and that perhaps in disadvantageous environmental conditions, such as those of the keys, females select communal egg-laying sites.

Seemingly, this species is common throughout its geographic range, since it is frequently observed despite difficulties in detecting it. The number of individuals at Bosque de La Habana varied with the year: ten lizards were captured in 100 min. of collection in December 1957, but only two in two weeks in December 1958.

In June 1990 a relative abundance of 5.76 individuals/hour was found in the coastal shrubwood of El Narigón, on the northeast coast of La Habana province; a relative abundance of three individuals/hour was found in the ecotone between the coastal shrubwood and the dry semideciduous forest in the same locality, on the same date. These figures are higher than those found in the coastal shrubwood (0.3) and semideciduous forest (0.4) of Península de Guanahacabibes during the same season of the year. Apparently the abundance of this species varies according to locality and environmental conditions.

Anolis paternus (plate 24)

Anolis angusticeps Barbour, 1916, *Ann. Carnegie Mus.* 10(2):302. *Anolis angusticeps paternus* Hardy, 1966, *Carib. J. Sci.* 6(1–2):25; holotype USNM 142156. Type locality: vicinity of Nueva Gerona, Isla de Pinos. *Anolis paternus:* Garrido, 1975, *Poeyana* 144:7.

Geographic Range

Isla de Cuba (south of Pinar del Río, from La Fe to Herradura); Isla de la Juventud (northern region).

Localities (map 26)

Pinar del Río: Herradura (Ruibal 1964); Ciudad Sandino; La Fé; San Waldo; La Arenera (Garrido 1975*d*); La Coloma (Garrido and Jaume 1984). *Isla de la Juventud:* Nueva Gerona (Barbour 1916); La Fe; Los Indios (Hardy 1966); Santa Bárbara (Schwartz and Thomas 1968); El Abra; Cayo Piedra; La Siguanea (Garrido 1975*d*).

Content

Polytypic endemic species; two subspecies are recognized:
  Anolis paternus paternus
Anolis angusticeps paternus Hardy, 1966, *Carib. J. Sci.* 6(1–2):25. Type locality: vicinities of Nueva Gerona, Isla de Pinos. Geographic range: Isla de la Juventud (to the north of Ciénaga de Lanier).
  Anolis paternus pinarensis
Anolis paternus pinarensis Garrido, 1975, *Poeyana* 144:8; holotype IZ 4073. Type locality: 5 km from Ciudad Sandino, Guane, Pinar del Río province, Cuba. Geographic range: sandy savannas in the southern part of Pinar del Río province, between La Fe and La Coloma, to Herradura.

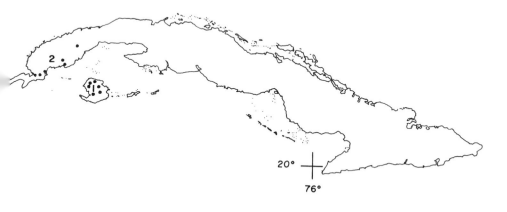

Map 26. Geographic distribution of *Anolis paternus.*

DESCRIPTION

Very small size. Mean snout-vent length from 44.1 to 45.8 mm in males and from 36.2 to 41.6 mm in females.

Greenish gray with dark gray transverse bands on tail and grayish brown transverse bands on limbs. Yellow labials; gray postlabial band to the ear opening, continuing to the groin. A pale yellow, lateral, longitudinal, narrow line is bordered with small black spots on its upper side. Yellow or reddish ventral region. Dewlap in males, carmine or yellow, with black streaks on the gular region and white inner scales. Brown iris.

Small, vertically oval ear opening. Smooth, circular, small dorsal scales; smooth, circular, small ventral scales, somewhat larger than dorsals. Enlarged postcloacal scales in males.

BIBLIOGRAPHIC SOURCES

Barbour (1916): coloration (as *A. angusticeps*). Barbour and Ramsden (1919): habitat data. Ruibal (1964): habitat data. Hardy (1966): original description; coloration; drawings of the dorsal scalation of the head (figs. 2 and 3), of the gular scalation (mental and postmental scales) (fig. 4A); habitat data. Schwartz and Thomas (1968): coloration; abundance. Garrido (1975*d*): coloration and description of metachromatic phases; habitat data; abundance. Garrido (1980*a*): reproductive data in captivity. Estrada and Silva Rodríguez (1984): morphometry; general index of locomotion. Silva Lee (1985): photograph of a specimen. Rodríguez Schettino (1996): habitat data.

MORPHOLOGICAL VARIATION

Metachromatism from greenish gray with dark gray crossbands on tail to brown with dark brown vermiculations on body and dark brown transverse bands on tail. Coloration varies slightly within each subspecies.

Geographic Variation

The two subspecies are described as follows:

Anolis paternus pinarensis (sandy savannas of Pinar del Río province): pale gray with some brown dorsal spots. Yellow ventral region. Gray and brown crossbands on tail. Pink carmine dewlap. Mean snout-vent length of 44.1 mm in males and of 36.2 mm in females.

A. p. paternus (Isla de la Juventud, northern region): greenish gray with dark brown crossbands. Two pale yellow longitudinal lines between forelimbs and hindlimbs. Pale yellow ventral region; orange in the subcaudal region. Greenish gray and dark gray transverse bands on tail. Yellow and carmine dewlap. Mean snout-vent length of 45.0 mm in males and of 41.6 mm in females.

Sexual Variation

Coloration is similar in both sexes. Smooth or slightly rugose head scales in males, quite rugose in females. Males are recognized by their dewlaps and their enlarged postcloacal scales. Measurements are shown in table 7.20.

Age Variation

Coloration of juveniles is similar to that of adults. Males are recognized by their enlarged postcloacal scales.

NATURAL HISTORY

The species lives in the sandy savannas of the southern part of Pinar del Río province and the northern part of Isla de la Juventud. It perches on trunks of palmettos and bushes and on fence posts; it is also found among pasture grasses and in urban zones. In general, it does not perch high above ground.

It is a diurnal species, apparently heliothermic since it frequently basks and is observed mainly during the warmest hours of the day.

It moves slowly, staying very close to branches and twigs; this, and its cryptic coloration, make detection difficult. When on tree trunks or bushes, it generally has the head upward and runs toward the crowns after staying motionless for a short time; when on fence posts or grasses, it hides beneath the herbaceous stratum. Based on the mean general index of locomotion (0.92), it is a crawler, suggesting that it is able to climb trees easily.

Table 7.20. Comparison of snout-vent length *(SVL)*, head length *(HL)*, and tail length *(TL)* between male and female adults of *Anolis paternus* (symbols as in table 7.1).

Sex	N	X	CV	m	M	t	p
				SVL			
Males	16	44.7	27.95	41.3	49.4	1.47	n.s.
Females	14	32.3	14.70	29.4	47.8		
				HL			
Males	16	16.0	28.19	12.9	16.2	1.39	n.s.
Females	14	12.9	15.67	9.8	16.0		
				TL			
Males	16	71.1	10.20	61.2	84.3	2.12	<0.05
Females	12	64.8	13.30	51.1	78.4		

Feeding in free life is unknown, but in captivity it has accepted fruit flies (*Drosophila* spp.) and ants.

Adult members of this species observed in captivity spent a mean time of 2 min. during fourteen copulations; the longest one lasted 4 min.

It is an abundant species throughout its geographic range and very abundant at Isla de la Juventud.

Anolis guazuma (plate 25)

Anolis guazuma Garrido, 1983, *Carib. J. Sci.* 19 (3–4):71; holotype CZACC 46128. Type locality: La Emajagua, Pico Turquino (about 600 m above sea level), Sierra Maestra, Santiago de Cuba, Cuba.

Geographic Range

Isla de Cuba (known only from the type locality).

Localities (map 27)

Santiago de Cuba: La Emajagua, southern slope of Pico Turquino (Garrido 1983); 1.5 km S of La Tabla, northern slope of Sierra Maestra (Díaz *et al.* 1996).

Content

Monotypic endemic species.

Description

Very small size. Mean snout-vent length of 44.1 mm in males and of 38.3 mm in females.

Pale brown with dark brown spots on body; six dark brown transverse bands on tail and many on limbs. Grayish brown labials, with very dark brown spots. A very dark brown spot behind the eye continues, joining with a dark spot on the neck. There is a dark brown, transverse interorbital band. White gular and ventral regions. Dewlap in males, dark orange with an ochraceous basal spot and white inner scales. Brown iris. Black oral cavity and throat.

Circular, very small ear opening. Prehensile tail, shorter than snout-vent length, very thick at the base, without caudal crest. Smooth, granular, and slightly prominent dorsal scales; smooth, flat, circular ventral scales, slightly imbricate on thorax and hexagonal nonimbricate at the abdomen. Smooth head scales; a concave depression between frontal ridges. Elongate and strongly imbricate tail scales. Enlarged postcloacal scales in males.

2 cm

Plate 1. *Leiocephalus carinatus.*

2 cm

Plate 2. *Leiocephalus cubensis.*

2 cm

Plate 3. *Leiocephalus raviceps.*

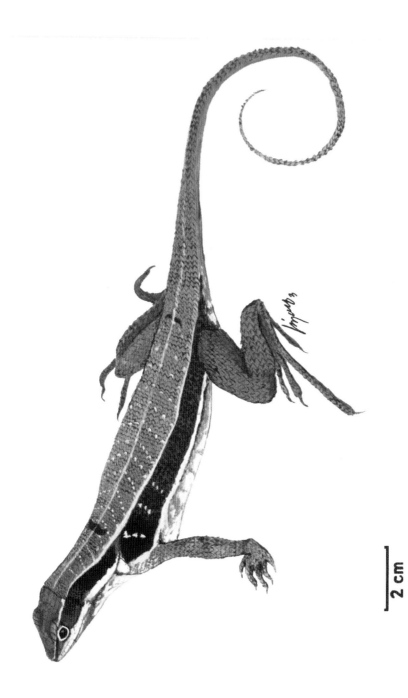

Plate 4. *Leiocephalus macropus.*

2 cm

Plate 5. *Leiocephalus stictigaster.*

Plate 6. *Leiocephalus onaneyi*.

Plate 7. *Cyclura nubila nubila* (adult male).

Plate 8. *Cyclura nubila nubila* (juvenile).

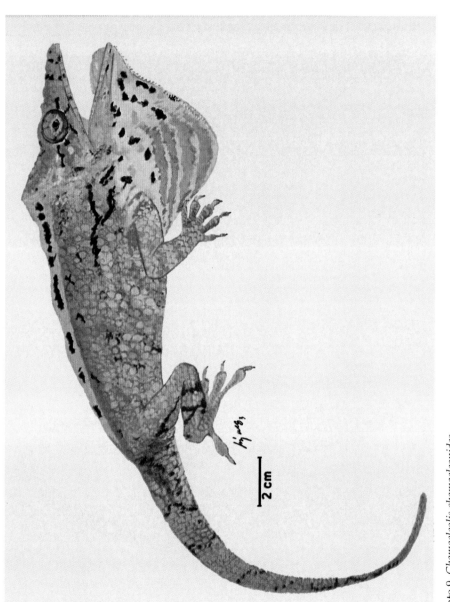

Plate 9. *Chamaeleolis chamaeleonides*.

Plate 10. *Chamaeleolis porcus.*

2 cm

Plate 11. *Chamaeleolis barbatus.*

Plate 12. *Anolis equestris*.

Plate 13. *Anolis equestris* (juvenile).

Plate 14. *Anolis luteogularis.*

Plate 15. *Anolis noblei.*

Plate 16. *Anolis smallwoodi.*

Plate 17. *Anolis baracoae.*

Plate 18. *Anolis pigmaequestris.*

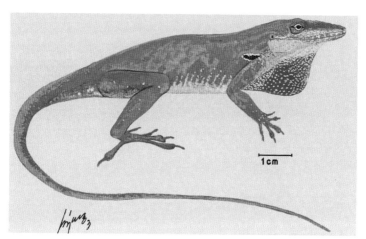

Plate 19. *Anolis porcatus.*

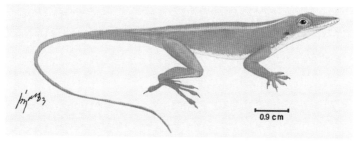

Plate 20. *Anolis porcatus* (female).

Plate 21. *Anolis isolepis.*

Plate 22. *Anolis allisoni.*

Plate 23. *Anolis angusticeps.*

Plate 24. *Anolis paternus.*

Plate 25. *Anolis guazuma.*

Plate 26. *Anolis loysiana.*

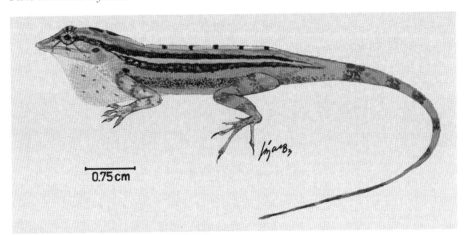

Plate 27. *Anolis argillaceus.*

Plate 28. *Anolis centralis.*

Plate 29. *Anolis lucius.*

Plate 30. *Anolis lucius* (female).

Plate 31. *Anolis argenteolus.*

Plate 32. *Anolis vermiculatus.*

Plate 33. *Anolis vermiculatus* (female).

Plate 34. *Anolis vermiculatus* (juvenile).

Plate 35. *Anolis bartschi.*

Plate 36. *Anolis alutaceus.*

Plate 37. *Anolis clivicola.*

Plate 38. *Anolis anfiloquioi.*

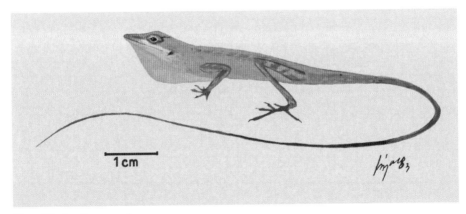

Plate 39. *Anolis cyanopleurus.*

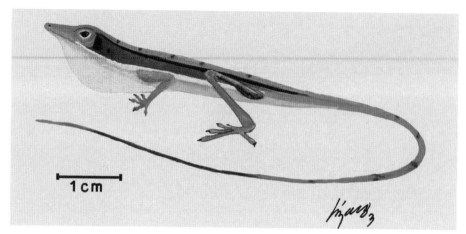

Plate 40. *Anolis cupeyalensis.*

Plate 41. *Anolis mimus.*

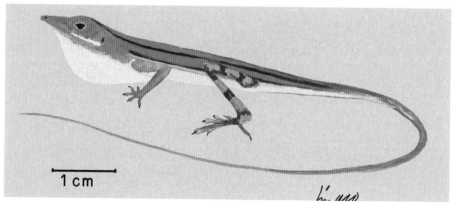

Plate 42. *Anolis fugitivus.*

Plate 43. *Anolis juangundlachi.*

Plate 44. *Anolis spectrum.*

Plate 45. *Anolis vanidicus.*

Plate 46. *Anolis sagrei.*

Plate 47. *Anolis bremeri.*

Plate 48. *Anolis homolechis.*

Plate 49. *Anolis homolechis* (female).

Plate 50. *Anolis quadriocellifer.*

Plate 51. *Anolis jubar.*

Plate 52. *Anolis mestrei.*

Plate 53. *Anolis allogus.*

0,9 cm

Plate 54. *Anolis ahli.*

Plate 55. *Anclis rubribarbus.*

1 cm

Plate 56. *Anolis inias.*

Plate 57. *Anolis ophiolepis.*

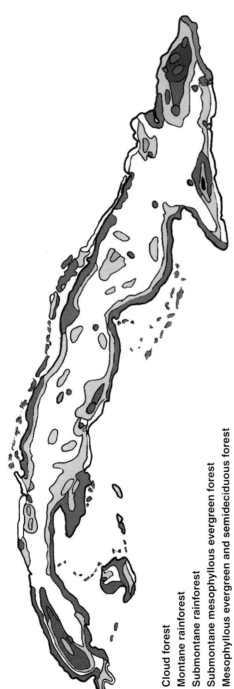

Cloud forest
Montane rainforest
Submontane rainforest
Submontane mesophyllous evergreen forest
Mesophyllous evergreen and semideciduous forest
Pine forest
Mogote complex vegetation
Swamp forest
Mangrove forest
Coastal and subcoastal microphyllus forest
Semidesert thorny shrubwood
Serpentine xeromorphus shrubwood
Cultures and pastures
Urban zones

Plate 58. Current vegetation of Cuba.

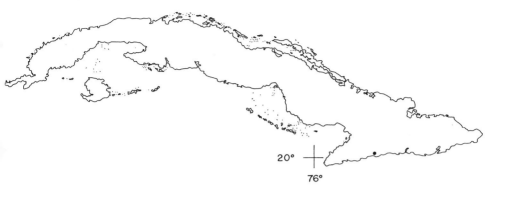

Map 27. Geographic distribution of *Anolis guazuma*.

BIBLIOGRAPHIC SOURCES

Garrido (1983): original description; morphology; habitat data; aggressive behavior between males. Rodríguez Schettino (1985): habitat data; abundance.

MORPHOLOGICAL VARIATION

Metachromatism from pale brown to dark brown.

Geographic Variation

Little individual variability within its limited geographic range.

Sexual Variation

Coloration and scalation are similar in both sexes. Females are pale grayish brown, with three well-defined reddish spots on back, ten brown crossbands on tail, and white rudimentary dewlap. Males are recognized by their dewlaps and their enlarged postcloacal scales. Differences in size are shown in table 7.21.

Age Variation

Juveniles are not known.

NATURAL HISTORY

The species lives on the thinnest twigs of guásimas (*Guazuma tomentosa*) at La Emajagua, Pico Turquino, at 600 m above sea level, and about 3 or 4 m

Table 7.21. Comparison of snout-vent length (SVL), head length (HL), and tail length (TL) between male and female adults of Anolis guazuma (symbols as in table 7.1).

Sex	N	X	CV	m	M	t	p
				SVL			
Males	7	44.1	9.15	38.9	48.5	2.80	<0.05
Females	3	38.3	4.08	36.6	39.7		
				HL			
Males	7	14.5	7.57	13.0	15.5	3.69	<0.01
Females	3	12.4	4.14	12.0	13.0		
				TL			
Males	4	38.4	8.27	35.2	42.8	0.78	n.s.
Females	2	35.8	14.80	32.0	39.5		

above ground. Males and females are found toward the peripheral ends of the twigs. It is a diurnal species, apparently heliothermic since it prefers high, peripheral twigs exposed to sun during the dry season.

It moves slowly, remaining very flattened against twigs; perhaps this is its principal defensive behavior. This and its cryptic coloration make detection difficult. Nevertheless, when danger is near it jumps to the ground and remarkably hides among grasses and beneath dead leaves. Its behavioral displays seem to be simple, based on the aggressive display observed between two males at the end of a twig of "guásima," during which one of them moved the head, nodded forward, and raised the anterior part of the body. The other male moved only the head. Both kept the dewlap partially extended. One of them opened the mouth, closed it as if swallowing something, and went back.

This species was common in its limited geographic range in March 1980. At least one pair was detected on each guásima. Nevertheless, no other individuals were found in other nearby trees or bushes, nor at other nearby place. Furthermore, no individuals could be found in May 1986, despite searches for them in the same trees where they had been seen before. To date, there is no explanation for such a change.

Anolis alayoni

Anolis alayoni Estrada and Hedges, 1995, Carib. J. Sci. 31(1–2):65; holotype MNHNCU 2746. Type locality: La Fangosa, Yateras, Guantánamo province, Cuba.

GEOGRAPHIC RANGE

Isla de Cuba (upland areas of Holguín and Guantánamo provinces).

LOCALITIES (MAP 28)

Holguín: Farallones de la Italiana, Levisa; Cayo Guan, Moa; Arroyo Culebra de Hacha, Moa; 3 km *E* of La Melba (Estrada and Hedges 1995). *Guantánamo:* La Fangosa, Yateras; between Cayo Fortuna and Riíto, Yateras; Arroyón, San Antonio del Sur; Cayo Fortuna, Yateras; Piedra La Vela; España Chiquita, Sierra de Canasta; Sumidero del Río Cuzco, El Salvador; La Poa, Baracoa; Arroyo Blanco, Baracoa; La Florida, Baracoa; Nibujón, Baracoa; base of Monte Iberia; Gran Tierra, Maisí; Yumurí, Maisí (Estrada and Hedges 1995).

CONTENT

Monotypic endemic species.

DESCRIPTION

Very small size. Mean snout-vent length of 42.2 mm in males and of 36.6 mm in females.

Pale grayish brown with black and brown spots on body; 4–5 brown or black X-shaped transverse dorsal bands. Gray and brown transverse bands on tail. Dorsal surface of head light brown; orbital region reddish brown;

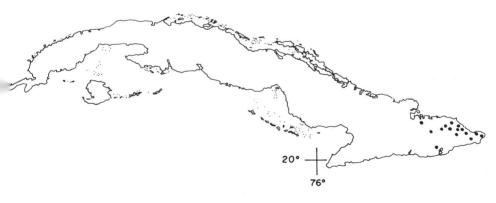

Map 28. Geographic distribution of *Anolis alayoni.*

yellowish lateral stripe from ear opening to hindlimb insertion. Yellow throat, with 5–6 brown bands on each side. Ventral region yellow, with brown spots; base of tail reddish. Dewlap in males, dark yellow.

Head long and narrow, with keeled, hexagonal, enlarged scales. Short limbs. Tail short, semiprehensile, and broad at the base. Smooth, granular dorsal scales; smooth, circular ventral scales, larger than dorsals. Enlarged postcloacal scales in males.

Bibliographic Sources

Estrada and Hedges (1995): original description; coloration; scalation; measurements; habitat data.

Morphological Variation

Metachromatism from pale grayish brown to dark reddish brown with almost the same pattern as in the light phase.

Geographic Variation

Little individual variation within its geographic range.

Sexual Variation

Coloration and scalation are similar in both sexes. Females have a yellow rudimentary dewlap. Males are recognized by their dewlaps and their enlarged postcloacal scales. Differences in size are shown in table 7.22.

Table 7.22. Comparison of snout-vent length (SVL), head length (HL), and tail length (TL) between male and female adults of *Anolis alayoni* (symbols as in table 7.1). Measurements after Estrada and Hedges (1995).

Sex	N	X	CV	m	M	t	p
				SVL			
Males	16	42.2	6.8	35.8	46.8	6.11	<0.01
Females	13	36.6	4.7	33.3	38.9		
				HL			
Males	16	12.6	6.3	10.9	13.8	10.46	<0.01
Females	13	10.2	4.6	8.9	10.7		
				TL			
Males	6	46.4	11.1	41.9	53.3	1.23	n.s.
Females	8	44.5	13.1	36.0	55.0		

Age Variation

Juveniles are not known.

NATURAL HISTORY

This species lives in dry or secondary rain forests and *"charrascales"* associated with pine forests, at elevations of 200 to 900 m asl. It occurs on twigs and tree trunks at heights of 2 or 3 m above the ground. At night it sleeps on high twigs.

Anolis garridoi

Anolis garridoi Díaz, Estrada, and Moreno, 1996, *Carib. J. Sci.* 32(1):54; holotype MNHNCU 4285. Type locality: Topes de Collantes, Sierra de Trinidad, Sancti Spíritus province, Cuba, approximately 700 m elevation.

GEOGRAPHIC RANGE

Isla de Cuba (Topes de Collantes).

LOCALITIES (MAP 29)

Sancti Spíritus: Topes de Collantes (Díaz *et al.* 1996).

CONTENT

Monotypic endemic species.

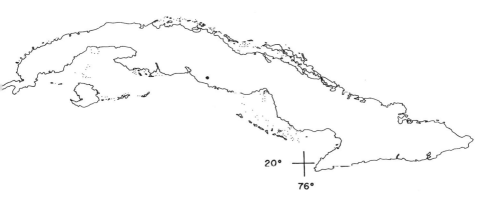

Map 29. Geographic distribution of *Anolis garridoi*.

Description

Very small size. Mean snout-vent length of 41.8 mm in males and of 36.8 mm in females.

Pale gray with black and brown spots on body; dark brown or black temporal blotch. There is a dark brown, transverse interocular band. A black or dark brown spot or X-shaped blotch over the sacral region. Yellowish white gular and ventral regions. Black oral cavity and inner throat. Dewlap in males, dark reddish orange, centrally grading to light yellow.

Head long and narrow, with smooth, enlarged scales. Short limbs. Prehensile tail, shorter than snout-vent length. Small, granular dorsal scales; ventrals larger than dorsals. Enlarged postcloacal scales in males.

Bibliographic Sources

Díaz *et al.* (1996): original description; coloration; scalation; measurements; habitat data; behavior.

Morphological Variation

Metachromatism from pale gray to reddish brown.

Geographic Variation

Little individual variation in its limited geographic range.

Sexual Variation

Coloration and scalation are similar in both sexes. Males are recognized by their enlarged poscloacal scales. Differences in size are shown in table 7.23.

Table 7.23. Comparison of snout-vent length *(SVL)*, head length *(HL)*, and tail length *(TL)* between male and female adults of *Anolis garridoi* (symbols as in table 7.1). Measurements after Díaz *et al.* (1996).

Sex	N	X	CV	m	M	t	p
			SVL				
Males	5	39.5	2.70	37.5	41.8	2.06	n.s.
Females	3	36.5	0.14	36.2	36.8		
			HL				
Males	5	12.2	5.10			3.75	<0.01
Females	3	10.8	0.11				

Age Variation

Juveniles are not known.

NATURAL HISTORY

This species lives in the hardwood forests of Topes de Collantes, on branches, twigs, and tree trunks, mainly in shaded places. Males perch at a mean of 1.78 m and females at a mean of 1.50 m above ground (from 0.3 to 3 m above ground). Individuals move slowly when walking and jumping, remaining motionless almost all the time.

During the observed aggressive display, the resident male raised nuchal and dorsal crests, compressed the body laterally, and started to bob the head and elevate the limbs. Then he increased the amplitude of the head bobs and pushups; during the last head bob, he extended the dewlap. The intruding male finally escaped as the resident male approached and tried to bite.

Species group *argillaceus* (splenial absent; short head and limbs).

Anolis loysiana (plate 26)

Anolis loysiana Duméril and Bibron, 1837, *Erp. Gén.* 4:100; holotype MNHN 2465. Type locality: Cuba.

GEOGRAPHIC RANGE

Cuba.

LOCALITIES (MAP 30)

Pinar del Rio: San Diego de los Baños (Barbour 1914); El Veral, Guanahacabibes; Soroa (Peters 1970); El Mulo (Martínez Reyes 1995); Rancho Mundito (CZACC); Pan de Guajaibón (L.R.S.). *La Habana and Ciudad de La Habana:* La Habana (Cocteau and Bibron 1838); Caobí; Nueva Paz (Garrido 1973f); Escaleras de Jaruco (González González 1989). *Matanzas:* Soplillar, Ciénaga de Zapata (Garrido 1980b). *Sancti Spíritus:* Sierra de Trinidad (CZACC). *Camagüey:* 15 km W of Camagüey (Ruibal 1964); Sierra de Najasa. *Holguín:* Cupeyal del Norte (Estrada *et al.* 1987); Sierra de Moa; Cananova (CZACC). *Granma:* Jiguaní (Barbour 1914). *Santiago de Cuba:* Pico Turquino (Garrido 1980a); El Cobre; La Gran Piedra (CZACC). *Guantánamo:* Bayate (Barbour and Ramsden 1919); springs of Jaguaní (Estrada *et al.* 1987); Monte Iberia; Felicidad (CZACC).

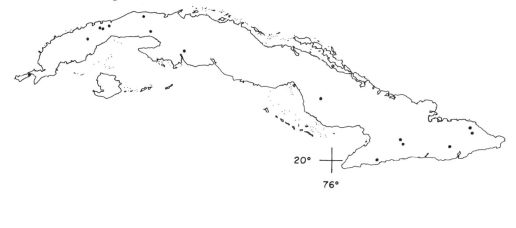

Map 30. Geographic distribution of *Anolis loysiana*.

Content

Monotypic endemic species.

Description

Very small size. Mean snout-vent length of 37.8 mm in males and of 35.9 mm in females.

Pale gray with brown transverse bands on limbs and tail. A brown, dorsolateral, longitudinal line from the posterior edge of the eye to the sacral region. A brown, lateral, longitudinal line from behind the ear opening to the groin; a brown, ill-defined, lateral, longitudinal line between forelimbs and hindlimbs. A brown, transverse, interorbital band; several brown radial lines from the eye. Several brown vertical lines on jaws. Dewlap in males, yellowish orange with reddish inner scales. Green iris.

Horizontally oval ear opening. Slightly dorsolaterally compressed body. Slightly concave central region of head. Smooth, circular, small dorsal scales; smooth, circular ventral scales, larger than dorsals. Smooth head scales. Many spine-like scales on body, limbs, and tail. Enlarged postcloacal scales in males.

Bibliographic Sources

Cocteau (1836): description. Duméril and Bibron (1837): original description; external morphology. Cocteau and Bibron (1838): external morphology; drawings of a specimen in dorsal view (pl. 14, fig. 1), of a specimen in ventral view (pl. 14, fig. 2), of the head in lateral view (pl. 14, fig. A), of the left forelimb (pl. 14, fig. B), of the ventral surface of the left forelimb (pl. 14, fig. C), of the dorsal surface of the left hindlimb (pl. 14, fig. D), of the ventral surface of the left hindlimb (pl. 14, fig. E). Gundlach (1867, 1880): external morphology; coloration; habitat data. Barbour (1914): coloration. Barbour and Ramsden (1916*b*): description; habitat data. Barbour and Ramsden (1919): scalation; coloration; drawings of a specimen in dorsal view (pl. 3, fig. 4), of the dorsal scalation of the head (pl. 7, fig. 4); habitat data. Barbour (1930*a, b*): habitat data; abundance. Alayo Dalmau (1955): coloration; drawing of the dorsal scalation of the head (pl. 3, fig. 1, no. 6). Etheridge (1960): osteology. Ruibal (1964): scalation; coloration; drawing of the interorbital scalation (fig. 9); habitat data; abundance. Schoener (1970*a*): interspecfic relationships in size. Williams and Rand (1977): coloration; ecomorphology. Coy Otero and Baruš (1979*a*): nematodes. Coy Otero and Lorenzo Hernández (1982): helminths. Estrada and Silva Rodríguez (1984): morphometry; general index of locomotion. Buide (1985): photograph of a specimen. Rodríguez Schettino (1985): habitat data; abundance; altitudinal distribution. Rodríguez Schettino and Martínez Reyes (1985): habitat data; abundance. Silva Lee (1985): photograph of a specimen; habitat data. Rodríguez Schettino *et al.* (1997): habitat data.

Morphological Variation

Metachromatism from pale gray with brown crossbands on tail, to gray with brown crossbands. Coloration varies a little individually and geographically.

Geographic Variation

Some lizards from Camagüey province have orange-red dewlaps, others rosaceous brown, and still others pinkish yellow. At Sierra Maestra, some individuals have very pale yellow dewlaps. However, the majority of the lizards in any locality have yellowish orange dewlaps.

Sexual Variation

Coloration is similar in both sexes. Males are recognized by their dewlaps and their enlarged postcloacal scales. Measurements are shown in table 7.24.

Age Variation

Juvenile coloration is similar to that of adults. Males have enlarged postcloacal scales.

Table 7.24. Comparison of snout-vent length *(SVL)*, head length *(HL)*, and tail length *(TL)* between male and female adults of *Anolis loysiana* (symbols as in table 7.1).

Sex	N	X	CV	m	M	t	p
				SVL			
Males	11	37.8	13.81	30.3	47.2	1.03	n.s.
Females	10	35.9	5.72	32.3	38.2		
				HL			
Males	11	12.5	6.13	11.4	13.3	3.95	<0.01
Females	10	11.4	5.14	10.4	12.4		
				TL			
Males	7	51.8	12.58	42.2	60.7	1.23	n.s.
Females	4	44.5	30.68	32.0	59.2		

NATURAL HISTORY

This species lives in evergreen and semideciduous forests, perching on trunks of trees with light colors and rugose bark; it takes shelter in fissures of trunks that offer a great deal of camouflage, such as *"ceiba"* (*Bombyx* sp.), *"guamá"* (*Lonchocarpus* sp.), royal palm (*Roystonea regia*), and "guásima" (*Guazuma tomentosa*). Individuals perched at a mean height above ground of 1.55 m at Península de Guanahacabibes, on trunks of trees in semideciduous forest over limestone outcrops, during both seasons of the year. The general index of locomotion (0.98) is very high, which indicates that is a crawler species with a great capacity to climb trees.

The species is diurnal, apparently nonheliothermic based on the places where it has been observed: in shadow in moist, dense forests. Its cryptic coloration makes it very difficult to find on tree trunks unless it is moving or displaying. It usually moves slowly except when being pursued. To escape, it runs rapidly upward on the trunk or jumps toward an adjacent tree.

Feeding in free life is unknown; it has accepted fruit flies (*Drosophila* spp.) and other small insects in captivity.

About reproduction it is known only that L. F. de Armas (pers. comm.) observed a pair copulating over a bare branch on the ground in September.

Seemingly, it is a scarce species: only one or two individuals are found on one tree, or some near trunks (generally of the same species). Nevertheless, because of the extraordinary mimetism of this lizard, it could seem more scarce than it actually is.

To date, two species of parasitic nematodes are known for *Anolis loysiana*: *Cyrtosomum sclopori* and *Oswaldocruzia lenteixeirai*.

Anolis argillaceus (plate 27)

Anolis argillaceus Cope, 1862, *Proc. Acad. Nat. Sci. Philadelphia* 14:176; holotype formerly in USNM, now lost. Type locality: Monte Verde, Oriente province, Cuba.

GEOGRAPHIC RANGE

Isla de Cuba (scattered populations from Camagüey province to Guantánamo province).

LOCALITIES (MAP 31)

Camagüey: Camagüey (Ruibal 1964); Santa Lucía; Puerto Tarafa (Garrido 1988). *Holguín:* Nuevo Mundo, Moa; Cayo Mambí (Garrido 1975*b*); Cupeyal del Norte; Sagua de Tánamo; Farallones de Moa; Calentura Arriba; La Melba (Estrada *et al.* 1987); Gibara (Garrido 1988); Guardalavaca (L.R.S.). *Granma:* Bayamo (CZACC). *Santiago de Cuba:* southern slope of Pico Turquino; Sierra Maestra (Garrido 1975*b*). *Guantánamo:* Monte Verde, Yateras (Cope 1862); Guantánamo; Baracoa (Gundlach 1867); Bayate; San Carlos (Barbour and Ramsden 1919); La Casimba, Maisí (Garrido 1967); Cupeyal (Peters 1970); Tabajó; Jauco; Cayo Güín; Punta de Maisí; Río Duaba (Garrido 1975*b*); Cayo Probado; springs of the Jaguaní River; Ojito de Agua; springs of the Yarey River; Cayo Fortuna; Alto del Yarey (Estrada *et al.* 1987); Boca de Boma (J. F. Milera, pers. comm.).

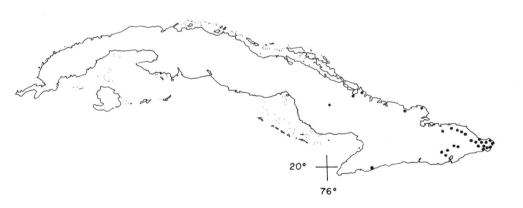

20°

76°

0   200 Km.

Map 31. Geographic distribution of *Anolis argillaceus*.

CONTENT

Monotypic endemic species.

DESCRIPTION

Very small size. Mean snout-vent length of 40.1 mm in males and of 36.0 mm in females.

Pale gray with brown crossbands on limbs and tail. Brown spots on the nuchal and dorsal crest. A brown, dorsolateral, longitudinal line from the posterior edge of the eye to the sacral region; a brown, ill-defined, lateral, longitudinal line between fore- and hindlimbs. A dark brown transverse interorbital band; several brown radial lines from the eye. Dewlap in males, pale orange with some red spots and white inner scales. Green iris.

Small, circular ear opening. Laterally compressed tail, vertically oval in transverse section. Smooth, circular, small dorsal scales; smooth, circular ventral scales, larger than dorsals. Smooth head scales. Enlarged postcloacal scales in males.

BIBLIOGRAPHIC SOURCES

Cope (1862): original description; morphology. Gundlach (1880): coloration in life. Barbour (1914): coloration. Barbour and Ramsden (1916b): habitat data. Barbour and Ramsden (1919): coloration; scalation; drawing of the dorsal scalation of the head (pl. 7, fig. 6); habitat data; behavior. Barbour (1930a, b): habitat data. Etheridge (1960): osteology. Ruibal (1964): coloration; scalation; drawings of the interorbital scalation (fig. 9), of the supraocular scalation (fig. 11); habitat data; abundance. Garrido (1967): habitat data. Williams (1969): habitat data. Peters (1970): coloration; external morphology; habitat data; reproductive data. Schoener (1970a): interspecific relationships in size. Coy Otero and Baruš (1973a): parasites. Garrido (1973d): coloration. Garrido (1975b): coloration in life; drawing of the head in lateral view (fig. 1, upper); photograph of the head in lateral view (fig. 2, left); habitat data; reproductive data. Williams and Rand (1977): coloration; ecomorphology. Coy Otero and Baruš (1979a, 1980): nematodes. Garrido (1980a): reproductive data in captivity. Coy Otero and Lorenzo Hernández (1982): helminths. Estrada and Silva Rodríguez (1984): morphometry; general index of locomotion. Schwartz and Henderson (1985): drawing of the lateral region of the head and of the forelimb (pl. 2, no. 9). Silva Lee (1985): two photographs of two specimens. Estrada (1987b): communal egg laying. Estrada et al. (1987): abundance.

Morphological Variation

Metachromatism from pale gray with brown transverse bands on tail, to dark gray with dark brown crossbands. Coloration varies individually, as does the disposition of the brown spots.

Geographic Variation

Little variation throughout its geographic range.

Sexual Variation

Coloration is similar in both sexes. Males are recognized by their dewlaps and their enlarged postcloacal scales. Measurements are shown in table 7.25.

Age Variation

Juvenile coloration is similar to that of adults. Males have enlarged postcloacal scales.

Natural History

This species lives in evergreen, semideciduous, microphyllous, and coniferous forests, and in the ecotone zones between these and open areas, also in serpentine shrubwoods. It perches on tree and bush trunks of rugose bark, or among twigs of the scrub layer. Generally it has the head downward. Based on the mean of the general index of locomotion (0.97), it is a crawler, which indicates that it has a great capacity to climb trunks. It moves

Table 7.25. Comparison of snout-vent length (SVL), head length (HL), and tail length (TL) between male and female adults of *Anolis argillaceus* (symbols as in table 7.1).

Sex	N	X	CV	m	M	t	p
			SVL				
Males	18	40.1	8.16	34.5	46.2	3.67	<0.01
Females	19	36.0	9.74	31.8	44.8		
			HL				
Males	18	12.6	7.23	10.7	14.1	3.61	<0.01
Females	19	11.4	9.61	10.0	14.3		
			TL				
Males	15	68.2	13.74	54.1	88.8	1.88	<0.01
Females	14	61.9	13.93	48.4	75.8		

slowly and its cryptic coloration makes its detection difficult. When escaping, it looks for shelter among bush twigs or hides in the interstices of tree barks; however, it does not try to climb trunks.

It is a diurnal species, apparently heliothermic. Males begin to leave their shelters under pine barks when sunbeams fall directly on the trunks.

The air temperature where this species was observed was 31.8°C, and that of the surface of the bush trunk was 32.0°C, at 2:30 P.M. in an evergreen forest at Baracoa.

The reproductive cycle is unknown; the majority of females collected at Cupeyal in January were gravid, but not the ones captured on the shores of the Toa River at Tabajó, Baracoa. A communal nest was found in a fragment of a semidecayed pine trunk at Yateras, Guantánamo, in April; it contained ten eggs and five shells recently eclosioned. Eggs were rosaceous white or yellowish white, and between 7.1 and 13.0 mm in length (x = 9.5 mm), with a mean weight of 0.3 mg. Six eggs eclosioned; the mean snout-vent length of the newly born lizards was 16.8 mm and their mean tail length, 30.1 mm. The mean time spent during precopulatory behavior in seven captive pairs was 17.6 sec., while that of copulation was 3.8 min.; the longest one lasted 5.25 min.

In general, one or two individuals are found on a tree or bush. However, because of its cryptic coloration it could seem more scarce than it actually is.

To date, three species of parasitic nematodes are known for *Anolis argillaceus: Cyrtosomum sclepori, Physalopteroides valdesi,* and *Porrocaecum* sp. larvae.

Anolis centralis (plate 28)

Anolis argillaceus centralis Peters, 1970, *Mitt. Zool. Mus. Berlin* 46(1):215; holotype ZMB 41616. Type locality: Victoria de Las Tunas, Oriente province, Cuba. *Anolis centralis:* Garrido, 1975, *Poeyana* 142:9.

GEOGRAPHIC RANGE

Isla de Cuba (northern and central parts of Camagüey, Las Tunas, and Holguín provinces; southern parts of Granma, Santiago de Cuba, and Guantánamo provinces).

LOCALITIES (MAP 32)

Ciego de Ávila: Cayo Coco (Martínez Reyes 1998). *Camagüey:* Camagüey (Ruibal 1964); Sierra de Cubitas; Playa Santa Lucía (Garrido 1975b); Nuevitas (Garrido 1988); Cayo Sabinal; Cayo Paredón Grande (Martínez Reyes 1998). *Las Tunas:* Las Tunas (Peters 1970); Manatí; Puerto Padre

(CZACC). *Holguín:* Gibara (Garrido 1975*b*); Banes (CZACC). *Granma:* Bayamo (Garrido 1975*b*); Birama (Garrido 1990); Cabo Cruz (CZACC). *Santiago de Cuba:* La Maya (Barbour and Ramsden 1919); Santiago de Cuba (Alayo Dalmau 1951); Pico Turquino; El Cobre; Alto Songo; La Gran Piedra; Puerto Boniato; La Socapa (Garrido 1975*b*); La Punta; Ocujal (Garrido and Jaume 1984); Siboney (Fong and del Castillo 1997); Baconao (L.R.S.). *Guantánamo:* Base Naval de Guantánamo (Lando and Williams 1969); Monte Líbano; Bayate; San Carlos; Baitiquirí; Macambo; Imías (Garrido 1975*b*); Sierra de la Canasta; Boca de Jaibo, Guantánamo (CZACC).

CONTENT

Polytypic endemic species; two subspecies are recognized:

Anolis centralis centralis

Anolis argillaceus centralis Peters, 1970, *Mitt. Zool. Mus. Berlin* 46(1):215. Type locality: Victoria de Las Tunas, Oriente province, Cuba. *Anolis centralis centralis:* Garrido, 1975, *Poeyana* 142:11. Geographic range: northern and central parts of Camagüey, Las Tunas, and Holguín provinces; northern part of Granma province.

Anolis centralis litoralis

Anolis centralis litoralis Garrido, 1975, *Poeyana* 142:12; holotype IZ 3472. Type locality: near Versalles, Santiago de Cuba, Oriente province, Cuba. Geographic range: southern parts of Granma, Santiago de Cuba, and Guantánamo provinces.

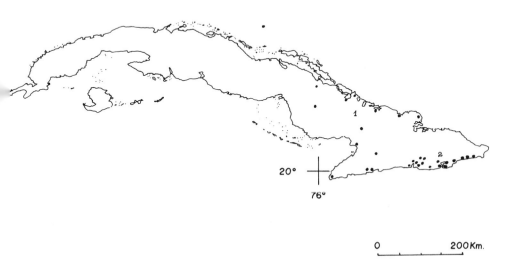

Map 32. Geographic distribution of *Anolis centralis*. 1, *A. c. centralis*; 2, *A. c. litoralis*.

DESCRIPTION

Very small size. Mean snout-vent length of 39.7 mm in males and of 35.0 mm in females.

Gray with brown transverse bands on tail and brown spots on limbs. A brown, dorsolateral, longitudinal line from the posterior edge of the eye to the sacral region. A pale brown, lateral, longitudinal line from the posterior edge of the ear opening to the groin. A brown, ill-defined, lateral longitudinal band between fore- and hindlimbs. Brown, ill-defined transverse interorbital bar; several brown radial lines from the eye. Dewlap in males, pale yellow, orange-red or pale orange, with white or yellow inner scales. Pale green iris.

Oblique oval ear opening, with a posterior fold of skin. Smooth, circular, small dorsal scales; smooth, circular ventral scales, larger than dorsals. Smooth head scales. Enlarged postcloacal scales in males.

BIBLIOGRAPHIC SOURCES

Alayo Dalmau (1955): coloration; drawing of the dorsal scalation of the head (pl. 3, fig. 1, no. 6). Schwartz and Ogren (1956): measurements. Ruibal (1964): coloration. Peters (1970): original description; coloration; external morphology; habitat data. Garrido (1975b): coloration in life; scalation; drawing of the head in lateral view (fig. 1, lower); photograph of the head in lateral view (fig. 2, right); habitat data. Coy Otero and Baruš (1979a, 1980): nematodes. Garrido (1980a): reproductive data in captivity. Coy Otero and Lorenzo Hernández (1982): helminths. Estrada and Silva Rodríguez (1984): morphometry; general index of locomotion. Silva Lee (1985): photograph of a specimen. Garrido (1988): courtship and mating patterns and aggressive behavior in captivity. Fong and del Castillo (1997): habitat data.

MORPHOLOGICAL VARIATION

Metachromatism from gray with brown transverse bands on tail, to dark gray with the crossbands dark brown. Coloration varies slightly.

Geographic Variation

The two subspecies are described as follows:

Anolis centralis centralis (central eastern region): the same color as the species. Red-orange dewlap with yellow inner scales. Mean snout-vent length of 40.3 mm in males and of 35.0 mm in females.

A. c. litoralis (southeastern region): the same color as the species. Yellow, orange, or pink dewlap with white inner scales. Mean snout-vent length of 39.7 mm in males and of 35.0 mm in females.

Sexual Variation

Coloration pattern is similar in both sexes. Males are recognized by their dewlaps and their enlarged postcloacal scales. Differences in size are shown in table 7.26.

Age Variation

Juvenile coloration is similar to that of adults. Males have enlarged postcloacal scales.

NATURAL HISTORY

This species lives in coastal and subcoastal forests, serpentine and semi-desert shrubwoods, and secondary scrubs. It perches on trunks and thin branches of bushes and trees in open areas or at the borders of forests, where the vegetation is not too dense. Mostly it dwells on sparse and spiny bushes, especially *"aroma"* (*Acacia farnesiana*) and *"marabú"* (*Cailliea glomerata*), although it also lives on trees of light bark, on fence posts, in tall grasses at the sides of roads and roadways, and on sea grape (*Coccoloba uvifera*). The mean height above ground at which eleven individuals were observed on the southern coast of Guantánamo province was 1.5 m. They were seen at heights from 0.7 to 3.0 m irrespective of sex, although only two males were detected at 3.0 m. The diameters of the branches and trunks they occupied varied from 1 to 6 cm, with a mean of 2 cm. The general index of locomotion calculated for this species (0.79) places it among crawlers, with a great capacity for climbing. It moves slowly; to escape, it runs short distances, stops, and runs away again.

It is a diurnal species, apparently nonheliothermic based on the shaded

Table 7.26. Comparison of snout-vent length (*SVL*), head length (*HL*), and tail length (*TL*) between male and female adults of *Anolis centralis* (symbols as in table 7.1).

Sex	N	X	CV	m	M	t	p
				SVL			
Males	20	40.0	10.62	31.2	47.2	3.81	<0.01
Females	20	34.5	11.65	28.6	46.0		
				HL			
Males	20	12.3	8.54	9.7	14.0	2.04	<0.05
Females	20	11.2	20.25	9.4	14.0		
				TL			
Males	20	63.0	21.35	45.0	82.7	0.36	n.s.
Females	17	62.2	17.70	35.9	87.2		

places where it spends most of its time, even though it inhabits open zones where heliophyle vegetation predominates. The body temperature of one male collected on a shaded bush at Baracoa in July 1985 was 33.4°C, while that of the air was 31.6°C.

Feeding is unknown in this species, although it has been observed gathering insects on branches of sea grapes.

About reproduction it is known only that courtship observed in captivity includes some sequences of bobbing and dewlap extension. The mean time spent during precopulatory behavior varied from 14.5 to 28 sec. according to the subspecies; during copulation the male positioned himself over the female, bit her neck, and stayed motionless. The mean copulatory time varied from 2.7 to 7.4 min; separation was deliberate and the male moved forward. Four females laid four eggs each, which measured from 6.6 to 7.4 mm in length and from 4.5 to 5.0 mm in width.

The aggressive displays of captive males included lateral placement, mouth opening, nuchal and dorsal crest raising, dewlap extension, bobbing, and extension and flexion of the four limbs during the 12 sec. before attacking each other.

Toward the easternmost region of this species' geographic range, populations are more numerous than toward the westernmost part, such as at Camagüey, where the species seems to be scarce. However, because of its cryptic coloration, it could seem less numerous in some localities than it actually is.

To date, only one species of parasitic nematode is known for *Anolis centralis: Cyrtosomum sclepori.*

Anolis pumilus

Anolis argillaceus: Collette, 1961, *Bull. Mus. Comp. Zool.* 125:140. *Anolis argillaceus:* Garrido, 1975, *Poeyana* 142:8, 18–19. *Anolis pumilus* Garrido, 1988, *Doñana, Acta Vertebrata* 15:46; holotype IZ 4470. Type locality: Bosque de La Habana, La Habana province, Cuba.

Geographic Range

Isla de Cuba (from Península de Guanahacabibes to Sancti Spíritus province); Isla de la Juventud; Santa María and Las Brujas keys in the Archipiélago de Sabana-Camagüey.

Localities (map 33)

Pinar del Río: Taco-Taco (Buide 1967); La Jaula, Península de Guanahacabibes (Peters 1970); El Fraile, Península de Guanahacabibes

(Rodríguez Schettino and Martínez Reyes 1985); Sierra de Santo Cristo del Valle; Sierra de Guane; Pica-Pica; Quiebrahacha (Garrido 1988). *La Habana and Ciudad de La Habana:* Bosque de La Habana (Collette 1961); Loma Colorada, Madruga; Santa Cruz del Norte (Garrido 1973*f*); Cojímar; El Cayuelo; Alamar (Garrido 1975*b*); Tapaste; Managua (Ortiz Díaz 1978); Jibacoa; Monte Barreto (Garrido 1988); Escaleras de Jaruco (González González 1989). *Matanzas:* Península de Hicacos (Buide 1967); Bacunayagua; Valle del Yumurí (Garrido 1988). *Villa Clara:* Cayo Santa María (Garrido 1973*d*); Cayo Las Brujas (Schwartz and Thomas 1975). *Sancti Spíritus:* Arroyo La Mariposa (Peters 1970); Sabanas de San Felipe (Garrido 1975*b*); Jobo Rosado, Yaguajay (Garrido 1988). *Isla de la Juventud:* Sierra de Casas (Garrido 1975*b*).

CONTENT

Monotypic endemic species.

DESCRIPTION

Very small size. Mean snout-vent length of 32.2 mm in males and of 33.6 mm in females.

Pale gray with brown transverse bands on limbs and tail. A pale brown longitudinal middorsal line that starts from the interparietal scale and ends on the tail. Two longitudinal dorsolateral pale yellow lines, from behind the ear opening to the sacrum. Several dark brown spots between longitudinal lines. A dark brown transverse interorbital bar. A dark brown triangular

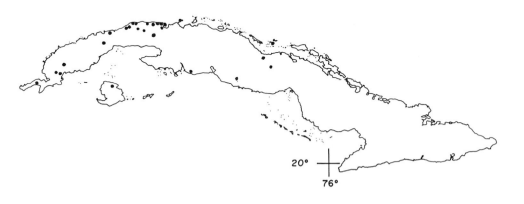

Map 33. Geographic distribution of *Anolis pumilus.*

spot over the nape, with the vortex backward, and a dark dot behind the vortex. A very dark brown suprascapular spot, the same size as the ear opening. Three brown radial lines centered in the eye. Grayish brown ventral region. Dewlap in males, pale orange with white inner scales. Green iris.

Small, circular ear opening. Smooth, granular dorsal scales; smooth, granular ventral scales, larger than dorsals. Supraorbital scales in contact. Enlarged postcloacal scales in males.

Bibliographic Sources

Collette (1961): habitat data; behavior. Garrido (1973d): habitat data. Garrido (1975b): coloration of a female; morphological variation. Llanes Echevarría (1978): coloration; measurements; habitat data; daily activity. Ortiz Díaz (1978): coloration; measurements; feeding. Rodríguez Schettino and Martínez Reyes (1985): habitat data; measurements; abundance; behavior. Garrido (1988): original description; coloration; scalation; morphological variation; photograph of a male (fig. 1); habitat data; daily activity; captivity conditions; longevity; reproductive data in captivity; courtship; mating; aggressive display between males. Rodríguez Schettino (1996): habitat data.

Morphological Variation

Metachromatism from pale gray with pale brown longitudinal lines, to gray with dark brown longitudinal lines. Coloration is not too variable.

Geographic Variation

Males from Jobo Rosado, Sancti Spíritus province, have yellow dewlaps. Lizards from the central region of the island of Cuba and the archipelagoes have ear openings somewhat larger than those from the western region.

Sexual Variation

Coloration pattern is the same for both sexes. Females are somewhat larger than males (table 7.27). Males are recognized by their dewlaps and their enlarged postcloacal scales.

Age Variation

Coloration of juveniles is similar to that of adults. The snout-vent length is about 20 mm. Males have enlarged postcloacal scales.

Natural History

This species lives in xeromorphous shrubwoods, coastal forest, and secondary vegetation, on trunks and thin limbs of bushes that are generally spiny and microphyllous, such as *"jía blanca"* (*Casearia aculeata*) and *"aroma"* (*Acacia farnesiana*). It perches at a mean height above ground of 0.75 m and no higher than 2 m.

It uses the holes of the "diente de perro" for shelter when escaping or going to rest. It is a diurnal species, apparently heliothermic based on the open places it inhabits. However, the majority of individuals were observed not under the sun but in the shadows of twigs and leaves of bushes, at a mean air temperature of 28.8°C. Most individuals have been found after 10:00 A.M., and at Península de Guanahacabibes were active between 5:00 P.M. and 7:00 P.M. during the wet season. Generally, two activity peaks are observed, one at 10:30 A.M. and the other at 2:30 P.M. This suggests that at noon, when air temperature and solar radiation are higher, it takes shelter in the shadows; therefore, it seems to be a behavioral thermoregulating species.

Based on the stomach contents of specimens collected at Tapaste, La Habana province, this species feeds on hymenopterans (mainly formicids), coleopterans, lepidopterans, dipterans, homopterans, and spiders. It has been fed in captivity with insects of several orders.

Males begin courtship with quick dewlap extensions and short nods, based on thirty-five matings observed in captivity. If the female is receptive, she nods; the male approaches the female and positions himself above her,

Table 7.27. Comparison of snout-vent length (*SVL*), head length (*HL*), and tail length (*TL*) between male and female adults of *Anolis pumilus* (symbols as in table 7.1).

Sex	N	X	CV	m	M	t	p
			SVL				
Males	7	32.2	5.95	28.4	34.2	1.06	n.s.
Females	11	33.6	9.32	26.1	39.2		
			HL				
Males	7	9.9	3.51	9.4	10.3	0.41	n.s.
Females	11	10.0	5.80	8.7	10.6		
			TL				
Males	3	62.6	3.87	60.0	64.8	1.26	n.s.
Females	6	57.3	11.95	50.6	64.6		

during a mean time of 12 sec. Then the male inserts his hemipenis without biting the female's neck; the longest time spent during copulation is 13 min. and the mean, 5 min. Separation is brusque: the male jumps to the side or behind the female. Eggs laid by captive females are white, have lengths from 5.0 to 7.6 mm and widths from 4.2 to 4.8 mm. The dates of oviposition varied from March to May.

When two males meet, both bob and extend the dewlaps two or three times. They place themselves at lateral positions to each other, raise the nuchal crests, and laterally compress their bodies. Their patterns of coloration look more defined and both males rise on their four limbs, keeping the dewlaps distended; they raise their heads and repeatedly bob. All parts of this ritual are repeated until one male begins to exhibit aggressive behavior.

Seemingly, this species is scarce, but its cryptic coloration may cause its abundance to be underestimated. It builds small populations, but very aggregate ones; a relative abundance of up to six individuals/hour has been calculated on the southern coast of Cabo Corrientes, Península de Guanahacabibes.

Series *lucius* (splenial absent; four lumbar vertebrae; six aseptate caudal vertebrae).

Species group *lucius* (two or three semitransparent scales in the lower eyelid).

Anolis lucius (coronel, gorrita; plate 29)

Anolis lucius Duméril and Bibron, 1837, *Erp. Gén.* 4:105; holotype MNHN 2466. Type locality: Cuba. *Anolis mertensi* Ahl, 1925, *Zool. Anz.* 62:86; holotype ZMB 27811. Type locality: Cuba.

Geographic Range

Isla de Cuba (from Escaleras de Jaruco, La Habana province, to Banes, Holguín province).

Localities (map 34)

La Habana: Madruga (Barbour 1914); Aguacate (Barbour and Ramsden 1919); Cueva de Don Martín (García Ávila *et al.* 1969); Jibacoa; Tapaste (Peters 1970); Sierra de Camarones; Arcos de Canasí; Jaruco (Garrido 1973f); Escaleras de Jaruco (Silva Taboada 1974); Cueva del Vaho; Cinco Cuevas (Silva Taboada 1988); El Narigón, Puerto Escondido (Alarcón Chávez *et al.* 1990); Punta Jíjira (L.R.S.). *Matanzas:* Coliseo (Gundlach 1867); Río Canímar (Buide 1967); Bacunayagua (Soto Ramírez 1994); San Miguel de los Baños (CZACC). *Villa Clara:* Sierra de Jumagua (Cochran 1934); Manicaragua;

Jibacoa (Schwartz and Ogren 1956); southeast of Caibarién (Garrido 1973c); Sagua la Grande (Menéndez *et al.* 1986). *Cienfuegos:* Guaos (Barbour and Ramsden 1919); Río Arimao (Barbour 1930a); Soledad (Smith and Willis 1955); Guajimico (Schwartz and Ogren 1956). *Sancti Spíritus:* Sierra de Jatibonico; Sancti Spíritus; Trinidad (Barbour 1914); Guanayara (Hardy 1957); Salto del Caburní (Rodríguez Schettino *et al.* 1979); Jobo Rosado, Yaguajay (Espinosa López, Sosa Espinosa, and Berovides Álvarez 1990); Caguanes (L.R.S.). *Ciego de Ávila:* Loma de Cunagua (Smith and Willis 1955). *Camagüey:* Guáimaro (Barbour and Ramsden 1919); Camagüey; Sierra de Cubitas (Ruibal 1961); Sierra de Najasa (Estrada 1987a). *Holguín:* Buenaventura (Smith and Willis 1955); Gibara; Guardalavaca (Garrido 1973e). *Granma:* Jiguaní (Alayo Dalmau 1955). *Santiago de Cuba:* Los Negros (Alayo Dalmau 1955).

CONTENT

Monotypic endemic species.

DESCRIPTION

Small size. Mean snout-vent length of 58.3 mm in males and of 51.5 mm in females.

Greenish gray, with pale brown transverse bands on limbs and tail. Four brown or reddish rhomboidal middorsal blotches. Two brown longitudinal lateral stripes. A brown postorbital band that bifurcates and continues with a semicircular shape to the upper and posterior part of the head. Brown labials and postlabial bands, to the ear opening, which continue with a

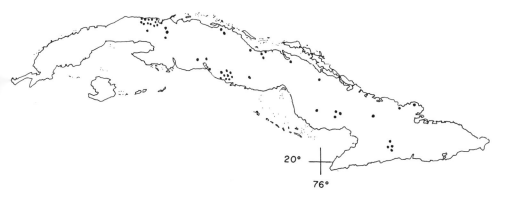

Map 34. Geographic distribution of *Anolis lucius*.

semicircular shape over the nuchal region. Ivory ground color between postorbital and postlabial bands. Yellow eyelids, the lower with three blue- and black-bordered semitransparent scales (fig. 21). White interparietal scale. Yellow ventral region. Dewlap in males, white with some gray semi- circular stripes at the base and six or eight yellow stripes to the posterior part. Brown iris.

Large, vertically oval ear opening. Vertical section of tail is oval in males and round in females. Smooth, circular, small dorsal scales; smooth, circu- lar ventral scales, somewhat larger than dorsals. Smooth head scales; large interparietal scale, with a conspicuous pineal eye. Enlarged postcloacal scales in males.

Six pairs of macrochromosomes and twelve pairs of microchromosomes (2n = 36).

Bibliographic Sources

Duméril and Bibron (1837): original description. Cocteau and Bibron (1838): coloration; scalation; external morphology; drawings of a specimen in dorsolateral view (pl. 12), of the dorsal scalation of the head (pl. 12, fig. 1), of the gular scalation (pl. 12, fig. 2), of cloacal and femoral scalation (pl. 12, fig. 3), of the ventral surface of the right forelimb (pl. 12, fig. 4), of the ventral

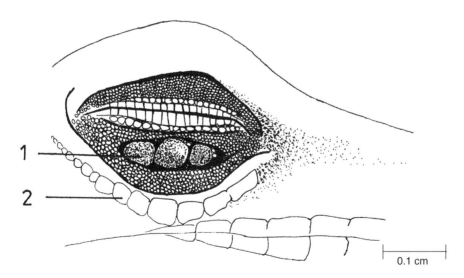

0.1 cm

Fig. 21. Ocular region of *Anolis lucius* with closed eye. *1,* semitransparent scales of the lower eyelid; *2,* suborbital semicircle.

surface of the right hindlimb (pl. 12, fig. 5). Gundlach (1867): habitat data; behavior. Bocourt (1874): (pl. 14, fig. 5). Gundlach (1880): coloration in life; habitat data; feeding; behavior. Barbour (1914): coloration; habitat data. Barbour and Ramsden (1916b): habitat data; vocalization. Barbour and Ramsden (1919): coloration; scalation; drawing of the dorsal scalation of the head (pl. 6, fig. 4); habitat data; abundance; vocalization. Dunn (1926): abundance; communal nests. Barbour (1930a, b): habitat data; abundance; egg features. Cochran (1941): external morphology. Alayo Dalmau (1955): coloration; drawing of the dorsal scalation of the head (pl. 6, fig. 1); habitat data; abundance. Koopman and Ruibal (1955): fossil remains. Smith and Willis (1955): external morphology; habitat data; behavior. Williams and Hecht (1955): scales of the lower eyelid; drawing of the scales of the lower eyelid (fig. 1); habitat data. Schwartz and Ogren (1956): external morphology; habitat data; behavior; abundance; vocalization. Allen and Neill (1957): external morphology; habitat data; abundance; behavior; vocalization; egg features; predators. Hardy (1957): juvenile coloration; photograph of a juvenile and a female (fig. 1); habitat data; abundance; communal egg laying; egg and hatchling features; reproductive data; mating. Etheridge (1960): osteology. Ruibal (1961): habitat data; behavior; abundance; thermic relations. Ruibal (1964): coloration; scalation; drawings of the interorbital scales (fig. 9), of the head in lateral view (fig. 10); habitat data; communal egg laying; vocalization. Gorman (1965): karyotype. Rand (1967): communal nests. Ruibal (1967): coloration; threatening behavior between males. Gorman and Atkins (1968): karyotype. Williams (1969): habitat data. Peters (1970): external morphology; vocalization; daily activity. Schoener (1970a): interspecific relationships in size. Garrido (1973c): habitat data. Coy Otero (1976): nematodes. Valderrama Puente et al. (1976): coloration in life; thermic relations. Williams and Rand (1977): coloration; habitat data. Llanes Echevarría (1978): coloration in life; habitat data; thermic relations; daily activity. Ortiz Díaz (1978): coloration in life; feeding. Coy Otero and Baruš (1979a): nematodes. Valderrama Puente (1979): morphometry; thermic relations; habitat data; abundance; reproductive cycle; territoriality. Espinosa López et al. (1979): electrophoretic pattern of general proteins and esterases; drawing of the pattern of general proteins (fig. 1), of the pattern of esterases (fig. 2). Garrido (1980a): reproductive behavior in captivity. Silva Rodríguez (1980): electrophoretic pattern of general proteins. González Martínez (1981): electrophoretic pattern of hemoglobin. Jiménez (1981): Leishmania. Mugica Valdés (1981): electrophoretic pattern of esterases of liver; photograph of the pattern of esterases (fig. 9). Mugica Valdés et al. (1982): electrophoretic pattern of esterases; drawing of the pattern of esterases (fig. 1g).

Rodríguez Schettino (1982): morphometry. Silva Rodríguez, Berovides Álvarez, and Estrada (1982): communal egg laying. Estrada and Silva Rodríguez (1984): morphometry; general index of locomotion. Garrido and Jaume (1984): coloration; habitat data. Buide (1985): photograph of a specimen. Schwartz and Henderson (1985): drawing of the lateral region of the head and of the forelimb (pl. 2, no. 8). Silva Lee (1985): coloration; habitat data; communal nests; photograph of a specimen. Menéndez *et al.* (1986): habitat data; thermic relations. Rodríguez Schettino and Valderrama Puente (1986): habitat data; thermic relations; sexual ratio. Espinosa López *et al.* (1987): electrophoretic pattern of plasmatic proteins. Estrada (1987*a*): habitat data. Granda (1987): morphometry; coefficients of arboreality. Marcellini and Rodríguez Schettino (1987): coloration; habitat data; thermic relations; behavior; sexual ratio. Quintana Ferrer (1987): morphometry. Socarrás *et al.* (1988): potential predators. Valderrama Puente and Rodríguez Schettino (1988): reproductive cycle; egg features; sexual ratio; habitat data; thermic relations. Rodríguez Schettino and Martínez Reyes (1989*a*): habitat data; thermic relations; reproductive cycle; egg features; population structure; density; biomass; feeding. Alarcón Chávez *et al.* (1990): habitat data; feeding; reproductive data; abundance. Espinosa López, Posada García *et al.* (1990): electrophoretic pattern of muscle; photograph of the electrophoretic pattern of general proteins (fig. 1B). Espinosa López, Sosa Espinosa, and Berovides Álvarez (1990): electrophoretic patterns of eight proteic systems. Espinosa López *et al.* (1991): electrophoretic patterns of eight proteic systems; relation with the anthropization and the stability of the habitat. Rodríguez Schettino and Martínez Reyes (1994): feeding.

MORPHOLOGICAL VARIATION

Metachromatism from greenish gray to pale brown, to dark gray, to pale brown, to dark gray, or dark brown; however, always with the ivory-colored stripes between labials and postlabial lines, the white interparietal scale, and the reddish spots over the dorsum. Coloration pattern slightly variable.

Genetically also little variation, as is seen in the low coefficient of variation of the number of bands of plasmatic proteins and other proteic systems detected by electrophoresis of plasma, muscle, and liver.

Geographic Variation

Little variation throughout its geographic range.

Sexual Variation

Females have a white dot, bordered with black, in the middle of each of the reddish spots (plate 30). Their tails are round in transverse section. Males are recognized by brown middorsal spots, tails laterally compressed or with a small caudal crest, dewlaps, and the enlarged postcloacal scales. Differences in size are shown in table 7.28.

Age Variation

Juveniles have the female coloration pattern, but more brilliant and with a white ventrolateral stripe from fore- to hindlimbs. The snout-vent length is from 25.9 to 39.9 mm. The transverse section of the tail is round. Subadults of both sexes have snout-vent lengths that vary between 40.0 and 44.9 mm and the coloration pattern of their respective sexes. These lizards have not attained sexual maturity.

NATURAL HISTORY

This species lives in coastal shrubwoods and semideciduous forests, in association with limestone soils and outcrops such as caves and cliffs, in lowlands, and in mountainous regions. It perches on rocks and walls of caves, from ground level to high sites, even hanging from cave ceilings. It is also found on complex tree trunks that are generally adjacent or on limestone outcrops, up to 3 m above ground, irrespective of sex or maturity.

However, means of height above ground calculated for different populations and dates of sampling coincided in an interval of 1 to 2 m except in the coastal shrubwood of the northern coast of La Habana province, where it

Table 7.28. Comparison of snout-vent length (SVL), head length (HL), and tail length (TL) between male and female adults of *Anolis lucius* (symbols as in table 7.1).

Sex	N	X	CV	m	M	t	p
				SVL			
Males	17	57.6	21.18	51.0	62.0	7.42	<0.01
Females	16	49.2	5.60	45.5	53.0		
				HL			
Males	17	17.2	11.00	13.0	19.6	4.57	<0.01
Females	16	14.8	6.13	13.4	16.0		
				TL			
Males	12	104.8	7.26	9.03	118.6	1.54	<0.10
Females	4	98.3	6.26	90.0	104.9		

was 0.75 m, and at Playa Jibacoa, where it was 0.79 m, during 1978. Height above ground seems to be related to vegetation structure, which in the first exceptional case is very reclined and in the second, sparse and low. According to the general index of locomotion (0.80), the species is a runner, while the coefficient of observed arboreality (0.28) is low, which indicates little ability to climb trees; these measurements are typical of petricolous and clivicolous species.

Males stay on their perches and defend them aggressively. During aggressive display, two captive males kept their dewlaps completely extended and protruded their tongues, which were inflamed and reddened. They raised their nuchal crests and forelimbs and bobbed. This process lasted 16 sec.

Individuals of this species move very quickly while hunting; to escape, they rapidly hide within crevices and hollows of stones and walls or run toward tree crowns. On some occasions, they stay immobile, flattened against the substrate, before escaping. They emit an acute, loud squeak when captured and when they slip away within the fissures of rocks.

Apparently, the semitransparent scales of the lower eyelid function like sunglasses to protect the eyes when the individual passes from dark to lighted places, such as when it leaves caves.

It is a diurnal, nonheliothermic species, with a marked preference for shaded sites. According to the available data, it can occasionally actively thermoregulate; however, in most cases it is a thermconformer species that maintains mean body temperatures from 24.0 to 30.0°C at the expense of mean air temperature, with no differences between sexes. Body temperature is usually higher than the temperatures of air and substrate and is positively correlated with both; it is higher during the wet season and increases with rising environmental temperature, both daily and seasonally. Means for different months varied from 20.3 to 30.6°C. Air temperature in places where this species has been found varied between 22.0 and 30.0°C and the temperature of substrate, from 22.0 to 28.0°C. The species has always been observed in warmer places during the wet season than during the dry season. The relative humidity varied between 70 and 80% in such places during the wet season and was maintained at above 70% during the dry season.

At night it takes refuge within fissures and under lapels of rocks or in the interstices of tree trunks. It begins to leave these nocturnal shelters after 8:00 A.M. during the wet season and after 9:00 A.M. during the dry season. The number of active individuals rapidly increases and stabilizes at about 10:00 A.M.; after this time it diminishes, slowly during the wet season and suddenly during the dry. However, during the wet season the number increases again in the afternoon: there is an observed bimodal pattern, in contrast with the single peak observed during the dry season. No individuals are active after 5:00 P.M. or 3:00 P.M. during the wet and dry season, respectively.

This is mainly a myrmecophagous species, although it also includes in its diet other hymenopterans, coleopterans, lepidopterans, dipterans, blatopterans, isopterans, insect larvae, spiders, and isopods. Occasionally individuals are saurophagous and even cannibalistic; they may even eat some fruits. The mean weight of the stomach contents was 0.049 g. Males eat somewhat larger prey than do females during the two seasons; during in the wet season, both sexes capture larger prey than in the dry season. In the population of Playa Jibacoa, means of prey length were 5.2 and 6.0 mm for males during the dry and wet season, respectively; for females these means were 3.5 and 4.1 mm, respectively.

Courtship and copulation patterns have been observed on several occasions among both free-living and captive individuals. The male positions himself at the right side of the female with the body slightly twisted outward, biting her neck; to insert the hemipenis, he bends the tail at a right angle toward the left, beneath the female's cloacal region. The female's tail remains extended backward with spasmodic waggings. The precopulatory phase lasts 26.8 sec. and copulation an average of 3.95 min.; the longest copulation period registered for captive individuals lasted 5.3 min. Members of a pair separate slowly when copulation ends.

Reproduction is cyclic. Gravid females appear in March and remain in the population until October; however, June, July, and August are the months during which the percentages of gravid females are highest. These are also the only months during which females have two oviductal eggs. Egg length varies between 9.0 and 16.6 mm, while mean egg length varies from 11.0 to 15.4 mm. The mean weights of two samples of oviductal eggs were 0.237 and 0.300 g.

Females lay several eggs, one each time, depositing them in communal sites for various females and seasons. The calculated reproductive effort is 0.68 and the reproductive potential, 0.69. Eggs are deposited in holes of cave walls, cliffs, or tree trunks. They are white with flexible, leathery shells and are strongly joined to the substrate and to other eggs. Eclosion occurs at one egg pole; afterward, the empty shells drop to the ground, forming groups of shells in which different stages of decay are in progress. These shell groups have been found on cave floors, inside or at the entrance, where the air temperature ranged from 26.0 to 29.0°C and the entrance orientation was from north to south.

An inverse relationship exists between size of fat bodies and reproduction. At the beginning of the year, medium-sized fat bodies are more frequent; during the months of greatest reproductive activity small ones predominate; and after October the percentage of large fat bodies increases. However, in females small fat bodies are more frequent because of the females' greater energy expenditure in egg production.

The sexual coefficient is 0.9 males/females and the ratio between adults and immatures (subadults and juveniles) is 1.4. Populations are composed of a greater percentage of adults from April to July, while from January to March and from September to December, subadults and juveniles predominate.

This is a gregarious species that establishes very numerous populations. The relative abundance in the coastal shrubwood of the northeastern littoral of La Habana province was 11.5 individuals/hour. The density of the population from Playa Jibacoa varied between 428 and 1,404 individuals/ha and that of the Jardín Botánico de Cienfuegos was 930 individuals/ha. The biomass at Playa Jibacoa varied between 0.9 and 4.9 kg/hour.

To date, seven species of parasitic nematodes are known for *Anolis lucius*: *Parapharyngodon cubensis, Cyrtosomum sclelopori, Physaloptera squamatae, Skrjabinoptera (Didelphysoma) phrynosoma, Oswaldocruzia lenteixeirai, Porrocacum* sp. larvae, and Physalopteridae gen. sp.

Furthermore, *Leishmania* sp. has been reported in the last portion of the intestine of *A. lucius*. The Barn Owl (*Tyto alba*) seems to be an important predator, as are giant anoles, which have been successfully fed with live individuals of this species.

Anolis argenteolus (plate 31)

Anolis argenteolus Cope, 1861, *Proc. Acad. Nat. Sci. Philadelphia* 13:213; holotype formerly in USNM, now lost. Type locality: Monte Verde, Oriente province, Cuba.

GEOGRAPHIC RANGE

Isla de Cuba (from Sierra de Najasa to Punta de Maisí).

LOCALITIES (MAP 35)

Camagüey: Santa Cruz del Sur (Ruibal 1964); Sierra de Najasa (Garrido and Jaume 1984). *Las Tunas:* El Yarey, Puerto Padre (CZACC). *Holguín:* Gibara; Guardalavaca (Garrido 1973e); Farallones de Moa; La Melba (Estrada *et al.* 1987); Levisa (Estrada 1988); Cayo Saetía; Mayarí; Bahía de Naranjo; Sagua de Tánamo (CZACC); Mayabe, Holguín (L.R.S.). *Granma:* Jiguaní;

Bueycitos (Barbour 1914); Belic (Barbour and Ramsden 1919); Puerto Portillo (Cochran 1934). *Santiago de Cuba:* Santiago de Cuba (Barbour 1914); El Cobre; San Luis (Stejneger 1917); Río Magdalena (Cochran 1934); Ocujal (Cooper 1958); Pico Turquino (Garrido 1980*a*); La Mula (Rodríguez Schettino 1985); Laguna de Baconao (Schwartz and Henderson 1985); Siboney (Fong and del Castillo 1997); Palma Soriano; Pinares de Mayarí (CZACC). *Guantánamo:* Monte Verde (Cope 1861); Baracoa (Gundlach 1867); Guantánamo (Alayo Dalmau 1955); Baitiquirí; Punta de Maisí (Peters 1970); Bayate; Boca de Jauco; Cupeyal (CZACC); Boca de Boma (J. F. Milera, pers. comm.); Playa Imías; Tortugilla (L.R.S.).

CONTENT

Monotypic endemic species.

DESCRIPTION

Small size. Mean snout-vent length of 55.5 mm in males and of 46.6 mm in females.

Gray with pale brown and purplish brown dots; brown transverse bands on limbs and tail. Some purplish brown vertical lines at the sides of the body. Two or three ivory-colored, semicircular bands on the posterior part of the head. Yellow eyelids, the lower with two blue, semitransparent scales, bordered with black. Pale gray interparietal scale. Yellow ventral region, with a midventral longitudinal reddish brown stripe. Dewlap in

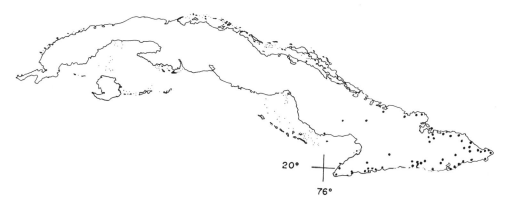

20°

76°

0   200 Km.

Map 35. Geographic distribution of *Anolis argenteolus*.

males, white with about eight purplish semicircular stripes, very thin and close to each other. Brown iris.

Small, vertically oval ear opening. Smooth, circular, small dorsal scales; smooth, circular vertical scales, somewhat larger than dorsals. Slightly keeled head scales. Large interparietal scale, with a conspicuous pineal eye. Somewhat enlarged postcloacal scales in males.

Six pairs of macrochromosomes and twelve pairs of microchormosomes (2n = 36).

BIBLIOGRAPHIC SOURCES

Cope (1861): original description. Gundlach (1867): habitat data. Bocourt (1874): pl. 14, fig. 14. Gundlach (1880): coloration in life; habitat data. Barbour (1914): coloration; habitat data. Barbour and Ramsden (1916b): habitat data. Stejneger (1917): coloration; drawings of the dorsal scales of the head (fig. 49), of the head in lateral view (fig. 50), of the tail in lateral view in the fifth verticil (fig. 51). Barbour and Ramsden (1919): coloration in life; scalation; drawing of the dorsal scales of the head (pl. 14, fig. 6); habitat data. Barbour (1930a, b): habitat data. Alayo Dalmau (1951): habitat data; abundance. Alayo Dalmau (1955): coloration; drawings of the dorsal scales of the head, of the head in lateral view, of the tail in lateral view in the fifth verticil (pl. 3, fig. 5); habitat data; abundance. Williams and Hecht (1955): scales of the lower eyelid; drawing of the scales of the lower eyelid (fig. 2). Schwartz and Ogren (1956): habitat data; behavior; measurements. Cooper (1958): habitat data. Etheridge (1960): osteology. Ruibal (1964): coloration; scalation; drawing of the interorbital scales (fig. 9); habitat data. Gorman and Atkins (1968): karyotype; photograph of the metaphase II (fig. 1A). Williams (1969): habitat data. Peters (1970): external morphology; photograph of the head in dorsal view (fig. 1d); habitat data; vocalization. Schoener (1970a, b): interspecific relationships in size. Coy Otero and Baruš (1979a): parasitic nematode. Garrido (1980a): reproductive data in captivity. Garrido (1983): habitat data. Rodríguez Schettino (1985): habitat data; abundance; altitudinal distribution. Schwartz and Henderson (1985): drawing of the lateral region of the head and of the forelimb (pl. 2, no. 7). Silva (1985): coloration; habitat data; abundance; behavior; photograph of a specimen. Estrada (1987a): habitat data. Estrada et al. (1987): habitat data. Granda Martínez (1987): morphometry; coefficient of arboreality. Quintana Ferrer (1987): morphometry. Estrada (1988): communal egg laying. Rodríguez Schettino and Martínez Reyes (1988, 1989b): reproduction; density; biomass; population structure. Socarrás et al. (1988): predators. Coy Otero (1989): trematode. Rodríguez Schettino and Martínez Reyes (1989a):

morphometry; habitat data; thermic relations; daily and seasonal activity; population structure; reproduction; density; biomass; feeding. Espinosa López, Sosa Espinosa, and Berovides Álvarez (1990): electrophoretic patterns of eight proteic systems. Rodríguez Schettino and Martínez Reyes (1991, 1996): feeding. Fong and del Castillo (1997): feeding. Rodríguez Schettino and Martínez Reyes (in press): reproductive cycle; population structure.

MORPHOLOGICAL VARIATION

Metachromatism from pale gray with brown dots, to gray with brown dots. Dewlap varies from white with purplish stripes, to white with pale or dark gray stripes. Coloration pattern varies slightly.

Geographic Variation

Slight variation throughout its geographic range.

Sexual Variation

Coloration pattern is similar in both sexes. Males are recognized by their dewlaps and their somewhat enlarged postcloacal scales; they are larger than females (table 7.29).

Age Variation

Juvenile coloration is similar to that of adults. Males have somewhat enlarged postcloacal scales. The minimal snout-vent length of newly born lizards is 21.4 mm; female juveniles grow to 35.9 mm with ovaries to 0.9 mm in length; male juveniles grow to 40.9 mm with testes to 1.9 mm in length.

Table 7.29. Comparison of snout-vent length (SVL), head length (HL), and tail length (TL) between male and female adults of Anolis argenteolus (symbols as in table 7.1).

Sex	N	X	CV	m	M	t	p
			SVL				
Males	52	55.5	3.8	48.6	59.8	24.12	<0.01
Females	62	46.6	3.9	42.0	51.0		
			HL				
Males	52	16.7	5.6	15.0	19.5	21.29	<0.01
Females	62	13.5	4.9	12.0	15.2		
			TL				
Males	44	113.1	6.5	91.4	126.6	15.51	<0.01
Females	53	90.5	7.7	74.7	103.5		

Male subadult snout-vent length reaches a maximum of 46.9 mm and testes length of 2.9 mm; female subadult snout-vent length reaches a maximum of 40.9 mm and ovary length of 2.9 mm. Both males and females have coloration patterns like those of adults but have not attained sexual maturity.

Natural History

This species lives in semideciduous forests, coastal shrubwoods, coffee plantations, and urban zones of the central eastern region of the country. It is very frequent in areas near rivers and coastal cliffs, and on trees of sea grape (*Coccoloba uvifera*).

It perches on tree trunks, limestone cliffs, and walls of human constructions. Individuals are usually found at heights above ground ranging from 1 to 2 m, although they can quickly climb very high. Males perch higher than females. The coefficient of observed arboreality (1.34) is high and indicates a great capacity to climb vertical surfaces. Generally, the species rapidly escapes to tree crowns or toward the highest parts of walls; occasionally, it jumps from one branch to another.

Apparently, males defend their territories only slightly or not at all; some heterosexual pairs have been found on the same tree, and even various males in different places of a trunk, without showing any aggressive display. Only in a few cases have advertising displays been observed, in which males bobbed and repeatedly extended the dewlaps without making any other threatening gestures.

It is a diurnal nonheliothermic species most frequently found in shaded places. Body temperature, which varies according to air and substrate temperatures, is maintained above and at the expense of both, between means of 27.0 and 30.0°C. Body temperature is higher during the wet season and increases in relation to environmental temperatures, both daily and seasonally. Mean air temperature in the places where the species was found varied from 26.0 to 29.0°C and that of the substrate, from 26.0 to 28.0°C. The relative humidity varied between 50 and 70% in the dry season and between 70 and 100% in the wet season.

Individuals begin to leave their nocturnal shelters—in the highest branches of the trees where they sleep—after 6:00 A.M. in both seasons. The number of active individuals rapidly increases and stabilizes at about 10:00 A.M. During the wet season this number diminishes between 11:00 A.M. and 2:00 P.M.; then it increases again, although there are fewer individuals than in the morning. During the dry season the number of active individuals is almost stable until 3:00 P.M., diminishing thereafter. Few individuals are active after 6:00 P.M. in either season, and at 7:00 P.M. none are observed outside the refuges.

The species feeds mainly on ants, although it also includes in its diet dipterans, coleopterans, hemipterans, lepidopterans, orthopterans, blatopterans, insect larvae (mainly of lepidopterans), and spiders. Males capture larger prey than females during both seasons. During the wet season, males and females eat larger prey than during the dry season. In the population at La Mula, on the southern coast of Santiago de Cuba province, the means of prey length were 4.2 and 4.0 mm for males during the wet and dry seasons, respectively, and 3.5 and 3.9 mm for females.

Reproduction is continuous throughout the year, although with a higher percentage of gravid females from March to September. Oviductal egg length varies between 10.0 and 15.8 mm, while monthly means are between 11.5 and 12.6 mm. Most gravid females have only one egg, but from March to September about 10% have two.

The male courtship pattern begins with bobs and short runs toward the female, along with several dewlap extensions. The precopulatory phase lasts 13.2 sec. as a mean, and copulation, 1.18 min. The longest copulation observed in captivity lasted 1.23 min. Both members of the pair separate hastily. Females lay several eggs a year, one each time. During the months of greatest reproductive activity, approximately eight days elapse between each egg deposition; during other months, about 20 days. Therefore, it has been calculated that a female produces 19 + 10 eggs, respectively, in one year of fertile life. The reproductive potential varies from 0.85 to 1.00, with a mean of 0.91. Eggs are laid in communal sites for several females and various dates, in holes of tree trunks and under lapels of limestone cliffs. They are white with leathery, flexible shells and a mean length of 12.0 mm.

Size of fat bodies is inversely related to reproductive activity. Males have medium-sized fat bodies during the entire year and females small ones, consistent with the greater energy expenditure implied by continuous egg production.

The sexual coefficient is 0.9 males/females, and the ratio between adults and immatures (subadults and juveniles) is 7.7. Populations are composed of a large proportion of adults throughout the year.

This is a gregarious species that establishes numerous populations. The population from La Mula ranged in population density from 82 to 200 individuals/ha; the biomass was between 0.19 and 0.40 kg/ha.

To date, one parasitic nematode, *Cyrtosomum sclepori*, and one parasitic trematode, *Turquinia cubensis*, are known for *Anolis argenteolus*. This species is subject to predation by *Anolis imias* on the southern coast of Guantánamo province.

Species group *vermiculatus* (without dewlap).

Anolis vermiculatus (lagarto caimán; plate 32)

Anolis vermiculatus Duméril and Bibron, 1837, *Erp. Gén.* 4:128; syntypes MNHN 2349 and 2407. Type locality: Cuba; restricted to Viñales, Pinar del Río, Cuba by Ruibal, 1964, *Bull. Mus. Comp. Zool.* 130(8):511. *Deiroptyx vermiculatus:* Fitzinger, 1843, *Syst. Rept.*:17. *Anolis vermiculatus:* Etheridge, 1960, *Univ. Microfilms*:94.

GEOGRAPHIC RANGE

Isla de Cuba (Sierra de los Órganos and Sierra del Rosario).

LOCALITIES (MAP 36)

Pinar del Río: Taco Taco; Río Santa Cruz (Gundlach 1880); Sierra de Guane (Barbour 1914); San Diego de los Baños; El Guamá, Pinar del Río (Stejneger 1917); Sumidero; Isabel Rubio (Barbour and Ramsden 1919); Rancho Mundito (Neill and Allen 1957); Viñales (Ruibal 1964); Nortey (Garrido 1976*b*); Río San Juan; El Taburete (González Bermúdez and Rodríguez Schettino 1982); Río San Marcos, Mil Cumbres (Rodríguez Schettino and Novo Rodríguez 1982); Soroa (Schwartz and Henderson 1985); El Mulo; La Cañada del Infierno (Martínez Reyes 1995); El Cuzco, Cabañas; Río Cuyaguateje; Luis Lazo (CZACC); Río Los Palacios, Seboruco; Pan de Guajaibón; Cajálbana; Sierra de la Güira; San Vicente, Viñales (L.R.S.)

CONTENT

Monotypic endemic species.

DESCRIPTION

Medium or intermediate size. Mean snout-vent length of 111.9 mm in males and of 77.3 mm in females.

Dark olivaceous brown with yellow and black vermiculations, and black, ill-defined transverse bands on limbs and tail. Yellow labials, with black spots. Yellow gular region, with a transverse skin fold. White interparietal scale. Pale brown ventral region. Dewlap in males, very small and not differentiated from the gular region. Blue or brown iris.

Vertically oval ear opening. Very reduced digital pads. Keeled, circular, very small dorsal scales; ventral scales as dorsals. Large interparietal scale with conspicuous pineal eye. Middorsal longitudinal crest of scales modified as spines. Somewhat enlarged postcloacal scales in males.

Six pairs of macrochromosomes and eleven pairs of microchromosomes (2n = 34).

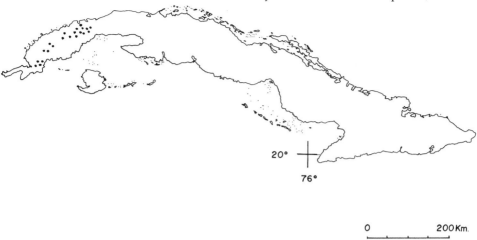

Map 36. Geographic distribution of *Anolis vermiculatus*.

BIBLIOGRAPHIC SOURCES

Duméril and Bibron (1837): original description; coloration; measurements. Cocteau and Bibron (1838): coloration in life; external morphology; scalation; drawings of a specimen in lateral view (pl. 10), of the dorsal scalation of the head (pl. 10, fig. 1), of the scales of the gular region (pl. 10, fig. 2), of the scales of the cloacal and femoral regions (pl. 10, fig. 3), of the ventral surface of the left hindlimb (pl. 10, fig. 4). Gundlach (1867, 1880): coloration in life; sexual variation; habitat data; behavior. Barbour and Ramsden (1916b): behavior. Stejneger (1917): drawings of the dorsal scales of the head (fig. 39), of the head in lateral view (fig. 40), of the tail in lateral view in the fifth verticil (fig. 41). Barbour and Ramsden (1919): coloration in life; scalation; drawing of the dorsal scales of the head (pl. 14, fig. 4); habitat data; behavior. Barbour (1930b): habitat data; behavior. Allen and Neill (1957): vocalization. Neill and Allen (1957): photograph of one male and of the head in lateral view; behavior; vocalization; feeding; aggressive display between males. Etheridge (1960): osteology. Ruibal (1964): coloration; scalation; drawing of the suborbital scales (fig. 17); habitat data; behavior; feeding; vocalization. Buide (1967): morphology of the gular region; behavior. Gorman and Atkins (1968): karyotype; photograph of the diakinesis (fig. 2); behavior. Williams (1969): habitat data; behavior. Peters (1970): external morphology; photographs of the head in dorsal view (fig. 1a), of the head in lateral view (fig. 3); vocalization; habitat data. Buide González *et al.* (1974): behavior; anthropic damage. Garrido (1976b): behavior; feeding; measurements of eggs laid in captivity. Schwartz (1978): habitat data; behavior; scalation; measurements. Coy Otero and Baruš (1979a): nematodes. Garrido (1980a): data of eggs laid in captivity. Rodríguez Schettino (1981a): feeding. Coy Otero and Rodríguez Schettino (1982): parasitic nematode. González Bermúdez and Rodríguez Schettino (1982): habitat data; thermic relations; behavior; feeding. Rodríguez Schettino and Novo Rodríguez (1982): habitat data; feeding; thermic relations. Garrido and Jaume (1984): behavior. Silva Lee (1984): habitat data; behavior; photograph of the head of one male in lateral view. Buide (1985): coloration; behavior; feeding; photograph of the head of one male in lateral view. Rodríguez Schettino and Novo Rodríguez (1985): habitat data; thermic relations; feeding; behavior; daily activity. Schwartz and Henderson (1985): drawing of the lateral region of the head and of the forelimb (pl. 3, no. 1). Silva Lee (1985): habitat data; territoriality; fights between males; sexual variation; vocalization; photographs of the head of one male in lateral view, of a fight between two males, of a female, and of two males. Granda Martínez (1987): morphometry; coefficients of arboreality. Quintana Ferrer (1987): morphometry. Rodríguez Schettino et al. (1987): habitat data; thermic relations; daily activity; abundance; behavior. Rodríguez Schettino and Martínez Reyes (1989a): habitat data; thermic relations; reproductive cycle; egg features; population structure; density; biomass; feeding. Espinosa López, Posada García et al. (1990): electrophoretic pattern of general proteins of muscle. Espinosa López, Sosa Espinosa, and Berovides Álvarez (1990): electrophoretic patterns of eight proteic systems. Rodríguez Schettino (1994): description of the nemasperms. Berovides Álvarez (1995): conservation. Rodríguez Schettino and Chamizo Lara (1997): conservation. Rodríguez Schettino et al. (1997): habitat data. Rodríguez Schettino and Lizana (1997): feeding; reproduction; habitat data; behavior.

Morphological Variation

Metachromatism from olivaceous brown with yellow and black vermiculations, to dark olivaceous brown with brown and black vermiculations. Coloration pattern varies slightly within each sex and age class. Age and sex are the main sources of variation in this species.

Geographic Variation

Little variation throughout its geographic range.

Sexual Variation

Females are olivaceous brown with black and yellow vermiculations and black transverse bands on the body, limbs, and tail; they have yellow labials and gular regions, both with white spots (plate 33). They are smaller than males (table 7.30). Males have canthal crests to the nostrils and somewhat enlarged postcloacal scales.

Age Variation

Snout-vent length of juveniles varies from 34.0 to 100.0 mm; they have female coloration pattern and flat head scales; males have small canthal crests that do not reach nostrils (plate 34).

NATURAL HISTORY

This species lives in the gallery forests of the Cordillera de Guaniguanico, Pinar del Río province, strictly associated with the banks of water courses. Males get around to a mean distance of 2.01 m off the banks; the mean distance for females is 0.93 m. Individuals perch on tree trunks, showing a preference for *"pomarrosa"* (*Syzigium jambos*), *"ocuje"* (*Calophyllum antillarum*), and royal palm (*Roystonea regia*). Males perch higher above ground or water than females and juveniles, which prefer the lower parts of trunks, tree roots, or rocks either at the water's edge or emerging from the water. Generally, the species is found between 1 and 2 m above ground or water, although some individuals have been seen at heights up to 6 m. The coefficient of observed arboreality (1.28) is high, which indicates a great capacity for climbing trees.

Adult males run away upward to tree crowns and rapidly jump into the water, where they hide under stones, roots, or bank lapels. On some occasions, they strongly swim under or over the water or stay underwater for several minutes (more than 30 min. in free life and up to 50 min. in captiv-

Table 7.30. Comparison of snout-vent length (SVL), head length (HL), and tail length (TL) between male and female adults of *Anolis vermiculatus* (symbols as in table 7.1).

Sex	N	X	CV	m	M	t	p
				SVL			
Males	21	111.9	6.86	98.4	124.5	10.55	<0.01
Females	12	77.3	7.23	69.6	84.9		
				HL			
Males	21	37.1	6.89	32.7	41.4	11.38	<0.01
Females	12	23.7	8.21	20.7	26.4		
				TL			
Males	15	237.1	10.02	193.2	268.4	6.33	<0.01
Females	10	164.8	9.55	129.5	184.3		

ity). Females and juveniles escape downward to the stream sides and hide among the vegetation. This species utters an acute squeak in free life and when it is captured.

Males are highly territorial; each individual lives on a different trunk, separated from other males by several meters and occupying the same perch. There are often aggressive displays between males fighting for territories and opportunities to court females. In such situations, both males stand up, raising their limbs and nuchal crests; extend their rudimentary dewlaps; and open their mouths. If they mutually attack, they bite each other on the neck, each trying to overturn his opponent; they fall into the water, still fighting, and are carried along by the current until one of them separates and escapes to the riverside. Occasionally the loser is pursued again, but in general the winner returns to his territory.

With the exception of juveniles, which behave heliothermically throughout the year, this is a diurnal nonheliothermic species. Usually, the greatest proportion of the population is found in shaded places, although in the coldest months the percentage of basking females increases. The mean body temperature of males (27.4°C) is higher than that of females (24.7°C) during the dry season, while in the wet season both sexes have a mean of 27.9°C. Means of air and substrate temperatures in the places where the species was found varied between 25.5 and 27.2°C and between 24.7 and 26.3°C, respectively. There is a positive significant correlation between body and air temperature, although the former is not higher than the latter during any season of the year. This correlation is observed daily and seasonally. The relative humidity in these gallery forests varies little throughout the year but fluctuates during the day from 65 to 90%.

Males have nocturnal refuges where they sleep on the highest branches of trees, while females hide in hollows and lapels dug in the mud and in dead leaves along stream sides. They begin to abandon these shelters after 7:00 A.M. and 8:00 A.M. during the wet and dry seasons, respectively. The number of active individuals rapidly increases, reaching a maximum at about 11:00 A.M. in both seasons; after this time the number diminishes, increasing slightly at 2:00 P.M. during the wet season. After 6:00 P.M. no individuals remain outside the refuges.

The species feeds mainly on insects (formicids, coleopterans, dipterans, lepidopterans, hemipterans, odonats, blatopterans, orthopterans, hemipterans, and larvae); also included in its diet are spiders, crustaceans (shrimps and land isopods), annelids, mollusks, fishes, amphibians, reptiles, and vegetable material (flowers and fruits). In general, it is an opportunistic feeder that consumes a great variety of trophic resources. Males eat larger prey than females during the dry season (X_m = 17.5 and X_f = 3.2 mm) and the wet season (X_m = 20.9 and X_f = 10.2 mm). In the dry season both sexes obtain smaller prey.

Reproduction is cyclic in this species. The first gravid females appear in populations in April, and in September they do not have oviductal eggs. During the months of egg production, most females have one oviductal egg but from 30 to 60% have two, which seems to compensate for the short duration of the reproductive period.

Oviductal egg length varies from 11.0 to 17.5 mm, while monthly means vary from 14.1 to 15.7 mm. The egg-laying sites are unknown; only one empty white shell has been found in the sand on the bank of the Manantiales River, in Soroa. Some females in captivity have deposited their eggs in the holes of rocks and attached to the rocks; the eggs were white and had a mean length of 19.5 mm. Other eggs laid in captivity, with a mean length of only 15.8 mm, were infertile.

Size of fat bodies corresponds with the energy expenditure that reproduction implies. Males have medium-sized fat bodies throughout the year, while females have small fat bodies during the months of egg production and large- or medium-sized ones in other months.

The sexual ratio is 1.2 males/female, while the ratio between adults and immatures (subadults and juveniles) is 2.6. Populations are composed of a larger proportion of adults, although subadults and juveniles are also observed every month. Population density at Soroa varies between 36 and 132 individuals/hour, and the biomass ranges from 0.43 to 3.67 kg/ha. This high variation in biomass is due to a mean body mass that is quite variable intraspecifically. Because of the low relative abundance of its populations and its limited ecological and geographical range, this species is considered vulnerable according to the IUCN Red List Categories.

To date, four species of parasitic nematodes are known for *Anolis vermiculatus: Parapharyngodon cubensis, Cyrtosomum sclepori, Piratuba digiticauda,* and *Physalopteridae* gen. sp. Potential predators on this species are the racer (*Alsophis cantherigerus*) and the freshwater snake (*Tretanorhinus variabilis*), both of which are very common and have a close spatial relationship with cayman lizards.

Anolis bartschi (plate 35)

Deiroptyx bartschi Cochran, 1928, *Proc. Biol. Soc. Washington* 41:169; holotype USNM 75805. Type locality: Baños de San Vicente, Pinar del Río province, Cuba. *Anolis bartschi:* Etheridge, 1960, *Univ. Microfilms:*93.

GEOGRAPHIC RANGE

Isla de Cuba (Sierra de los Órganos).

LOCALITIES (MAP 37)

Pinar del Río: San Vicente (Cochran 1928); Valle de Viñales (Buide 1967); Isabel Rubio (Schwartz 1968); Cueva del Indio, Viñales (Zajicek and Mauri Méndez 1969); Sierra de Guane; Pica Pica (Peters 1970); Cueva de los Portales, Los Palacios (Garrido and Jaume 1984); Sierra de la Güira (L.R.S.).

CONTENT

Monotypic endemic species.

DESCRIPTION

Small or medium size. Mean snout-vent length of 70.9 mm in males and of 59.1 mm in females.

Pale olivaceous brown with yellow, blue and orange vermiculations and brown and yellow ill-defined transverse bands on limbs and tail. Blue hands and feet. Blue central part of the tail. Yellow labials with gray spots. Yellow upper eyelid, blue lower one. A blue postocular spot to behind the ear opening, where it becomes darker, purple. Rosaceous yellow gular region, with a transverse skin fold. Brown interparietal scale. Very small dewlap, not differentiated from the gular region. Brown iris.

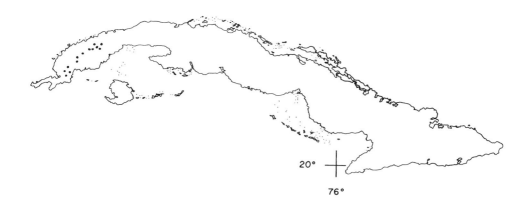

20°

76°

0   200 Km.

Map 37. Geographic distribution of *Anolis bartschi*.

Very small, vertically oval ear opening. Very long hindlimbs. Smooth, circular, small dorsal scales; ventral scales as dorsals. Large interparietal scale, with conspicuous pineal eye. Somewhat enlarged postcloacal scales in males.

Six pairs of macrochromosomes and twelve pairs of microchromosomes (2n = 36).

BIBLIOGRAPHIC SOURCES

Cochran (1928): original description. Barbour (1930b): abundance. Etheridge (1960): osteology. Ruibal (1964): coloration; scalation; habitat data. Buide (1967): morphology of the gular region; habitat data; abundance. Gorman and Atkins (1968): karyotype. Schwartz (1968): habitat data. Zajicek and Mauri Méndez (1969): hemoparasites; drawing of red cells parasitized by *Hemogregarina* sp. (fig. 2 A-C). Williams (1969): habitat data. Coy Otero (1970): parasitic nematode. Peters (1970): external morphology; habitat data; vocalization; photograph of the head in dorsal view (fig. 1b). Coy Otero and Baruš (1979a): nematodes. Garrido (1982b): habitat data. Novo and Rodríguez Schettino (1982): habitat data; thermic relations; feeding. Silva Rodríguez, Berovides Álvarez, and Estrada (1982): communal egg laying. Estrada and Novo Rodríguez (1984b): communal egg laying; features of the eggs and newly born lizards. Estrada and Silva Rodríguez (1984): morphometry; general index of locomotion. Buide (1985): habitat data; photograph of a specimen. Estrada and Novo Rodríguez (1986b): reproduction. Estrada and Novo Rodríguez (1986c): habitat data. Novo Rodríguez and Estrada (1986): reproduction. Estrada and Novo Rodríguez (1987): thermic relations. Granda Martínez (1987): morphometry; coefficients of arboreality. Quintana Ferrer (1987): morphometry. Rodríguez Schettino and Martínez Reyes (1989a): habitat data; thermic relations; reproductive cycle; egg features; population structure; density; biomass; feeding. Espinosa López, Posada García *et al.* (1990): electrophoretic pattern of general proteins of muscle. Espinosa López, Sosa Espinosa, and Berovides Álvarez (1990): electrophoretic patterns of eight proteic systems. Rodríguez Schettino and Martínez Reyes (1992): feeding. Rodríguez Schettino (1994): description of the nemasperms.

MORPHOLOGICAL VARIATION

Metachromatism from pale green to pale olivaceous brown, to dark olivaceous brown with black and yellow vermiculations. Coloration pattern varies slightly. Sex is the main source of variation in this species.

Geographic Variation

Slightly variable throughout its geographic range.

Sexual Variation

Coloration pattern is similar in both sexes. Females have a middorsal longitudinal white stripe, with three yellow spots; the blue postocular spot is ill defined and extends only to the anterior part of the ear opening. Males are recognized by their more extended and intense blue spots, their enlarged postcloacal scales, and their larger size (table 7.31).

Age Variation

Juvenile coloration is similar to that of adult females. Newly born lizards have a mean snout-vent length of 20.5 mm.

NATURAL HISTORY

This species lives in the "mogotes" of Sierra de los Órganos, in places where the vegetation shades the rocky substrate. Individuals perch on rocks and limestone cliffs and, exceptionally, on trunks of bushes adjacent to the "mogotes," at a height above ground between 1 and 3 m (2 m as a mean), although some individuals can climb up to 6 m. No differences in perching habits have been found between males and females.

The coefficient of observed arboreality in this species ranges from 0 to 0.09, which indicates its poor ability to climb trees; this is typical of petricolous species. Likewise, the calculated general index of locomotion is 0.76; that is, the species is a runner.

Table 7.31. Comparison of snout-vent length *(SVL)*, head length *(HL)*, and tail length *(TL)* between male and female adults of *Anolis bartschi* (symbols as in table 7.1).

Sex	N	X	CV	m	M	t	p
			SVL				
Males	25	70.9	4.6	61.8	75.1	13.42	<0.01
Females	31	59.1	5.5	52.5	63.6		
			HL				
Males	25	22.0	5.8	18.2	24.5	14.43	<0.01
Females	31	17.6	5.6	15.7	19.3		
			TL				
Males	15	137.6	13.5	106.1	168.8	3.11	<0.01
Females	24	123.5	7.9	101.8	139.7		

Females escape more rapidly than males. Both sexes escape toward rock crevices, but they suddenly stop before moving farther inside. They emit acute squeaks when they are captured and also in free life.

This species is not very territorial; individuals of both sexes can be observed in proximity to each other on cliffs and cave walls. However, males may perform advertising displays, even aggressive displays, to defend their shelters. In such cases, males raise onto their limbs, extend their rudimentary dewlaps, and open their mouths. If they approach to attack, they bite each other jaws to jaws.

It is a diurnal nonheliothermic species that is found in shaded sites. Both sexes maintain similar mean body temperatures, not only in the dry (24.3°C) but also in the wet season (27.2°C), and the latter is higher than the former. Air temperature in places where this species is found is higher in the wet (26.6°C) than in the dry season (24.2°C), the same as substrate temperatures (26.3 and 23.8°C, respectively). Body temperature does not differ from environmental temperatures during the dry season but is higher in the wet season. There is a significant positive correlation between body and environmental temperatures in both seasons of the year and also daily. The relative humidity is generally high, varying during the day from 46 to 90% in the dry season and from 70 to 100% in the wet season.

Members of this species sleep inside crevices, holes, and lapels of cliffs and cave walls. They begin to leave these shelters after 9:00 A.M. during the dry season and after 7:00 A.M. during the wet season. The number of active individuals rapidly increases until about 10:00 A.M.; then it diminishes, increasing again at 5:00 P.M. during the wet season and at 2:00 P.M. during the dry. After 7:00 P.M. and 5:00 P.M. in the wet and dry seasons, respectively, no individuals are observed outside the shelters.

This species feeds mainly on insects (formicids, coleopterans, dipterans, lepidopterans, blatopterans, and larvae), although it also consumes considerable numbers of gastropods as well as spiders, isopods, chilopods, diplopods, and vegetable material (fruit and flowers). Males eat larger prey than females in the dry season ($X_m = 6.60$; $X_f = 3.76$ mm) as well as in the wet season ($X_m = 6.41$; $X_f = 5.13$ mm).

Reproduction is cyclic in this species. The first gravid females appear in populations in February, and in October they have neither oviductal eggs nor developing ovarian follicles. Only 15% of ovigerous females have two oviductal eggs. Oviductal egg length ranges between 10.0 and 17.0 mm, while monthly means vary from 13.7 and 14.6 mm. Eggs are laid in communal sites, under lapels of cliffs and walls of "mogotes," and have mean lengths ranging from 15.5 to 18.0 mm. The reproductive potential is low: from 0.61 to 0.87 ($X = 0.76$).

Size and weight of fat bodies correspond with the energy expenditure of reproduction. At the beginning and end of the year, both sexes have medium-sized fat bodies; during the months of egg production, they only have small ones.

The sexual ratio is 0.9 males/female and the ratio between adults and immatures (subadults and juveniles) is 9.9. This indicates the higher proportion of adults that constitute the populations: from 77 to 100% in February and April, respectively.

Population density at Cueva del Indio varies from 180 to 470 individuals/ha, with biomass ranging from 1.01 to 3.08 kg/ha. In general, populations are numerous in every place; thus the species can be classified as gregarious.

To date, protozoa of the genus *Hemogregarina* are known to parasitize red cells of *Anolis bartschi*. In addition, five species of parasitic nematodes are known: *Cyrtosomum scelopori, Parapharyngodon cubensis, Skrjabinoptera (Didelphysoma) phrynosoma, Oswaldocruzia lenteixeirai,* and *Porrocaecum* sp. larvae.

Series *alutaceus* (splenial absent; five lumbar vertebrae; seven anterior aseptate caudal vertebrae).

Species group *alutaceus* (very small size; long head, body, and tail).

Anolis alutaceus (plate 36)

Anolis alutaceus Cope, 1861, *Proc. Acad. Nat. Sci. Philadelphia* 13:212; syntype MCZ 10932. Type locality: Monte Verde, Oriente province, Cuba. *Anolis alutaceus alutaceus:* Barbour, 1937, *Bull. Mus. Comp. Zool.* 82(2):124. *Anolis alutaceus saltatus* Peters, 1970, *Mitt. Zool. Mus. Berlin* 46(1):217; holotype ZMB 41868. Type locality: Arroyo La Mariposa, Sierra de Trinidad, 4 km NW of Topes de Collantes, Las Villas province, Cuba.

GEOGRAPHIC RANGE

Isla de Cuba; Isla de la Juventud.

LOCALITIES (MAP 38)

Pinar del Río: San Diego de los Baños (Stejneger 1917); Cabo de San Antonio; El Cayuco (Manuel Lazo) (Schwartz and Garrido 1968); Soroa (Peters 1970); Sierra de los Órganos; Cabezas (Schwartz and Garrido 1971); Pica Pica; Viñales; Bahía Honda; Cayajabos (Garrido 1980a); El Salón, Sierra del Rosario (Silva Rodríguez and Estrada 1982); El Veral (Rodríguez Schettino and Martínez Reyes 1985); El Mulo, Sierra del Rosario (Martínez Reyes

1995); Mil Cumbres (L.R.S.). *La Habana and Ciudad de La Habana:* Madruga (Barbour 1914); Bosque de La Habana (Collette 1961); Canasí; Arana; Jaruco; Nueva Paz (Garrido 1973*f*); Guanajay; Tapaste; San Antonio de los Baños (Garrido 1980*a*); Escaleras de Jaruco (González González 1989); Jardín Botánico Nacional (Quesada Jacob *et al.* 1991). *Matanzas:* Matanzas (Barbour and Ramsden 1916*a*); San Miguel de los Baños (Garrido 1980*a*); Soplillar (Garrido 1980*b*). *Cienfuegos:* Cienfuegos (Barbour 1914); Soledad (Dunn 1926). *Villa Clara:* Santa Clara (Barbour and Ramsden 1916*a*); Cubanacán (Garrido and Estrada 1989). *Sancti Spíritus:* Mayajigua (Barbour and Ramsden 1919); Arroyo La Mariposa, Topes de Collantes (Peters 1970); Sierra de Trinidad (Schwartz and Garrido 1971); Jatibonico (Garrido 1980*a*); Río Caburní; Pico de Potrerillo (Rodríguez Schettino *et al.* 1979). *Camagüey:* Sierra de Cubitas (Barbour and Ramsden 1919); Sierra de Najasa (Schwartz and Garrido 1971). *Ciego de Ávila:* Loma de Cunagua (Garrido and Estrada 1989). *Holguín:* Nuevo Mundo; Monte Iberia; Cupeyal del Norte; Moa (Garrido 1980*a*); Farallones de Moa; Arroyo Bueno; El Palenque; La Melba (Garrido and Estrada 1989). *Granma:* Jiguaní (Barbour 1914); Maffo (Schwartz and Garrido 1971); Moa (Garrido 1980*a*). *Santiago de Cuba:* Sierra Maestra (Barbour and Ramsden 1916*a*); Santiago de Cuba (Alayo 1951); Sierra del Cobre; Sierra de la Gran Piedra; Pico Turquino (Schwartz and Garrido 1971). *Guantánamo:* Monte Verde (Cope 1861); Monte Líbano (Barbour 1914); Tabajó; Cupeyal (Peters 1970); Maisí; Duaba Arriba; Yateras (Garrido 1980*a*). *Isla de la Juventud:* Nueva Gerona (Barbour 1916); Mogotes de Santa Isabel (Garrido 1980*a*).

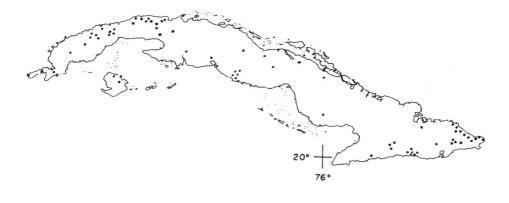

Map 38. Geographic distribution of *Anolis alutaceus.*

CONTENT

Monotypic endemic species.

DESCRIPTION

Very small size. Mean snout-vent length of 34.1 mm in males and of 32.9 mm in females.

Dark brown with pale brown middorsal longitudinal line. Brown limbs and tail. Ivory-colored labial stripe, extending to the ear opening. An ivory-colored longitudinal lateral stripe from the sides of the neck to the groin. A black angular spot over the head, and a black postocular spot. Grayish white or yellowish ventral region. Dewlap in males, pale yellow with white inner scales. Blue iris.

Circular, small ear opening. Long, slender hindlimbs. Smooth, small dorsal scales, with a middorsal longitudinal zone of eight rows of smooth, large scales; smooth, circular ventral scales, larger than dorsals. Enlarged postcloacal scales in males.

BIBLIOGRAPHIC SOURCES

Cope (1861): original description. Gundlach (1867): morphology. Gundlach (1880): habitat data; three coloration patterns. Barbour (1916): scalation. Barbour and Ramsden (1916a): abundance; habitat data. Barbour and Ramsden (1919): coloration; drawing of the dorsal scales of the head (pl. 9, fig. 1); habitat data; abundance; feeding; egg laying. Dunn (1926): egg-laying sites. Barbour (1930a, b): habitat data; abundance. Alayo Dalmau (1955): coloration; drawing of the dorsal scales of the head (pl. 3, fig. 4); abundance. Collette (1961): morphology of the subdigital lamellae; drawing of the subdigital lamellae of the third left toe (fig. 2a); habitat data; behavior; abundance; thermic relations; peritoneal pigment. Ruibal (1961): habitat data. Williams (1961): scalation; habitat data. Ruibal (1964): drawing of the dorsal scales of the head (fig. 12); habitat data; behavior. Buide (1967): habitat data. Rand (1967): egg-laying sites. Garrido and Schwartz (1968): habitat data; behavior. Peters (1970): photograph of two males and two females in dorsal view (fig. 4 a–d). Schwartz and Garrido (1971): two coloration patterns. Baruš and Coy Otero (1974): parasites. Coy Otero and Baruš (1979a): nematodes. Garrido (1980a): coloration of both sexes at different localities; drawing of the head and anterior part of the body (fig. 3 B); habitat data; courtship, mating, and aggressive display between males in captivity. Garrido (1980b): dewlap coloration; habitat data. Estrada and Silva Rodríguez (1984): morphometry; general index of locomotion. Rodríguez

Schettino (1985): habitat data. Rodríguez Schettino and Martínez Reyes (1985): habitat data; behavior. Schwartz and Henderson (1985): drawing of the lateral region of the head and of the forelimb (pl. 2, no. 10). Silva Lee (1985): photograph of a specimen. Garrido and Estrada (1989): measurements. Quesada Jacob *et al.* (1991): abundance. Rodríguez Schettino (1996): habitat data.

Morphological Variation

Without metachromatism. Coloration is individually quite variable. In all populations there are two coloration patterns: in the first, the middorsal longitudinal line is separated from the body color by a very dark brown longitudinal line; the other has an ill-defined middorsal longitudinal line and an angular spot on the head.

Geographic Variation

Individual variations are seen in all populations and are not geographic variants.

Sexual Variation

Both sexes have similar coloration patterns. Males are recognized by their dewlaps and their enlarged postcloacal scales. Measurements are shown in table 7.32.

Age Variation

Juveniles are similar to adults. Males have enlarged postcloacal scales.

Table 7.32. Comparison of snout-vent length *(SVL)*, head length *(HL)*, and tail length *(TL)* between male and female adults of *Anolis alutaceus* (symbols as in table 7.1).

Sex	N	X	CV	m	M	t	p
			SVL				
Males	9	34.1	7.07	28.2	37.5	1.15	n.s.
Females	8	32.9	4.71	28.2	36.0		
			HL				
Males	9	10.8	7.68	9.2	11.4	2.25	<0.05
Females	8	10.1	4.29	9.3	10.7		
			TL				
Males	6	89.6	6.92	52.3	99.0	3.67	<0.01
Females	8	76.1	9.35	53.1	87.2		

NATURAL HISTORY

The species lives in shrubwoods; semideciduous, evergreen, and submontane rain forests; and secondary vegetation. Individuals perch on bush twigs and leaves in the herbaceous stratum; on some occasions, they are also found on stones, walls, and cliffs. Usually, they do not climb to heights of more than 1 m above ground. Individuals have been found within deep holes of the "diente de perro" at Península de Guanahacabibes, but also on the herbaceous stratum there. In addition, they have been observed in caves such as Santo Tomás, Pinar del Río province, and Cueva del Aura, Pico Turquino.

Members of this species hide on twigs and leaves, but they quickly jump or shelter inside holes of rocks when pursued. According to the calculated general index of locomotion (0.62), this is a jumping species, which is consistent with its behavior in free life and in captivity. Males defend their territories, at least in captivity; they laterally place and extend their dewlaps, flex their four limbs, and raise their nuchal crests. After a pause, this process is repeated until the resident male attacks the other with his mouth open.

It is a diurnal nonheliothermic species, based on the shaded sites where it is found; it was observed at a mean air temperature of 28.5°C at Península de Guanahacabibes. Body temperatures measured in captivity were high: between 30.6 and 32.4°C in males and between 30.4 and 31.0°C in females. Most individuals are active in the first hours of the morning; at noon they shade, avoiding the warmest time. They sleep on long leaves of bushes and grasses at night.

The species feeds on ants on shrub twigs. In captivity it has accepted small insects, such as fruit flies (*Drosophila* spp.) and ants.

Females lay only one egg at a time, under dead leaves and semidecayed bark, in bromeliad leaves, among grasses, or under stones. One egg laid by a captive female measured 7.8 x 5.0 mm. Courtship behavior, as observed in captivity, is complex; generally, it includes body balancing, raising and vibration of the head, slight nodding, dewlap extension, and flexion. It begins when the male places himself laterally in front of the female; then he nods quickly and extends the dewlap, which is maintained thus the entire time. The male flexes his forelimbs ten or twelve times and repeats the process. During courtship, the female slightly nods; if receptive, she extends the rudimentary dewlap and remains immobile as the male bites her neck and begins copulation.

This species establishes numerous populations throughout its geographic range. It is more easily observed on rainy days.

To date, three species of parasitic nematodes are known for *Anolis alutaceus: Skrjabinodon anolis, Skrjabinoptera (Didelphysoma) phrynosoma*, and *Porrocaecum* sp. larvae.

Anolis clivicola (plate 37)

Anolis clivicolus Barbour and Shreve, 1935, *Occ. Papers Boston Soc. Nat. Hist.* 8:251; holotype MCZ 39664. Type locality: Loma del Cardero, Pico Turquino, 4,000–6,000 ft., Oriente province, Cuba. *Anolis alutaceus clivicolus:* Barbour, 1937, *Bull. Mus. Comp. Zool.* 82(2):124. *Anolis clivicola:* Schwartz and Garrido, 1971, *Carib. J. Sci.* 11(1–2):12.

Geographic Range

Isla de Cuba (Sierra Maestra, from 800 m above sea level, at Sierra del Turquino; Peladero, Sierra del Cobre).

Localities (map 39)

Santiago de Cuba: Pico Turquino (Barbour and Shreve 1935); Peladero, Sierra del Cobre (Schwartz and Garrido 1971); Río Palma Mocha (CZACC).

Content

Monotypic endemic species.

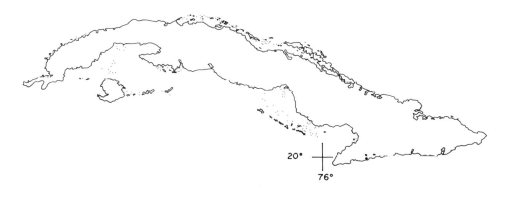

20°

76°

0   200 Km.

Map 39. Geographic distribution of *Anolis clivicola*.

DESCRIPTION

Very small size. Mean snout-vent length of 45.6 mm in males and of 39.6 mm in females.

Greenish gray with small, isolated green scales and a gray middorsal longitudinal line. Pale brown limbs and tail. White labial stripe extending to the ear opening. Yellow eyelids. Grayish ventral region, with some red scales. Dewlap in males, ochraceous yellow with green inner scales. Green iris.

Small, circular ear opening. Long, slender hindlimbs. Smooth, circular, small dorsal scales, with a middorsal longitudinal zone composed of eight rows of keeled, imbricate scales, larger than dorsals; smooth or weakly keeled ventral scales. Enlarged postcloacal scales in males.

BIBLIOGRAPHIC SOURCES

Barbour and Shreve (1935): original description; measurements. Barbour (1937): habitat data. Williams (1961): scalation. Ruibal (1964): drawing of the dorsolateral scalation of the body (fig. 13). Schoener (1970a, b): interspecific relations in size. Schwartz and Garrido (1971): scalation. Garrido (1980a): coloration; habitat data. Estrada and Silva Rodríguez (1984): morphometry; general index of locomotion. Rodríguez Schettino (1985): habitat data; abundance.

MORPHOLOGICAL VARIATION

Without metachromatism. Coloration is slightly variable.

Geographic Variation

Little variation throughout its limited geographic range.

Sexual Variation

Both sexes have similar coloration patterns. Females are yellowish brown with isolated gray or black spots. Males are recognized by their dewlaps and their enlarged postcloacal scales. Differences in size are shown in table 7.33.

Age Variation

Juvenile coloration is similar to that of adults but with broader gray middorsal line. Males have enlarged postcloacal scales.

NATURAL HISTORY

This species lives in the shrubby and herbaceous vegetation of rain and cloud forests, at more than 800 m above sea level, perching on trunks of

Table 7.33. Comparison of snout-vent length (SVL), head length (HL), and tail length (TL) between male and female adults of *Anolis clivicola* (symbols as in table 7.1).

Sex	N	X	CV	m	M	t	p
			SVL				
Males	9	45.6	7.52	29.8	49.4	4.54	<0.01
Females	10	39.6	5.53	36.3	44.1		
			HL				
Males	9	14.0	4.61	9.7	15.0	4.58	<0.01
Females	10	12.6	5.56	11.5	13.9		
			TL				
Males	8	98.0	4.19	61.5	103.2	5.97	<0.01
Females	6	81.7	7.53	52.1	88.2		

trees, bushes, and arboreous ferns. Males perch on trunks and branches up to 6 m above ground; females and juveniles perch at lesser heights, between the herbaceous stratum and leaf litter, and occasionally on roots. When on trunks or branches, individuals escape quickly, jumping from branch to branch; however, if surprised, they run on the ground to escape. According to the calculated general index of locomotion (0.66), this is a jumping species, which is consistent with its free-life behavior.

It is a diurnal, seemingly nonheliothermic species based on the shaded places where it has been observed. The majority of individuals are detected during the warmest hours of the day; before and after this period, they are sheltered.

The population living at Pico Turquino is abundant from approximately 800 to 1,500 m above sea level; at higher altitudes, it becomes scarce.

Anolis anfiloquioi (plate 38)

Anolis anfiloquioi Garrido, 1980, *Poeyana* 201:17; holotype IZ 4283. Type locality: hill in front of the house of A. Suárez, La Poa, Sabanilla, Oriente province, Cuba.

GEOGRAPHIC RANGE

Isla de Cuba (vicinity of Levisa; Baracoa; Guantánamo).

LOCALITIES (MAP 40)

Holguín: Levisa (Garrido 1980*a*); Los Tibes; Loma Blanca; El Yayal (Navarro and Peña 1995). *Guantánamo:* Baracoa; Lajas; Laguna Lajial; Boca del Jaibo; El Palmar (Garrido 1980*a*).

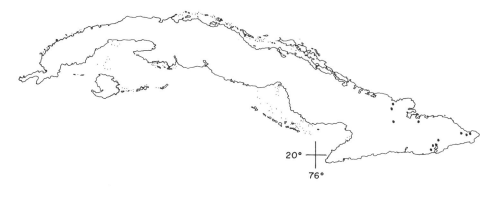

Map 40. Geographic distribution of *Anolis anfiloquioi*.

CONTENT

Monotypic endemic species.

DESCRIPTION

Very small size. Mean snout-vent length of 36.3 mm in males and of 35.1 mm in females.

Grayish brown, with an orange spot on the scapular region. Pale brown limbs and tail, with some ill-defined brown transverse bands on limbs. White labial stripe, extending to the scapular region, bordered with a black upper line. Grayish white gular region. Gray ventral region, with dark gray dots. Black transverse bands or a black central portion on tail. Dewlap in males, ochraceous with white inner scales. Pale brown iris.

Small, circular ear opening. Long, slender hindlimbs. Smooth, circular, small dorsal scales, with a middorsal longitudinal zone consisting of eight rows of keeled scales, somewhat larger than dorsals; smooth or weakly keeled ventral scales, smaller than dorsals. Enlarged postcloacal scales in males.

BIBLIOGRAPHIC SOURCES

Garrido (1980*a*): original description; drawing of the head and the anterior part of the body in lateral view (fig. 3A); habitat data; behavior. Estrada and Silva Rodríguez (1984): morphometry; general index of locomotion. Silva Lee (1985): photograph of a specimen. Garrido and Estrada (1989): measurements.

MORPHOLOGICAL VARIATION

Without metachromatism. Coloration is slightly variable individually and geographically.

Geographic Variation

Individuals collected at Levisa, north of Holguín province, have some meristic and color differences, but they are not considered a distinct subspecies.

Sexual Variation

Both sexes have similar coloration patterns and do not differ significantly in size (table 7.34). Males are recognized by their dewlaps and their enlarged postcloacal scales.

Age Variation

Juveniles are not known.

NATURAL HISTORY

This species lives in shrubwoods, semideciduous and gallery forests, and secondary vegetation in the easternmost region of the country. Males perch on branches and leaves of small, sparse bushes and grasses, at a height above ground near 0.5 m—higher than females and juveniles, which also live on the ground.

Individuals flatten on leaves and branches in free life and in captivity. When escaping, they quickly jump among leaves and hide in the herba-

Table 7.34. Comparison of snout-vent length (SVL), head length (HL), and tail length (TL) between male and female adults of *Anolis anfiloquioi* (symbols as in table 7.1).

Sex	N	X	CV	m	M	t	p
				SVL			
Males	11	36.3	10.32	30.9	40.5	0.85	n.s.
Females	9	35.1	7.29	25.2	38.2		
				HL			
Males	11	11.7	6.96	10.3	12.8	2.53	<0.05
Females	9	10.8	6.49	8.3	12.2		
				TL			
Males	5	101.5	6.47	42.3	107.3	0.47	n.s.
Females	6	93.0	8.67	43.2	103.9		

ceous stratum. According to the calculated general index of locomotion (0.58), this is a jumping species, which is consistent with its behavior in free life and in captivity .

Males actively defend their territories, at least in captivity against *Anolis alutaceus,* with head bobbing and nodding, dewlap extension, limb flexion, and upward arching of the tail.

This is a diurnal species, seemingly nonheliothermic based on the shaded sites where it is found.

Feeding in free life is unknown. The species has accepted small insects in captivity.

Anolis inexpectata

Anolis inexpectata Garrido and Estrada, 1989, *Rev. Biol.* 3:59; holotype MNHN 291. Type locality: Farallones de Moa, Holguín, Cuba.

GEOGRAPHIC RANGE

Isla de Cuba (south of the Sagua de Tánamo and Moa districts, and north of the Yateras district).

LOCALITIES (MAP 41)

Holguín: Farallones de Moa; Cupeyal del Norte; Calentura Arriba; La Melba; Arroyo los Gatos; Arroyo Bueno (Garrido and Estrada 1989). *Guantánamo:* Ojito de Agua; springs of the Yarey River; springs of the Jaguaní River (Garrido and Estrada 1989).

CONTENT

Monotypic endemic species.

DESCRIPTION

Very small size. Mean snout-vent length of 34.4 mm in males and of 30.6 mm in females.

Olivaceous brown head and anterior dorsolateral region of body; brown posterior dorsolateral region. A pale brown middorsal longitudinal line, bordered on both sides by a brown line. Dark brown postocular spot. White labial stripe extending to the suprascapular region, bordered above by a fine brown line. Dark and pale brown transverse bands on tail. Grayish white ventral region. Dewlap in males, grayish yellow or greenish gray. Blue iris.

Oval, oblique ear opening, pointed upward. Frontal ridges higher than canthals, with a groove between them. A frontal depression. Smooth, circu-

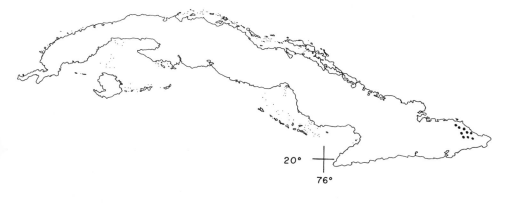

Map 41. Geographic distribution of *Anolis inexpectata*.

lar, small dorsal scales; smooth, circular ventral scales, the same size as or somewhat larger than dorsals. One row of scales between supraorbital semicircles.

BIBLIOGRAPHIC SOURCES

Garrido and Estrada (1989): original description; morphometry; coloration; morphological variation; habitat data; general index of locomotion.

MORPHOLOGICAL VARIATION

Without metachromatism. Coloration varies a little individually and geographically. The postocular spot is reddish, instead of black, in some individuals, as is the posterior part of the thigh.

Geographic Variation

Color variations are not geographic, but within populations.

Sexual Variation

Coloration pattern is similar in both sexes. Males and females differ in size (table 7.35) and in the number of ventral and canthal scales. Males have higher frontal ridges, and the groove between ridges and the frontal depression is deeper than in females.

Age Variation

Juveniles are not known.

Table 7.35. Comparison of snout-vent length *(SVL)*, head length *(HL)*, and tail length *(TL)* between male and female adults of *Anolis inexpectata* (symbols as in table 7.1).

Sex	N	X	CV	m	M	t	p
				SVL			
Males	16	34.4	5.0	29.0	36.5	4.13	<0.05
Females	15	30.6	9.0	24.5	35.0		
				HL			
Males	16	10.3	8.0	8.5	11.3	3.20	<0.05
Females	15	9.1	9.0	7.2	10.2		
				TL			
Males	12	87.5	7.0	76.0	96.0	3.70	<0.05
Females	10	75.7	11.0	62.0	92.0		

NATURAL HISTORY

This species lives in coniferous forests and rain forests of the Cuchillas de Moa, on trunks of bushes and ferns ranging from 1 to 2 cm in diameter, with a mean height above ground of 1.5 m.

The general index of locomotion is 0.59 for males and 0.58 for females, indicating that this is a jumping species. Apparently, it is a diurnal nonheliothermic species based on the shaded and moist places where it has been found.

Anolis macilentus

Anolis macilentus Garrido and Hedges, 1992, *Carib. J. Sci.* 28(1,2):22; holo-type MNHNCU 2721. Type locality: Río Pai, Monte Líbano, southern slope of Meseta del Guaso, 650 m, Guantánamo province, Cuba.

GEOGRAPHIC RANGE

Isla de Cuba (known only from the type locality).

LOCALITIES (MAP 42)

Guantánamo: Río Pai, Monte Líbano (Garrido and Hedges 1992).

CONTENT

Monotypic endemic species.

DESCRIPTION

Very small size. Snout-vent length from 34 to 41 mm in males and from 30 to 36 mm in females.

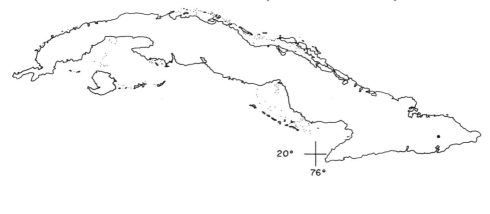

20°

76°

0 _____ 200 Km.

Map 42. Geographic distribution of *Anolis macilentus*.

Pale brown. White labial stripe extending to the suprascapular region, bordered above by a narrow dark brown or black stripe. Black postocular spot. Beige frontal depression; dark brown occipital region. Black, ill-defined spots on sides between limbs. Grayish beige ventral region, with isolated pink scales. Dewlap in males, grayish brown, small. Deep blue iris.

Keeled, rhomboidal dorsal scales; smooth, rhomboidal ventral scales, somewhat larger than dorsals. One or two rows of scales between supraorbital semicircles. Enlarged postcloacal scales in males.

BIBLIOGRAPHIC SOURCES

Garrido and Hedges (1992): original description; coloration; habitat data; photographs of the holotype (figs. 1 A and 2); drawing of the dorsal region of the head (fig. 3B).

MORPHOLOGICAL VARIATION

Little variation in this species.

Geographic Variation

Varies slightly in its limited geographic range.

Sexual Variation

Coloration pattern is similar in both sexes. Males are somewhat larger than females and are recognized by their dewlaps and their enlarged postcloacal scales.

Age Variation

Juveniles are not known.

NATURAL HISTORY

This species lives in the Río Pai gallery forest at 650 m above sea level, in shrubby and herbaceous vegetation. Individuals are found on twigs and stems of grasses, from 0.5 to 2 m above ground, where they remain immobile. When escaping, they jump down among the grasses, where they do not move.

It is a diurnal species, apparently nonheliothermic based on the shaded places where it is found. The greatest activity was observed in the middle of the morning.

Anolis vescus

Anolis vescus Garrido and Hedges, 1992, *Carib. J. Sci.* 28(1,2):25; holotype MNHNCU 2729. Type locality: Palmarito, 4.7 km N de Los Calderos, Municipio Imías, Guantánamo province, 700 meters, Cuba.

GEOGRAPHIC RANGE

Isla de Cuba (known only from the type locality).

LOCALITIES (MAP 43)

Guantánamo: Palmarito, Sierra del Purial (Garrido and Hedges 1992).

CONTENT

Monotypic endemic species.

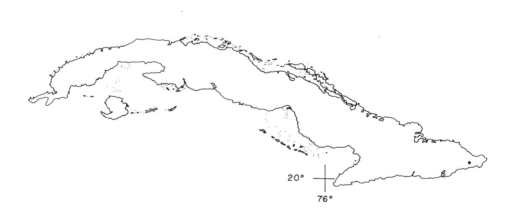

Map 43. Geographic distribution of *Anolis vescus.*

DESCRIPTION

Very small size. Snout-vent length of 41 mm in the only known male and from 33 to 37 mm in females.

Brown. White labial stripe extending to the forelimbs, bordered above by a narrow dark stripe. Dark brown snout, eyelids, and occipital region. Black suprascapular spot. Many dark spots on venter. A wide black band in the middle of the tail. Dewlap in males, pale grayish brown, small, but long. Dark metallic gray iris.

Keeled, rhomboidal dorsal scales; smooth, rhomboidal ventral scales, larger than dorsals. One or two rows of scales between supraorbital semi-circles. Enlarged postcloacal scales in males.

BIBLIOGRAPHIC SOURCES

Garrido and Hedges (1992): original description; coloration; habitat data; drawing of the dorsal surface of the head (fig. 3A).

MORPHOLOGICAL VARIATION

Only one male and four females are known.

Geographic Variation

The few specimens known have no geographic variations.

Sexual Variation

Females have a greenish shade on the dorsal surface of the head. Males are larger than females and have dewlaps and enlarged postcloacal scales.

Age Variation

Juveniles are not known.

NATURAL HISTORY

This species lives in shrubby and herbaceous vegetation at Palmarito, adjacent to a banana plantation, at 700 m above sea level.

Anolis alfaroi

Anolis alfaroi Garrido and Hedges, 1992, *Carib. J. Sci.* 28(1,2):27; holotype MNHNCU 2725. Type locality: 2 km *N* of La Munición, Yateras district, Guantánamo province, Cuba, 730 m.

GEOGRAPHIC RANGE

Isla de Cuba (known only from the type locality).

LOCALITIES (MAP 44)

Guantánamo: La Munición, Yateras (Garrido and Hedges 1992).

CONTENT

Monotypic endemic species.

DESCRIPTION

Very small size. Mean snout-vent length from 32 to 36 mm in males and from 28 to 33 mm in females.

Brown with six or seven vertical rows of small, yellowish spots on both sides of body. Five suffused dark spots on back. Dark brown dorsal surface of the head and loreal region. Dark brown interocular spot, V-shaped, with the vortex backward. White labial stripe extending to the forelimbs. Dewlap in males, pale gray with white inner scales, very small. Bluish iris.

Keeled, rhomboidal dorsal scales; keeled, rhomboidal ventral scales, somewhat larger than dorsals. One or two rows of scales between supraorbital semicircles. Enlarged postcloacal scales in males.

BIBLIOGRAPHIC SOURCES

Garrido and Hedges (1992): original description; coloration; habitat data; photograph of a male (fig. 2A).

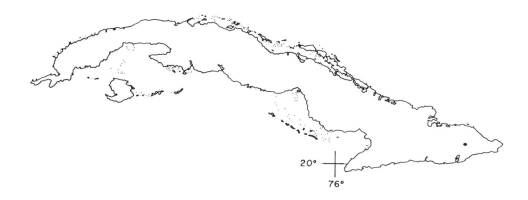

Map 44. Geographic distribution of *Anolis alfaroi.*

MORPHOLOGICAL VARIATION

Little variation in coloration and scalation.

Geographic Variation

Little variation throughout its limited geographic range.

Sexual Variation

Scales around the interparietal are smooth in males, while some of them are keeled in females. Female ventral scales are larger and less keeled than those of males, which are recognized by their dewlaps and their enlarged postcloacal scales.

Age Variation

Juveniles are not known.

NATURAL HISTORY

This species lives in a coniferous forest at 780 m above sea level. Individuals perch on stems of grasses and bushes at a height of 1 m above ground.

The species is diurnal. When escaping, individuals jump among grasses or hide on stems. They sleep at night on stems of ferns, grasses, and bushes.

Anolis cyanopleurus (plate 39)

Anolis cyanopleurus Cope, 1861, *Proc. Acad. Nat. Sci. Philadelphia* 13:211; syntypes USNM 62068–70. Type locality: Monte Verde, Cuba; restricted to La Prenda, Monte Verde, Yateras, Oriente province, Cuba (Garrido 1975*c*).

GEOGRAPHIC RANGE

Isla de Cuba (easternmost end of the island, from the vicinities of Cupeyal, Felicidad, and Bayate, to Maisí, in Guantánamo province).

LOCALITIES (MAP 45)

Holguín: Cupeyal del Norte (Estrada *et al.* 1987). *Guantánamo:* Monte Verde (Cope 1861); Tabajó; Baracoa; Cupeyal (Peters 1970); Bayate; La Munición; Duaba Arriba; Cuchillas de Guajimero; Sierra de Imías; Sierra del Maquey; Sierra del Purial; Río Ovando; Maisí (Garrido 1975*c*); Cayo Probado; La Melba; springs of the Jaguaní River (Estrada *et al.* 1987); springs of the Yarey River (Garrido and Estrada 1989).

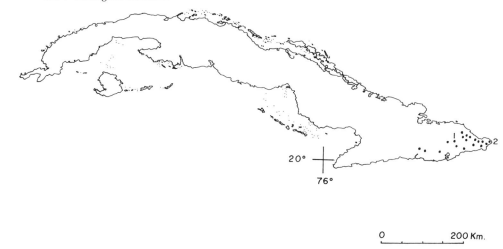

Map 45. Geographic distribution of *Anolis cyanopleurus. 1, A. c. cyanopleurus; 2, A. c. orientalis.*

CONTENT

Polytypic endemic species; two subspecies are recognized:

Anolis cyanopleurus cyanopleurus

Anolis cyanopleurus Cope, 1861, *Proc. Acad. Nat. Sci. Philadelphia* 13:211. Type locality: Monte Verde, Cuba, restricted to La Prenda, Monte Verde, Yateras, Oriente province, Cuba (Garrido 1975c). *Anolis cyanopleurus cyanopleurus:* Garrido, 1975, *Poeyana* 143:8. Geographic range: vicinities of Cupeyal; Bayate; Guantánamo; Duaba Arriba; Cuchillas del Toa; Sierra del Maquey; Sierra del Purial; Sierra de Imías; Baracoa.

Anolis cyanopleurus orientalis

Anolis cyanopleurus orientalis Garrido, 1975, *Poeyana* 143:16; holotype IZ 1564. Type locality: vicinity of Punta de Maisí, Baracoa, Oriente province, Cuba. Geographic range: Maisí and Río Ovando.

DESCRIPTION

Very small size. Mean snout-vent length between 35.9 and 38.6 mm in males and between 32.0 and 32.1 mm in females.

Green or pale brown, with a pale brown middorsal longitudinal line that has six brown dots on its borders. Greenish brown limbs, with pale brown, ill-defined transverse bands. White labial stripe extending to the ear opening. A black angular spot behind eyes. Greenish yellow ventral region. Olivaceous brown tail, with its center black. Dewlap in males, yellowish green with the inner scales of the same color. Green iris.

Oblique, oval ear opening. Long, slender hindlimbs. Keeled, circular, large dorsal scales, with a middorsal longitudinal zone consisting of six or seven rows of keeled, larger scales; keeled, circular ventral scales, larger than dorsals. Keeled, striated head scales. Enlarged postcloacal scales in males.

BIBLIOGRAPHIC SOURCES

Cope (1861): original description; coloration. Bocourt (1874): drawing of the dorsal scales of the head (pl. 16, fig. 29). Gundlach (1880): description; abundance. Barbour and Ramsden (1919): coloration; drawings of a specimen (pl. 4, fig. 1), of the scalation of the central portion of the tail (pl. 6, fig. 6), of the head in lateral view (pl. 8, fig. 1) and dorsal view (pl. 8, figs. 2 and 3); measurements. Alayo Dalmau (1955): drawings of the head in dorsal and lateral views (pl. 3, fig. 7, nos. 1–3), of the tail scales (pl. 6, fig. 6); habitat data; abundance. Williams (1961): scalation. Ruibal (1964): description; measurements; habitat data. Peters (1970): coloration of both sexes; photograph of a female and a male in dorsal view (fig. 5, a and b). Schoener (1970a): interspecific relationships in size. Garrido (1975c): coloration of both sexes; drawing of mental scales (fig. 1 D); photographs of the gular scales (fig. 2, upper left), of the dorsal scales (fig. 3, center); habitat data. Coy Otero and Baruš (1979a): nematode. Garrido (1980a): reproductive data in captivity. Estrada and Silva Rodríguez (1984): morphometry; general index of locomotion. Estrada *et al.* (1987): habitat data.

MORPHOLOGICAL VARIATION

Without metachromatism. Coloration is quite variable individually and geographically.

Geographic Variation

The two subspecies are described as follows:
Anolis cyanopleurus cyanopleurus (Guantánamo province, except the vicinities of Maisí): green, with ivory-colored middorsal line, bordered on both sides by dark brown lines and about six white dots along the center. White labial stripe extending to the armpit. Yellowish green dewlap. Bluish green iris. Six rows of scales in the middorsal line. Mean snout-vent length of 35.9 mm in males and of 32.1 mm in females.
A. c. orientalis (Maisí): pale brown with some dark brown spots on back. White labial stripe extending to the ear opening. Bluish white dewlap. Blue iris. Seven rows of large, keeled scales in the middorsal line. Mean snout-vent length of 38.6 mm in males and of 32.0 mm in females.

Sexual Variation

Coloration pattern is similar in both sexes, which differ slightly in size (table 7.36). Males are recognized by their dewlaps and their enlarged postcloacal scales.

Age Variation

Coloration of juveniles is similar to that of adults. Males have enlarged postcloacal scales.

NATURAL HISTORY

This species lives in semideciduous and evergreen forests of the eastern region. Individuals perch on bushes and grasses, preferably in places near streams and rivers.

Members of this species stay immobile on twigs and grasses; when escaping, they quickly jump from twig to twig. According to the calculated general index of locomotion (0.64), this is a jumping species, which is consistent with its free-life behavior.

It is a diurnal species, apparently nonheliothermic based on the shaded sites where it is found. The majority of individuals are observed during the first hours of the day and in the afternoon, although on rainy and cool days they can be observed throughout the daylight hours.

About reproduction little is known, only that females from the population at Río Duaba were gravid in June. The precopulatory phase for captive adults lasted 96 sec.; the mean duration of copulation was 22.1 min. and the longest copulation lasted 33 min.

Table 7.36. Comparison of snout-vent length *(SVL)*, head length *(HL)*, and tail length *(TL)* between male and female adults of *Anolis cyanopleurus* (symbols as in table 7.1).

Sex	N	X	CV	m	M	t	p
				SVL			
Males	9	35.9	9.78	29.8	39.5	2.65	<0.05
Females	9	32.1	8.10	26.1	36.2		
				HL			
Males	9	11.2	8.70	9.4	12.3	4.89	<0.01
Females	9	9.5	4.51	8.7	10.1		
				TL			
Males	8	92.6	9.35	60.2	106.9	3.00	<0.01
Females	9	79.9	10.84	70.3	96.7		

Populations are composed of few individuals; however, the species' cryptic coloration and slow movements can make it appear scarce.

To date, one species of parasitic nematode is known for *Anolis cyanopleurus: Cyrtosomum sclepori.*

Anolis cupeyalensis (plate 40)

Anolis cyanopleurus cupeyalensis Peters, 1970, *Mitt. Zool. Mus. Berlin* 46(1):225; holotype ZMB 41059. Type locality: Cupeyal, Sierra de Maquey (south of the Sagua de Tánamo district). *Anolis cupeyalensis:* Garrido, 1975, *Poeyana* 143:20.

Geographic Range

Isla de Cuba (San Felipe, Jatibonico; Pinares de Mayarí; Cupeyal; Yateras).

Localities (map 46)

Sancti Spíritus: San Felipe, Jatibonico (Garrido 1975c). *Holguín:* Pinares de Mayarí (Garrido 1975c); Cupeyal del Norte (Estrada *et al.* 1987). *Guantánamo:* Cupeyal (Peters 1970); Yateras (Garrido 1975c); springs of the Yarey River; Ojito de Agua (Estrada *et al.* 1987).

Content

Monotypic endemic species.

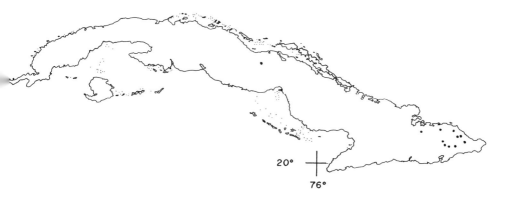

Map 46. Geographic distribution of *Anolis cupeyalensis.*

DESCRIPTION

Very small size. Mean snout-vent length of 30.6 mm in males and of 28.5 mm in females.

Pale brown, with olivaceous brown head; very pale brown middorsal longitudinal line, bordered on both sides by brown lines. White labial stripe extending to groin. A pale brown midventral longitudinal line from the gular region to the vent. Dewlap in males, pale yellow. Greenish blue iris.

Circular, small ear opening. Long, slender hindlimbs. Keeled, circular, large dorsal scales, with six rows of larger, keeled scales in the middorsal longitudinal zone; keeled, circular ventral scales, larger than dorsals. Enlarged postcloacal scales in males.

BIBLIOGRAPHIC SOURCES

Peters (1970): original description; photograph of a female and a male (fig. 5c and d); habitat data. Garrido (1975c): coloration of both sexes; drawing of the mental scales (fig. 1c); photograph of the gular scales (fig. 2, upper right); habitat data. Alayón García (1976): feeding in captivity. Estrada *et al.* (1987): habitat data.

MORPHOLOGICAL VARIATION

Without metachromatism. Coloration is slightly variable.

Geographic Variation

Individuals from the population of San Felipe vary slightly in coloration pattern.

Sexual Variation

Coloration pattern is similar in both sexes. Size differences are shown in table 7.37. Males are recognized by their dewlaps and their enlarged postcloacal scales.

Age Variation

Juveniles are not known.

NATURAL HISTORY

This species lives in grasslands along river and stream sides at the Macizo de Sagua-Baracoa. Individuals perch on stems and leaves of tall grasses and crawler ferns, in shaded places near the ground. They remain immobile in the grasses; when escaping, they rapidly jump among leaves.

The species is diurnal, seemingly nonheliothermic. The majority of individuals are observed during the first hours of the day and in the afternoon,

Table 7.37. Comparison of snout-vent length (SVL), head length (HL), and tail length (TL) between male and female adults of Anolis cupeyalensis (symbols as in table 7.1).

Sex	N	X	CV	m	M	t	p
			SVL				
Males	10	30.6	3.43	28.9	32.1	4.20	<0.01
Females	9	28.5	3.85	27.1	30.5		
			HL				
Males	10	9.9	3.88	9.2	10.4	4.27	<0.01
Females	9	9.2	3.11	8.7	9.8		
			TL				
Males	9	83.1	9.88	68.5	93.8	3.11	<0.01
Females	9	69.9	13.90	55.5	83.7		

avoiding direct sunlight; during rainy and cool days they are found all the time.

Feeding in free life is unknown; captive individuals of this species prefer spiders.

Anolis mimus (plate 41)

Anolis cupeyalensis montanus Garrido, 1975, *Poeyana* 143:24; holotype IZ 3917. Type locality: La Gran Piedra, Santiago de Cuba, Oriente province, Cuba. *Anolis montanus* Garrido, 1975, *Poeyana* 143:57. *Anolis mimus*: Schwartz and Thomas, 1975, *Carnegie Mus. Nat. Hist. Special Publ.* 1:93.

GEOGRAPHIC RANGE

Isla de Cuba (Sierra del Cobre; Sierra de la Gran Piedra; Santa María del Loreto).

LOCALITIES (MAP 47)

Santiago de Cuba: Sierra Maestra (Gundlach 1880); Sierra del Cobre; Sierra de la Gran Piedra; Santa María del Loreto; La Maya (Garrido 1975c).

CONTENT

Monotypic endemic species.

DESCRIPTION

Very small size. Mean snout-vent length of 33.4 mm in males and of 32.6 mm in females.

Dark brown with a gray middorsal longitudinal line bordered on both

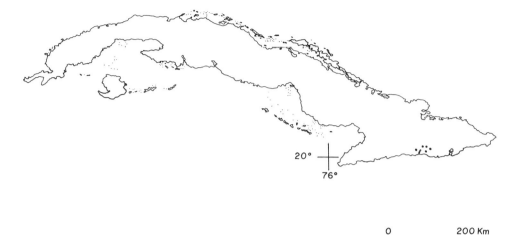

Map 47. Geographic distribution of *Anolis mimus*.

sides by very dark brown discontinuous lines. Greenish forehead. Pale brown limbs. White labial stripe exending to the anterior border of the ear opening. Yellow eyelids. Pale gray gular and ventral regions; in some individuals, with a white longitudinal line along the center. Dark brown tail with a pale brown central region. Dewlap in males, pale yellow with concolor inner scales. Green iris.

Small, circular ear opening. Long, slender hindlimbs. Keeled, circular, small dorsal scales, with six rows of larger, keeled scales in the middorsal line; keeled, circular ventral scales, larger than dorsals. Enlarged postcloacal scales in males.

BIBLIOGRAPHIC SOURCES

Gundlach (1880): coloration in life. Garrido (1975c): original description; coloration of both sexes; habitat data. Socarrás *et al.* (1988): predator.

MORPHOLOGICAL VARIATION

Without metachromatism. Coloration varies a little individually and geographically.

Geographic Variation

Slightly variable throughout its geographic range.

Sexual Variation

Both sexes have a similar coloration pattern; females have black spots and short dashes on the dorsum and do not differ in size from males (table 7.38).

Males are recognized by their dewlaps and their enlarged postcloacal scales.

Age Variation

Juveniles are not known.

NATURAL HISTORY

This species lives in grasslands along river and stream sides, in coniferous and submontane rain forests. Individuals perch on bushes and grasses. They are found among the tall grasses and sparse vegetation of felled forests at La Gran Piedra. Remaining immobile on twigs and leaves, they quickly jump and hide under leaves when escaping.

The species is diurnal, apparently nonheliothermic based on the shaded places where it is found. The majority of individuals are observed during the first hours of the day and in the afternoon. During rainy days, they can be seen all the time. Populations of this species are numerous throughout its geographic range.

Anolis allogus has been reported as a predator on *A. mimus* in free life.

Anolis fugitivus (plate 42)

Anolis fugitivus Garrido, 1975, *Poeyana* 143:28; holotype IZ 3854. Type locality: 2 km south of the Aserrío de Nuevo Mundo, Oriente province, Cuba.

GEOGRAPHIC RANGE

Isla de Cuba (vicinity of Nuevo Mundo).

Table 7.38. Comparison of snout-vent length *(SVL)*, head length *(HL)*, and tail length *(TL)* between male and female adults of *Anolis mimus* (symbols as in table 7.1).

Sex	N	X	CV	m	M	t	p
				SVL			
Males	6	33.4	10.17	27.7	37.0	0.43	n.s.
Females	7	32.6	10.34	27.3	36.7		
				HL			
Males	6	9.8	7.45	8.9	10.6	0.95	n.s.
Females	7	9.4	8.52	8.0	10.3		
				TL			
Males	6	77.4	9.60	64.1	85.1	0.24	n.s.
Females	6	76.4	9.73	50.8	88.5		

LOCALITIES (MAP 48)

Holguín: Nuevo Mundo, Moa (Garrido 1975*c*).

CONTENT

Monotypic endemc species.

DESCRIPTION

Very small size. Mean snout-vent length of 33.5 mm in males and of 31.3 mm in females.

Pale brown with very pale brown middorsal longitudinal line, which has some white dots. White labial stripe extending to the groin. Dewlap in males, pale yellow. Greenish blue iris.

Small, circular ear opening. Long, slender hindlimbs. Keeled, circular, large dorsal scales, with six rows of larger, keeled scales in the middorsal line; keeled, circular ventral scales, the same size as dorsals. Enlarged postcloacal scales in males.

BIBLIOGRAPHIC SOURCES

Garrido (1975*c*): original description; habitat data; behavior; drawing of the mental scales (fig. 1A); photograph of the gular scales (fig. 3, left). Estrada and Silva Rodríguez (1984): morphometry; general index of locomotion.

MORPHOLOGICAL VARIATION

Without metachromatism. Coloration is slightly variable.

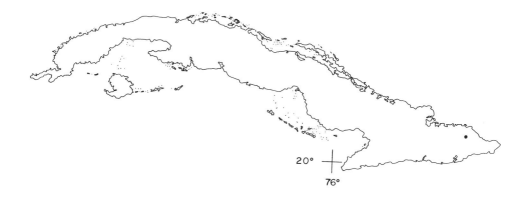

20°

76°

0   200 Km.

Map 48. Geographic distribution of *Anolis fugitivus*.

Geographic Variation

Slightly variable throughout its limited geographic range.

Sexual Variation

Both sexes have similar coloration patterns and do not differ in size (table 7.39). Males are recognized by their dewlaps and their enlarged postcloacal scales.

Age Variation

Juveniles are not known.

NATURAL HISTORY

The species lives in grasslands along river and stream sides. Individuals perch on stems and leaves of tall grasses, in shaded places near the ground. Seemingly, they prefer the *"cortadera"* grasses of the genus *Arthrostylidium*. They remain immobile on leaves of grasses; when escaping, they quickly jump and hide under leaves. According to the calculated general index of locomotion (0.67), this is a jumping species, which is consistent with its free-life behavior.

It is a diurnal, apparently nonheliothermic species based on the shaded sites of the grasslands where it is usually observed. The majority of individuals are found during the first hours of the day and in the afternoon, avoiding direct sunlight. During rainy and cool days, they can be seen all the time.

Apparently the reproductive cycle is short, since females collected in January, February, and June were not gravid.

Table 7.39. Comparison of snout-vent length *(SVL)*, head length *(HL)*, and tail length *(TL)* between male and female adults of *Anolis fugitivus* (symbols as in table 7.1).

Sex	N	X	CV	m	M	t	p
				SVL			
Males	4	33.5	6.84	30.6	36.2	1.69	n.s.
Females	5	31.3	4.95	29.7	33.3		
				HL			
Males	4	10.5	6.22	9.6	11.1	2.13	<0.10
Females	5	9.7	5.60	8.9	10.4		
				TL			
Males	2	80.8	4.55	78.2	83.4	0.31	n.s.
Females	4	79.3	7.43	74.1	87.2		

Anolis juangundlachi (plate 43)

Anolis cyanopleurus: Gundlach, 1880 *(part.), Contr. Erp. Cubana:*48. *Anolis juangundlachi* Garrido, 1975, *Poeyana* 143:34; holotype IZ 3755. Type locality: Finca Ceres (Los Montes), 4 km north of Carlos Rojas, Matanzas province, Cuba.

Geographic Range

Isla de Cuba (known only from the type locality).

Localities (map 49)

Matanzas: vicinity of Carlos Rojas (Gundlach 1880).

Content

Monotypic endemic species.

Description

Very small size. Mean snout-vent length of 33.2 mm in males and of 29.6 mm in females.

Pale brown with brown middorsal longitudinal line that has isolated black dots. A dark brown dorsolateral longitudinal line on both sides of the middorsal line, from behind the eye to the sacrum. Yellow labial stripe extending to the groin. Two black dots behind the ear opening. A brown lateral longitudinal line. Brown limbs; hindlimbs with orange on thigh and isolated black dots. Pale brown ventral region with a dark brown midven-

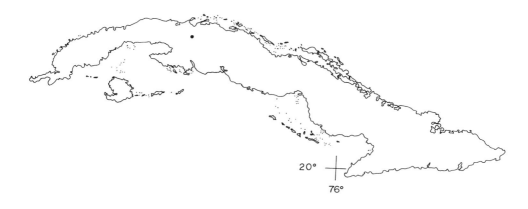

Map 49. Geographic distribution of *Anolis juangundlachi*.

tral longitudinal line, from the gular region to the vent. Pale brown tail, with its central part dark gray. Dewlap in males, pale yellow. Blue iris.

Vertically oval, small ear opening. Long, slender hindlimbs. Keeled, circular, small dorsal scales, with eight rows of larger keeled scales in the middorsal line; keeled, lanceolate, imbricate ventral scales, larger than dorsals. Enlarged postcloacal scales in males.

BIBLIOGRAPHIC SOURCES

Gundlach (1880): coloration. Garrido (1975c): original description; drawing of the mental scales (fig. 1B); photographs of the gular scales (fig. 2, right down), of the middorsal zone (fig. 3, right); habitat data; behavior. Garrido (1980a): reproductive behavior in captivity.

MORPHOLOGICAL VARIATION

Metachromatism from pale brown to dark brown. Coloration is slightly variable.

Geographic Variation

Little variable throughout its limited geographic range.

Sexual Variation

Females are yellowish brown with a pale brown middorsal longitudinal line bordered by yellow, very narrow lines, and these by dark brown lines, on both sides. Differences in size are shown in table 7.40. Males are recognized by their dewlaps and their enlarged postcloacal scales.

Table 7.40. Comparison of snout-vent length (SVL), head length (HL), and tail length (TL) between male and female adults of Anolis juangundlachi (symbols as in table 7.1).

Sex	N	X	CV	m	M	t	p
				SVL			
Males	12	33.2	4.54	30.0	35.8	3.41	<0.01
Females	6	29.6	10.26	25.2	34.3		
				HL			
Males	12	10.9	8.24	9.6	12.5	2.58	<0.05
Females	6	9.8	7.21	8.7	10.8		
				TL			
Males	5	90.3	12.07	78.9	105.3	1.01	n.s.
Females	3	83.4	6.18	78.4	88.7		

Age Variation

Juvenile coloration is similar to that of adult females, but with a white labial stripe, a larger number of longitudinal lines, and two black sacral dots. Males have enlarged postcloacal scales.

Natural History

This species lives in grasslands in the gallery forest of the stream that crosses Los Montes, Carlos Rojas. Individuals perch on stems and leaves of grasses, in shaded places near the ground. They remain immobile on leaves and stems; when escaping, they quickly jump and hide under leaves. The species is diurnal, apparently nonheliothermic based on the shaded sites where it is found. The majority of active individuals are observed at noontime.

The precopulatory phase lasted 202 sec. in captivity; the mean duration of copulation was 16.7 min., and the longest copulation lasted 22 min. When copulation ends, both members of the pair hastily separate.

The only known population of this species was found to be abundant during the census taken in 1982; nowadays it is very scarce because of extensive modification of its habitat due to Hurricane Lili in 1996.

Anolis spectrum (plate 44)

Anolis spectrum W. Peters, 1863, *Monastb. Akad-wiss. Berlin:* 136; syntypes ZMB 421 a and b. Type locality: Cuba, restricted to the vicinity of Matanzas and Cárdenas by Gundlach 1875, *An. Soc. Española Hist. Nat.* 4:358; further restricted to the "mogotes" at San Miguel de los Baños, 500 m from the swimming pool at San Miguel, before arriving in Río Paredones, Matanzas province, Cuba, by Garrido and Schwartz 1972, *Proc. Biol. Soc. Washington* 85(45):512. *Anolis spectrum sumiderensis* G. Peters, 1970, *Mitt. Zool. Mus. Berlin* 46(1):226; holotype ZMB 41783. Type locality: Valle de Pica Pica near Sumidero, Pinar del Río province, Cuba.

Geographic Range

Isla de Cuba (isolated populations in the western and central parts of the country).

Localities (map 50)

Pinar del Río: Pica Pica (Buide 1967). *Ciudad de La Habana:* Fontanar (Silva Rodríguez, Estrada, and Garrido 1982). *Matanzas:* Matanzas; Cárdenas

(Gundlach 1867); San Miguel de los Baños (Garrido and Schwartz 1972); Carlos Rojas (Garrido 1975c). *Cienfuegos:* Sierra de Trinidad (CZACC). *Sancti Spíritus:* Jobo Rosado, Yaguajay (Garrido and Jaume 1984).

CONTENT

Monotypic endemic species.

DESCRIPTION

Very small size. Mean snout-vent length of 37.7 mm in males and of 35.6 mm in females.

Very pale brown with about seven black dots on the middorsal longitudinal line. A lateral, longitudinal, very narrow brown line extending from behind the eye to the groin. Another lateral, longitudinal, very narrow brown line from behind the ear opening to the groin. Two brown angular spots on head, with the vortex backward. Brown radial lines on the upper eyelid. A black spot on shoulders. A dark brown transverse bar over another, larger, orange bar on sacrum. Brown limbs with an orange spot on elbow and dark brown transverse bands on hindlimbs. Pale brown tail with black central part and small black transverse bands to the end. Dark grayish brown gular and ventral regions. Dewlap in males, orange brown with pale brown inner scales. Pale brown iris.

Vertically oval, small ear opening. Long, slender hindlimbs. Keeled, circular, small dorsal scales, with ten rows of keeled, larger scales in the mid-

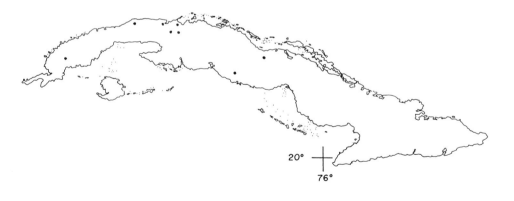

20°

76°

0   200 Km.

Map 50. Geographic distribution of *Anolis spectrum.*

dorsal longitudinal line; keeled, circular ventral scales, larger than dorsals. Enlarged postcloacal scales in males.

BIBLIOGRAPHIC SOURCES

W. Peters (1863): original description. Bocourt (1874): drawing of the dorsal scales of the head (pl. 16, fig. 24). Gundlach (1880): coloration in life. Barbour and Ramsden (1919): external morphology; measurements; drawing of the dorsal scales of the head (pl. 7, fig. 5). Barbour (1930a): abundance. Alayo Dalmau (1955): drawing of the dorsal scales of the head (pl. 3, fig. 1, no. 5). Williams (1961): external morphology; scalation; habitat data. Buide (1967): habitat data; abundance. G. Peters (1970): description; photographs of the ventral scalation (fig. 7a and b), of a specimen in dorsal and ventral views (fig. 6 a–d). Schoener (1970): interspecific relations in size. Garrido and Schwartz (1972): external morphology. Garrido (1980a): reproductive data in captivity.

MORPHOLOGICAL VARIATION

Metachromatism from very pale brown to brown. Coloration varies a little individually and geographically.

Geographic Variation

Slightly variable throughout its geographic range.

Sexual Variation

Females have four or five yellow transverse bands on the ventral surface of the first portion of the tail and two dots on vent, ochraceous orange. They do not differ in size from males (table 7.41), which are recognized by their dewlaps and their postcloacal scales.

Age Variation

Coloration of juveniles is similar to that of adults. Males have enlarged postcloacal scales.

NATURAL HISTORY

This species lives in semideciduous and secondary forests, on small bushes in the herbaceous stratum. Individuals perch on leaves of bushes and grasses, no higher than 1 m above ground; some individuals are also observed in leaf litter. They remain immobile on twigs and leaves, where they are almost imperceptible. When escaping, they quickly jump and hide un-

Table 7.41. Comparison of snout-vent length *(SVL)*, head length *(HL)*, and tail length *(TL)* between male and female adults of *Anolis spectrum* (symbols as in table 7.1).

Sex	N	X	CV	m	M	t	p
			SVL				
Males	12	37.7	7.52	32.1	42.1	1.52	n.s.
Females	6	35.6	6.35	32.5	39.3		
			HL				
Males	12	12.0	8.60	9.7	13.6	1.84	n.s.
Females	6	11.1	5.70	10.3	11.7		
			TL				
Males	7	98.2	8.51	44.1	108.6	1.05	n.s.
Females	6	84.3	6.52	76.2	91.2		

der leaves, although some individuals escape by running. According to the calculated general index of locomotion (0.64), this is a jumping species, which is consistent with its free-life behavior.

The species is diurnal, seemingly nonheliothermic based on the shaded sites where it is found.

About reproduction little is known. The precopulatory phase in captive adults lasted 30 sec.; the mean duration of copulation was 11.7 min., and the longest copulation lasted 19 min. When copulation ends, both members of the pair hastily separate. One female laid a white egg measuring 6.5 x 4.1 mm.

Apparently populations are not numerous; however, it is possible that the species seems more scarce than it actually is because of its cryptic coloration.

Anolis vanidicus (plate 45)

Anolis spectrum: Schwartz and Ogren, 1956 *(part.),* Herpetologica 12(2):98. *Anolis vanidicus* Garrido and Schwartz, 1972, *Proc. Biol. Soc. Washington* 85(45):515; holotype AMNH 78400. Type locality: 4 km *W*, 12 km *N* of Trinidad, Las Villas province, Cuba.

Geographic Range

Isla de Cuba (south of Cienfuegos and Sancti Spíritus provinces; Santiago de Cuba).

Localities (map 51)

Cienfuegos: Sierra de Trinidad (Dunn 1926); Soledad; Buenos Aires (Garrido and Schwartz 1972). *Sancti Spíritus:* Topes de Collantes (Garrido and Schwartz 1972); Río Caburní; Pico de Potrerillo (Rodríguez Schettino *et al.* 1979). *Santiago de Cuba:* Santiago de Cuba (Schwartz and Ogren 1956).

Content

Polytypic endemic species; two subspecies are recognized:

Anolis vanidicus vanidicus

Anolis vanidicus vanidicus Garrido and Schwartz, 1972, *Proc. Biol. Soc. Washington* 85(45):515. Type locality: 4 km *W*, 12 km *N* of Trinidad, Las Villas province, Cuba. Geographic range: Sierra de Trinidad.

Anolis vanidicus rejectus

Anolis vanidicus rejectus Garrido and Schwartz, 1972, *Proc. Biol. Soc. Washington* 85(45):517; holotype ChM 55.1–63. Type locality: 2 mi. (3.2 km) *N* of Santiago de Cuba, Oriente province, Cuba. Geographic range: known only from the type locality.

Description

Very small size. Mean snout vent length of 34.3 mm in males and of 33.4 mm in females.

Olivaceous brown with a pale brown, ill-defined middorsal longitudinal line that has some black spots on its borders. Pale brown or white labial stripe extending to the anterior edge of the ear opening. A black angular spot on the dorsal region of the head. Yellow ventral region. Pale brown tail

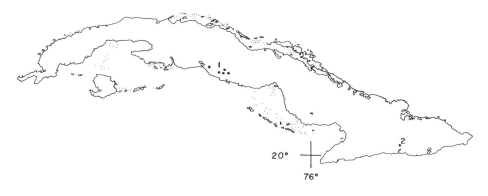

Map 51. Geographic distribution of *Anolis vanidicus. 1, A. v. vanidicus; 2, A. v. rejectus.*

with black central region. Dewlap in males, greenish ocher, with yellowish white inner scales. Green iris.

Small, circular ear opening. Long, slender hindlimbs. Keeled, circular, large dorsal scales, with seven or nine rows of keeled, larger scales in the middorsal longitudinal line; keeled, circular ventral scales, larger than dorsals. Enlarged postcloacal scales in males.

Bibliographic Sources

Schwartz and Ogren (1956) *Anolis spectrum* (*sic.*): coloration; measurements; habitat data. Ruibal (1964) *Anolis spectrum* (*sic.*): description; measurements; drawings of the dorsal scales of the head (fig. 14), of the dorsolateral scales (fig. 15); abundance. Garrido and Schwartz (1972): original description; habitat data. Estrada and Silva Rodríguez (1984): morphometry; general index of locomotion.

Morphological Variation

Without metachromatism. Coloration varies a little individually.

Geographic Variation

The two subspecies are described as follows:

Anolis vanidicus vanidicus (Sierra de Trinidad): the same as the species. The scales of the middorsal line are larger than those of the other subspecies. Mean snout-vent length of 34.3 mm in males and of 33.4 mm in females.

A. v. rejectus (Santiago de Cuba): the same as the species, but with smaller scales of the middorsal line. Snout-vent length of one female was 38.0 mm.

Sexual Variation

The pale brown middorsal line is more defined in females than in males; males and females do not differ in size (table 7.42). Males are recognized by their dewlaps and their enlarged postcloacal scales.

Age Variation

Juvenile coloration is similar to that of adults. Males have enlarged postcloacal scales.

Natural History

This species lives in evergreen forests and rain forests, as well as in pine, banana, and coffee plantations and in coniferous and gallery forests. It is found on grasses or on the ground. Individuals remain immobile on leaves;

Table 7.42. Comparison of snout-vent length *(SVL)*, head length *(HL)*, and tail length *(TL)* between male and female adults of *Anolis vanidicus* (symbols as in table 7.1).

Sex	N	X	CV	m	M	t	p
				SVL			
Males	10	34.3	6.73	31.2	37.7	1.13	n.s.
Females	9	33.4	3.37	31.8	35.4		
				HL			
Males	10	11.1	7.56	10.0	11.9	2.60	<0.05
Females	9	10.4	11.64	54.8	11.1		
				TL			
Males	6	85.3	10.35	40.3	97.8	1.40	n.s.
Females	6	77.9			88.2		

when escaping, they quickly jump to distance themselves, then jump again or hide among the vegetation. According to the calculated general index of locomotion (0.63), this is a jumping species, which is consistent with its free-life behavior.

The species is diurnal, apparently nonheliothermic based on the shaded sites where it is found.

Feeding in free life is unknown; captive individuals have accepted small insects.

The precopulatory phase in captive adults lasted 82 sec. in captivity; the mean duration of copulation was 14.3 min., and the longest copulation lasted 35 min. When copulation ends, both members of the pair quickly separate. Two eggs measured 8.2 x 5.0 and 8.2 x 4.3 mm.

It is found in isolated, numerous colonies at Sierra de Trinidad, which stay in the same places all the time.

Section beta (caudal autotomic vertebrae with transverse processes).

Series *sagrei* (splenial absent; two inscriptional ribs attached to the corresponding dorsal ribs and two floating free in the myocommata; a Y-shaped parietal crest; three or four lumbar vertebrae; five anterior aseptate caudal vertebrae).

Species group *sagrei* (short head; stocky body; long tail).

Anolis sagrei (chino, torito; Cuban brown lizard; plate 46)

Anolis sagrei Duméril and Bibron, 1837, *Erp. Gén.* 4:149; syntypes MNHN 24230, MNHN 6797, MCZ 2171. Type locality: Cuba, restricted to La Habana, La Habana province, Cuba, by Ruibal 1964, *Bull. Mus. Comp. Zool.* 130(8):490. *Anolis de la sagra:* Cocteau and Bibron, 1838, *Hist. Fís. Pol. Nat. Isla de Cuba* 4:82. *Dactyloa sagrei* Fitzinger, 1843, *Syst. Rept.*:67. *Dracontura catenata* Gosse, 1850, *Ann. Mag. Nat. Hist.* 2(6):346; holotype BMNH 1946.8.29.21. Type locality: Bluefields, Westmoreland Parish, Jamaica. *Anolis sagrae:* Gundlach, 1875, *An. Soc. Española Hist. Nat.* 4:356. *Anolis stejnegeri* Barbour, 1931, *Copeia* 3:88; holotype MCZ 29907. Type locality: Key West, Monroe County, Florida. *Anolis sagrei mayensis* Smith and Burger, 1949, *Anal. Inst. Biol.* 20:407; holotype UIMNH 4170. Type locality: Panlao, Campeche, México.

GEOGRAPHIC RANGE

North America (Florida and Florida Keys); Central America (Atlantic coast of México; Isla de Cozumel; Belice; Bay Islands); Isla de Cuba (islandwide); Isla de la Juventud; Archipiélago de los Colorados; Archipiélago de los Canarreos; Archipiélago de Sabana-Camagüey; Archipiélago de los Jardines de la Reina; Jamaica; Cayman Islands; Swan Island; Bahama Islands.

LOCALITIES (MAP 52)

Pinar del Río: Bahía Honda (Garman 1887); Pinar del Río (Barbour 1914); San Diego de los Baños (Stejneger 1917); Península de Guanahacabibes; Isabel Rubio; San Juan y Martínez (Garrido and Schwartz 1968); San Vicente (Zajicek and Mauri Méndez 1969); Cayo Juan García (Varona and Garrido 1970); La Coloma; Consolación del Sur; Taco Taco; Nortey (Garrido 1972); El Salón, Sierra del Rosario (Silva Rodríguez and Estrada 1982); Cayo Inés de Soto, Archipiélago de los Colorados (Estrada and Novo Rodríguez 1984*a*); Las Peladas; Las Terrazas; Soroa (Martínez Reyes 1995). *La Habana and Ciudad de La Habana:* Habana (Garman 1887); Guanajay; Mariel (Stejneger 1917); Bosque de La Habana (Collette 1961); Guanímar; Nueva Paz; Sierra de Camarones (Garrido 1972); Jaruco; Santa Cruz del Norte; Madruga (Garrido 1973*f*); Santiago de las Vegas (Llanes Echevarría 1978); San Antonio de los Baños (Armas and Alayón García 1987); Cueva de Sandoval, Vereda Nueva (Silva Taboada 1988); Jardín Botánico Nacional (García Rodríguez 1989); Escaleras de Jaruco (González González 1989); Atabey, C. de La Habana (Martínez Reyes 1989); El Narigón, Puerto Escondido (Hechevarría *et al.* 1990); Playa Caimito; Quivicán (L.R.S.). *Matanzas:* Matanzas (Garman 1887); Quemados (Stejneger 1917); Península de Hicacos (Buide 1966); Península de Zapata (Zajicek and Mauri Méndez 1969); Playa Larga; Playa Girón; Caleta Buena (Martínez Reyes 1994*c*); Bacunayagua (Soto Ramírez 1994); San Miguel de los Baños (CZACC). *Villa Clara:* Caibarién (Garman 1887); Manicaragua (Schwartz and Ogren 1956);

Cayo Santa María (Garrido 1972); Cayo Francés; Cayo Tío Pepe; Cayo Monos de Jutía; Cayo Lanzanillo (Garrido 1973*d*); Cayo Caimán del Faro; Cayo Las Brujas; Cayo Fragoso; Cayo Conuco (Martínez Reyes 1998). *Cienfuegos:* Soledad (Dunn 1926); Cayo Macho de Tierra (Cooper 1958); Rancho Luna (L.R.S.). *Sancti Spíritus:* Caguanes; Mayajigua (Garrido 1972); Topes de Collantes; Salto del Caburní; Finca Cudina (Rodríguez Schettino *et al.* 1979); San Felipe, Jatibonico (CZACC). *Ciego de Ávila:* Cayo Coco (Garrido 1976*a*); Cayo Guillermo (Martínez Reyes 1998). *Camagüey:* Camagüey (Barbour 1914); Sierra de Cubitas (Barbour and Ramsden 1919); Cayos de las Doce Leguas (Cochran 1934); 7.5 mi. *SW* of Banao (Schwartz and Henderson 1985); Cayo Guajaba (Garrido *et al.* 1986); Cayo Sabinal; Cayo Paredón Grande (Martínez Reyes 1998); Santa Cruz del Sur (L.R.S.). *Las Tunas:* Las Tunas (Díaz Castillo *et al.* 1991). *Holguín:* Cananova; Sagua de Tánamo; Gibara; Banes; Holguín; Potosí (Garrido 1972); Mayarí; Nicaro (CZACC). *Granma:* Bueycito (Schwartz and Ogren 1956); Embalse Leonero (Montañez Huguez *et al.* 1985); Birama; keys at Golfo de Guacanayabo (Garrido 1990); Cabo Cruz; Pilón (CZACC). *Santiago de Cuba:* Santiago de Cuba (Barbour 1914); El Cobre; San Luis (Stejneger 1917); Ocujal; La Gran Piedra (Garrido 1972); Las Cuevas; La Mula (Rodríguez Schettino 1985); Siboney (Fong and del Castillo 1997). *Guantánamo:* Guantánamo (Barbour 1914); Maisí (Garrido 1967); Cupeyal (Peters 1970); Baracoa (CZACC); Yumurí del Sur (L.R.S.). *Isla de la Juventud:* Nueva Gerona (Stejneger 1917); Cayo Cantiles (Garrido and Schwartz 1969); Cayo Campo (Estrada and Rodríguez 1985); Cayo Matías (Acosta Cruz *et al.* 1985); Siguanea; Cayo Piedra; Punta del Este; Cocodrilo (CZACC).

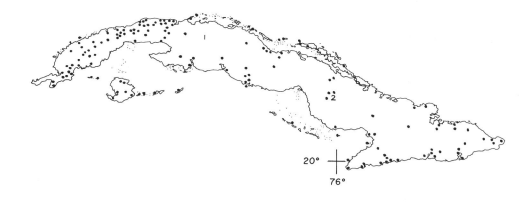

Map 52. Geographic distribution of *Anolis sagrei* in Cuba. *1*, *A. s. sagrei*; *2*, *A. s. greyi*.

CONTENT

Polytypic endemic species; five subspecies are recognized, two of them living in Cuba:

Anolis sagrei sagrei

Anolis sagrei Duméril and Bibron, 1837, *Erp. Gén.* 4:149. Type locality: Cuba, restricted to La Habana, La Habana province, Cuba (Ruibal 1964). *Anolis sagrei sagrei*: Barbour, 1937, *Bull. Mus. Comp. Zool.* 82(2):126. Geographic range: North America (Florida and Florida Keys); Central America (Atlantic coast of México; Isla de Cozumel; Belice; Bay Islands); Isla de Cuba (islandwide, except in the vicinity of Camagüey city); Isla de la Juventud; Archipiélago de los Colorados; Archipiélago de Sabana-Camagüey; Archipiélago de los Canarreos; Archipiélago de los Jardines de la Reina; Jamaica; Cayman Islands; Bahama Islands.

Anolis sagrei greyi

Anolis greyi Barbour, 1914, *Mem. Mus. Comp. Zool.* 44(2):287; holotype MCZ 7890. Type locality: Puerto Príncipe (Camagüey), Camagüey province, Cuba. *Anolis sagrei greyi*: Buide, 1967, *Torreia* 1:32. Geographic range: vicinity of the type locality.

DESCRIPTION

Very small or small size. Mean snout-vent length between 48.2 and 50.4 mm in males and between 35.3 and 39.1 mm in females.

Pale brown or brown with yellow, longitudinally arranged spots, dots, and lines. Pale and dark brown transverse bands on tail in some individuals. Pale and dark brown, ill-defined crossbands on limbs. Terra-cotta head; dark brown radial lines centered in eye. Yellow eyelids and labials. Very pale brown ventral region, with brown dots. Dewlap in males, red, ochraceous, or dark orange, yellow edged and with yellow or brown inner scales. Green or brown iris.

Small, circular ear opening. Nuchal, dorsal, and caudal crests in males. Vertically oval section of the tail. Keeled, circular, large dorsal scales, with a middorsal longitudinal zone of keeled, imbricate scales larger than dorsals; keeled, circular, imbricate ventral scales, larger than dorsals. Keeled head scales. Keeled or multicarinate supracarpal, supratarsal, and supradigital scales. Enlarged postcloacal scales in males.

Seven pairs of macrochromosomes and seven pairs of microchromosomes (2n = 28).

BIBLIOGRAPHIC SOURCES

Duméril and Bibron (1837): original description. Cocteau and Bibron (1838): coloration; scalation; description of the juvenile; drawings of a male in lateral view (pl. 13, fig. 1), of a female in dorsal view (pl. 13, fig. 2), of the dorsal scalation of the head (pl. 13, fig. a), of the head in lateral view (pl. 13, fig. b), of the gular scales (pl. 13, fig. c), of the femoral and cloacal regions (pl. 13, fig. d), of the body scales (pl. 13, fig. e), of the dorsal surface of the left hindlimb (pl. 13, fig. f), of the ventral surface of the right hindlimb (pl. 13, fig. g). Gundlach (1867, 1880): coloration; habitat data; abundance; metachromatism. Barbour (1914): coloration; scalation. Barbour and Ramsden (1916b): habitat data; abundance. Stejneger (1917): coloration; drawings of the dorsal scales of the head (fig. 52), of the head in lateral view (fig. 53), of the tail in lateral view in the fifth verticil (fig. 54); habitat data; behavior. Barbour and Ramsden (1919): coloration; scalation; drawings of the dorsal scales of the head (pl. 7, fig. 3; pl. 14, fig. 7); habitat data; abundance. Dunn (1926): habitat data; abundance. Barbour (1930a, b): abundance. Cochran (1934): coloration. Evans (1938): behavior; photographs of the aggressive display between two males (figs. 3–8), between two females (figs. 9 and 10). Pérez Vigueras (1940): trematode. Alayo Dalmau(1955): coloration; scalation; drawings of the dorsal scales of the head (pl. 3, fig. 1, no. 3; pl. 4, fig. 1, no. 52), of the head in lateral view (pl. 4, fig. 1, no. 53), of the tail in lateral view in the fifth verticil (pl. 4, fig. 1, no. 54); habitat data; behavior; abundance. Schwartz and Ogren (1956): measurements; habitat data; abundance. Allen and Neill (1957): morphology of the tail. Underwood and Williams (1959): coloration. Collette (1961): coloration; morphometry; drawing of the subdigital lamellae of the third toe (fig. 2e); habitat data; behavior; abundance. Ruibal (1961): habitat data; behavior; thermic relations; daily activity; egg-laying sites. Ruibal and Williams (1961b): habitat data. Ruibal (1964): coloration; scalation; habitat data; abundance. Ruibal and Ernst (1965): structure of the subdigital lamellae; electronic microphotographs of the digital setae (pl. 1, figs. 5 and 7; pl. 5, fig. 14); habitat data; abundance. Garrido (1967): habitat data. Ruibal (1967): coloration; territorial behavior in free life and captivity. Garrido and Schwartz (1968): coloration. Gorman and Atkins (1968): karyotype. Garrido and Schwartz (1969): habitat data; abundance. Williams (1969): habitat data; abundance. Zajicek and Mauri Méndez (1969): hemoparasites. Coy Otero (1970): helminths. Schoener (1970a, b): interspecific relationships in size. Varona and Garrido (1970): coloration; habitat data; abundance. Garrido (1972): coloration; scalation; photograph of the head in lateral view (fig. 2); measurements; abundance; dispersion. Garrido (1973b, d 1975c): habitat data; abundance; coloration. Ruiz (1977): behavior in captivity. Valderrama Puente (1977): reproductive data. Williams and Rand (1977): coloration; ecomorphology. Llanes Echevarría (1978): coloration; habitat data; thermic relations. Ortiz Díaz (1978): coloration; feeding. Coy Otero and Baruš (1979a, b): nematodes. Espinosa López et al. (1979):

electrophoretic pattern of general proteins (fig. 1), of esterases (fig. 2). Berovides Álvarez and Sampedro Marín (1980): habitat data; daily activity; feeding. Garrido (1980*a*): reproductive data in captivity. González Suárez (1980): electrophoretic pattern of hemoglobin. Silva Rodríguez (1980): electrophoretic pattern of general proteins. De Smet (1981): karyotype; photographs of the metaphase (fig. 1a), of meiotic cells (fig. 2a), of the metaphase II (fig. 2b), of the idiogram (fig. 6). Mugica Valdés (1981): electrophoretic patterns of esterases; photograph of the pattern of esterases (fig. 6, right). González Martínez (1982): histostructure of the liver. Mugica Valdés *et al.* (1982): electrophoretic pattern of esterases; drawing of the pattern of esterases (fig. 1c). Sampedro Marín *et al.* (1982): coloration; habitat data; feeding. Lieb *et al.* (1983): biochemical polymorphism. Estrada and Silva Rodríguez (1984): morphometry; general index of locomotion. García de Francisco (1984*a*): liver histology. García de Francisco (1984*b*): gastric mucose histology. Buide (1985): habitat data; two photographs of two specimens. Espinosa López, Benítez *et al.* (1985): electrophoretic patterns of esterases of muscle. Regalado (1985): tonic immobility. Rodríguez Schettino (1985): habitat data. Schwartz and Henderson (1985): drawings of the lateral region of the head and the forelimb (pl. 2, nos. 5 and 6). Silva Lee (1985): aggressive display; cannibalism; two photographs of two specimens. Tokarz (1985): behavior in captivity. Armas and Alayón García (1986): feeding. Estrada and Novo Rodríguez (1986*a*): habitat data. Armas and Alayón García (1987): predator. Espinosa López *et al.* (1987): electrophoretic pattern of general proteins of plasma. Quintana Ferrer (1987): morphometry. García Rodríguez (1989): habitat data; feeding; reproduction; abundance; photograph of a specimen. Manójina *et al.* (1989): predator. Martínez Reyes (1989): coloration patterns. Alarcón Chávez *et al.* (1990): abundance. Burnell and Hedges (1990): phylogenetic relationships. Espinosa López, Posada García *et al.* (1990): electrophoretic patterns of general proteins of muscle. Espinosa López, Sosa Espinosa, and Berovides Álvarez (1990): electrophoretic patterns of eight proteic systems. Quesada Jacob *et al.* (1991): habitat data; abundance. Regalado and Garrido (1993): behavior. Martínez Reyes (1994*c*): abundance. Rodríguez Schettino (1994): description of the nemasperms. Rodríguez Schettino (1996): habitat data. Fong and del Castillo (1997): habitat data. Sanz *et al.* (1997): histology.

MORPHOLOGICAL VARIATION

Metachromatism from pale brown to dark brown with yellow dots and lines; sometimes to very dark brown. The inner scales of the dewlap vary from pale yellow to dark brown, independently of changes in body coloration. Coloration is quite variable individually and geographically.

Some individuals in the same population have red dewlaps, others ocher, and others dark orange. In addition, it is possible to distinguish at least five dorsal color patterns: brown with yellow reticulations; brown with a middorsal yellow line; reddish brown; reddish brown with yellow reticulations; and reddish brown with a yellow middorsal line.

In some specimens, seven pairs of macro- and seven pairs of microchromosomes have been found, while others have had twenty macro- and nine microchromosomes (2n = 28 + 1).

Geographic Variation

The two Cuban subspecies are described as follows:

Anolis sagrei sagrei (Cuban territory, except Camagüey City): the same as the species, with green iris and canthal scales separated by two or three rows of scales. Mean snout-vent length of 50.4 mm in males and of 39.1 mm in females.

A. s. greyi (Camagüey City): the same as the species but with ochraceous dewlap, brown iris and frontal ridges separated by one row of scales or by two small scales. Mean snout-vent length of 48.2 mm in males and 35.3 mm in one female.

Individuals from the archipelagoes have orange dewlaps. Mean snout-vent length of males is 50.6 mm and that of females 39.5 mm.

Intermediate features between *A. sagrei* and *A. bremeri*, such as the shape of the rostral scale in females and size, have been observed in several individuals from localities south of Pinar del Río province.

Sexual Variation

Females have pale brown middorsal rhombuses, bordered with dark brown, and a middorsal longitudinal line, pale brown or yellow. Females are smaller than males (table 7.43), which are recognized by their dewlaps and their enlarged postcloacal scales.

Age Variation

Juvenile coloration is similar to that of adult females. Males have enlarged postcloacal scales. Juveniles also differ in gastric features: mucose gastric cells have larger nuclei, the gastric folds are taller, and there is a higher level of gastric activity than in adults.

NATURAL HISTORY

This species lives in almost all vegetation zones of Cuba, in plantations and urban areas, although it is more frequent in places modified by humans. It is found at forest edges or within vegetation in open areas. Individuals

Table 7.43. Comparison of snout-vent length *(SVL)*, head length *(HL)*, and tail length *(TL)* between male and female adults of *Anolis sagrei* (symbols as in table 7.1).

Sex	N	X	CV	m	M	t	p
				SVL			
Males	106	50.2	8.26	35.7	58.1	15.94	<0.01
Females	54	38.6	12.11	35.3	47.8		
				HL			
Males	25	13.8	7.02	12.1	15.8	10.03	<0.01
Females	21	11.4	5.10	10.3	12.5		
				TL			
Males	16	85.3	10.27	58.8	107.3	11.47	<0.01
Females	20	54.3	13.71	46.3	70.3		

perch on trunks of trees and bushes and on fences, not only wooden ones but also those made of concrete or wire, up to 1.5 m above ground. Usually, females and juveniles are found at lesser heights than males; they also use the herbaceous stratum, dead leaves, and stones.

When escaping, individuals of this species run toward the herbaceous stratum, hiding among leaves, stones, and grasses. When on trunks, they perch with their heads downward; after climbing to a certain height they turn around, with the head again in a downward position. According to the calculated general index of locomotion (0.82), this is a runner species, which is consistent with its free-life behavior.

This species exhibits tonic immobility in free life as well as experimentally, seemingly as a response to threats by possible predators. Usually, during the immobility phase, the tail is slowly moved in a zigzag fashion; this may be a behavioral pattern that attracts predators toward the least vulnerable part of the body.

It is a diurnal heliothermic species that frequently basks in open places, raising its body temperature by means of direct sunlight and through heat conduction from the substrate. In July and August the majority of individuals from Sierra de Cubitas maintained a body temperature of 34.0°C (mean 33.1°C), while the mean air temperature was 31.9°C. A significant positive correlation was found between both temperatures. Mean air temperature at several localities of La Habana province was 30.5°C and mean relative humidity 67.2%. In general, air temperature is high in places where this species is found, even in the coldest months of the year, since it uses basking in sunny sites and retreating to shade as an efficient mechanism of thermoregulation.

Individuals are active from the first hours of the day until the afternoon.

Partially active individuals have been observed under stones and other shelters between 8:30 A.M. and 10:00 A.M. On some occasions the number of active lizards increases, reaching a maximum level between 11:30 A.M. and 3:30 P.M., although sometimes fluctuations in daily activity are not observed.

This is a highly territorial species. If an individual is on its perch and another individual of the same species and sex approaches, the resident reacts by extending the dewlap, compressing the body laterally, and attacking, forcing the intruder to draw away. The resident darkens to dark brown or black, while the intruder pales. If aggression is accomplished, both individuals bite each other with their jaws until one of them quickly escapes. Captive males defend their territories, although with movements that are not so stereotyped. Resident males, defending against other males of the same species, bob their heads, flex their four limbs, wag their tails, and extend their dewlaps. In front of *A. allisoni*, some of these lizards escape and others defend their territories. Two captive males of *A. sagrei* returned to their terraria after escaping.

In addition, it has been experimentally determined that size is important in establishing dominance: the largest males defend their perches most successfully; they perch at greater heights; usually, they are the first to invade others' territories; and they initiate more challenges, few of which end in aggression.

The species feeds mainly on insects, although spiders, isopods, and mollusks have also been found in digestive tracts. The most frequently eaten insect orders are Hymenoptera (mainly formicids), Coleoptera, Lepidoptera, Diptera, and Dyctioptera. Males eat larger prey than females, and stomach content mass is greater in the largest individuals. Diet components are consistent with the types of invertebrates most frequently found in anthropic places; thus this opportunistic feeder serves as a biological control agent for such human-associated species. Individuals of this species kept in captivity have been fed with insects of the orders Diptera, Homoptera, Lepidoptera, and Isoptera; they exhibited a preference for winged termites of the genera *Neotermes* and *Nasutitermes*. In addition, they ate the remains of their own moltings. Another preferred food was the larvae of *Tenebrio molitor*; occasionally, adult males capture immatures of this coleopteran as well.

Reproductive activity is cyclic, with a notable increase during the wet season. The percentage of adult males with testes greater than 3 mm in length diminishes beginning in September, but it is still higher than the percentage of males with testes of lesser lengths. However, females with oviductal eggs have been found only between March and October. The mean length of these oviductal eggs is 7.7 mm. Fat body mass is much less than 0.1 g between April and September, that is, during the egg production period. Eggs are individually deposited in humus or leaf litter or under stones, only one per female in each egg-laying site. According to some histological tests, males may have sperm as early as March.

In free life courtship begins with dewlap extension in males, followed by head bobbing in receptive females. Behavior in captivity is similar. However, *A. sagrei* males exhibit an aggressive response to *A. allisoni* females that includes head bobbing, flexion, gorge swelling, raising of the head, and dewlap extension. After courtship, the precopulatory phase lasts between 24 and 39 sec.; the mean duration of copulation is between 2.6 and 3.8 min., with the longest copulation lasting 12 min. In addition, copulation has been observed between *A. sagrei* and *A. b. bremeri,* with a mean duration of 3.5 min. and a maximum of 5.1 min., from which offspring were not obtained.

Populations are numerous but contain scattered individuals. One lizard of each sex occupies a territory, in which may be established a family composed of one male and several females, one of which dominates. Juveniles and hatchlings may stay within this territory for at least several weeks before scattering to look for new territories. A relative abundance of 3.8 individuals/hour has been reported for the coastal shrubwood at El Narigón, on the northeastern coast of La Habana province; at the Jardín Botánico Nacional, between 8.4 and 83.3 individuals/hour have been reported for this kind of ecosystem. In Península de Zapata, relative abundance varied between 13.0 and 20.0 individuals/hour, depending on the locality (Playa Larga, Playa Girón, or Caleta Buena). These differences indicate that the frequency of *A. sagrei* at distinct places and dates varies, possibly due to characteristics of the habitat, climatic conditions, or both.

To date, protozoa of the genus *Hemogregarina* are known in red cells of *A. sagrei,* as well as two species of parasitic trematodes: *Urotrema wardi* and *U. scabridum;* eleven species of parasitic nematodes: *Cyrtosomum scelopori, Skrjabinoptera (Didelphysoma) phrynosoma, Parapharyngodon cubensis, Ozolaimus monhystera, Atractis opeatura, Porrocaecum* sp. larvae, *Physaloptera squamatae, Abbreviata* sp. larvae, *Trichospirura teixeirai, Oswaldocruzia lenteixeirai,* and Physalopteridae gen. sp.; and a mite, *Amblyomma torrei.*

The following species may be considered predators of this lizard: the Zapata Wren (*Ferminia cerverai*), in whose stomach contents were found remains of *A. sagrei;* the Sparrow Hawk (*Falco sparverius*), which is fed in captivity with live lizards supplied for this purpose, or which captures lizards that occasionally approach the cages; and the Cuban Blackbird (*Dives atroviolaceus*) and the Greater Antillean Grackle (*Quiscalus niger*), which catch *A. sagrei* in city gardens and parks.

Some captive reptiles, such as *A. porcatus, A. luteogularis, Leiocephalus cubensis, Antillophis andreai, Alsophis cantherigerus,* among others, have eaten *A. sagrei;* thus they are potential predators in free life as well. In fact, *L. cubensis* has been observed at the cemetery of La Habana City swallowing *A. sagrei.* Futhermore, in the enclosure of the Instituto de Ecología y Sistemática, hutias (*Capromys pilorides*) accepted these lizards when they were offered as a feeding experiment; such behavior indicates that this mainly phytophagous mammal may be a predator of lizards in natural conditions, as is Almiquí (*Solenodon cubanus*), which eats *A. sagrei* in captivity. On the other hand, the spider *Argiope trifasciata* has been observed eating *A. sagrei* in the wild.

Anolis bremeri (plate 47)

Anolis bremeri Barbour, 1914, *Mem. Mus. Comp. Zool.* 44(2):288; holotype MCZ 7889. Type locality: Herradura, Cuba.

Geographic Range

Isla de Cuba (southern coastal zone of Pinar del Río province); Isla de la Juventud (north of Ciénaga de Lanier).

Localities (map 53)

Pinar del Río: Herradura (Barbour 1914); La Coloma; 9 km *SW* of Pinar del Río; San Waldo (Garrido and Schwartz 1968); Guane (Peters 1970); La Fe; Alonso de Rojas; Los Palacios (Garrido 1972). *Isla de la Juventud:* Sierra de Caballos; Playa Bibijagua; Sierra de Casas; Peladero (road, between Nueva Gerona and Santa Bárbara); Sierra de la Guanábana; La Fe; Siguanea; La Reforma (Garrido 1972); Cayo Piedra; Carapachipey (CZACC).

Content

Polytypic endemic species; two subspecies are recognized:
  Anolis bremeri bremeri
Anolis bremeri Barbour, 1914, *Mem. Mus. Comp. Zool.* 44(2):288. Type locality: Herradura, Pinar del Río province, Cuba. *Anolis bremeri bremeri:* Garrido, 1972, *Carib. J. Sci.* 12 (1–2):62. Geographic range: from the vicinity of Manuel Lazo and La Fe, to localities south of Taco Taco.
  Anolis bremeri insulaepinorum
Anolis bremeri insulaepinorum Garrido, 1972, *Carib. J. Sci.* 12(1–2):63; holo-

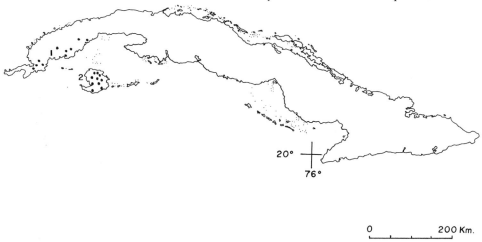

Map 53. Geographic distribution of *Anolis bremeri. 1, A. b. bremeri; 2, A. b. insulae-pinorum.*

type IZ 1626. Type locality: Hotel Colony, La Siguanea, Isla de Pinos. Geographic range: Isla de la Juventud, north of the Ciénaga de Lanier.

DESCRIPTION

Very small, small, or medium sized. Mean snout-vent length of 60.4 mm in males and between 44.4 and 45.2 mm in females.

Dark brown with yellow and black spots arranged in ill-defined longitudinal lines. Yellow eyelids. Brown labials with yellow spots. Dewlap in males, dark red or ocher, yellow edged and with black inner scales. Brown iris.

Vertically oval, large ear opening. Tip of the snout pointed upward in females. Small nuchal and dorsal crests in males, without caudal crest. Vertically oval section of the tail. Keeled, circular, small dorsal scales, with three or four rows of keeled, larger scales; keeled, circular ventral scales, larger than dorsals. Strongly keeled head scales. Keeled supracarpal, supratarsal, and supradigital scales. Enlarged postcloacal scales in males. Long, upward-directed rostral region in females.

BIBLIOGRAPHIC SOURCES

Barbour (1914): original description; habitat data. Barbour and Ramsden (1919): coloration; scalation; drawing of the dorsal scales of the head (pl. 6, fig. 5). Read and Amrein (1953): nematode. Alayo Dalmau (1955): coloration; drawing of the dorsal scales of the head (pl. 3, fig. 2). Ruibal (1964): coloration; measurements. Buide (1967): coloration. Garrido and Schwartz

(1968): coloration; chromatic variations. Coy Otero (1970): nematode. Garrido (1972): coloration; scalation; photograph of the head in lateral view (fig. 1). Williams and Rand (1977): coloration. Coy Otero and Baruš (1979a): nematodes. Garrido (1980a): reproductive data in captivity. González Suárez (1980): electrophoretic pattern of hemoglobin. Mugica Valdés (1981): electrophoretic pattern of esterases; photograph of the pattern (fig. 6, left). Rodríguez Schettino (1981b): features of the nemasperms. Mugica Valdés *et al.* (1982): electrophoretic pattern of esterasas; drawing of the pattern of esterases (fig. 1d). Estrada and Silva Rodríguez (1984): morphometry; general index of locomotion. Schwartz and Henderson (1985): drawing of the lateral region of the head and of the forelimb (pl. 2, no. 4). Silva Lee (1985): photographs of one male and one female; habitat data. Quintana Ferrer (1987): morphometry. Espinosa López, Posada García *et al.* (1990): electrophoretic pattern of general proteins of muscle. Espinosa López, Sosa Espinosa, and Berovides Álvarez (1990): electrophoretic patterns of eight proteic systems. Regalado and Garrido (1993): behavior. Rodríguez Schettino (1996): habitat data.

MORPHOLOGICAL VARIATION

Metachromatism from dark brown and yellow lines to very dark brown with pale brown lines. Coloration is quite variable individually and geographically.

Geographic Variation

The two subspecies are described as follows:

Anolis bremeri bremeri (Pinar del Río): brown with four yellow lateral longitudinal lines and spots. Ocher dewlap with yellow edge and black inner scales. Mean snout-vent length of 60.4 mm in males and of 44.4 mm in females.

A. b. insulaepinorum (Isla de la Juventud, northern region): brown with ivory-colored discontinuous longitudinal lines. Dark terra-cotta dewlap, with yellow edge and black inner scales. Mean snout-vent length of 60.4 mm in males and of 45.2 mm in females.

Intermediate features between *A. sagrei* and *A. bremeri*, such as the shape of the rostral scale in females and size, have been observed in several individuals from localities south of Pinar del Río province.

Sexual Variation

Females have pale brown middorsal rhombuses, bordered with dark brown, and a pale brown or yellow middorsal line. Rostral scale is long and surpasses mental scales. Males are recognized by their dewlaps, their enlarged postcloacal scales, and their larger size (table 7.44).

Age Variation

Juvenile coloration is similar to that of adult females. Males have enlarged postcloacal scales.

NATURAL HISTORY

This species lives in sandy savannas of Pinar del Río province and Isla de la Juventud, where palms of the genera *Thrynax* and *Sabal* are abundant. Individuals perch on trunks of palms, trees of the genus *Eucalyptus,* and bushes. Males perch higher than females and juveniles. The latter are more frequently found on fence posts and bushes and on the ground, while males climb palm trunks up to the leaves.

When escaping, individuals run toward the herbaceous stratum and hide among grasses and stones. According to the calculated general index of locomotion (0.65), this is a runner species, which is consistent with its free-life behavior. A diurnal heliothermic species, it usually basks on trunks and on the ground.

Feeding in free life is unknown. Individuals in captivity have been fed with insects such as fruit flies (*Drosophila* spp.), cockroaches (*Periplaneta americana*), and wood lice.

The mean duration of copulation between captive adults of *A. b. insulaepinorum* was 64.9 min., with the longest copulation lasting 127 min. Between individuals of *A. b. bremeri* and *A. sagrei,* the mean was 3.5 min. and

Table 7.44. Comparison of snout-vent length *(SVL)*, head length *(HL)*, and tail length *(TL)* between male and female adults of *Anolis bremeri* (symbols as in table 7.1).

Sex	N	X	CV	m	M	t	p
				SVL			
Males	27	60.4	6.98	50.6	70.8	13.69	<0.01
Females	28	44.9	9.35	33.1	52.3		
				HL			
Males	27	17.4	6.26	15.5	20.1	12.59	<0.01
Females	28	13.8	7.66	11.3	15.8		
				TL			
Males	16	108.7	10.53	62.8	130.6	7.92	<0.01
Females	17	83.4	7.65	46.7	92.3		

the longest copulation lasted 5.1 min.; no offspring were obtained. An *A. bremeri* captive female laid a white egg measuring 8.4 x 5.1 mm.

Some populations are numerous.

To date, eight species of parasitic nematodes are known for *Anolis bremeri: Parapharyngodon cubensis, Cyrtosomum scelopori, Porrocaecum* sp. larvae, *Physaloptera squamatae, Skrjabinoptera* (*Didelphysoma*) *phrynosoma, Oswaldocruzia lenteixeirai,* and Physalopteridae gen. sp.

Anolis homolechis (plate 48)

Xiphosurus homolechis Cope, 1864, *Proc. Acad. Nat. Sci. Philadelphia* 16:169; holotype BMNH 1946.8.5.78. Type locality: unknown, restricted to La Habana, La Habana province, Cuba, by Ruibal and Williams 1961*b*, *Bull. Mus. Comp. Zool.* 125(8):228. *Anolis homolechis:* Boulenger, 1885, *Cat. Lizards Brit. Mus.* 2:28. *Anolis muelleri* Ahl, 1924, *Zool. Archiv. f. Naturgesch* 90:247; holotype ZMB 4178. Type locality: Cuba. *Anolis calliurus* Ahl, 1924, *Zool. Archiv. f. Naturgesch* 90:249; holotype ZMB 9014. Type locality: Cuba. *Anolis cubanus* Ahl, 1925, *Zool. Anz.* 62:87; holotype ZMB 27810. Type locality: Cuba. *Anolis patricius* Barbour, 1929, *Proc. New England Zool. Club* 11:37; holotype MCZ 28759. Type locality: Mina Piloto, Sagua de Tánamo, Oriente province, Cuba.

GEOGRAPHIC RANGE

Isla de Cuba; Isla de la Juventud; Archipiélago de los Canarreos; Cayos de San Felipe.

LOCALITIES (MAP 54)

Pinar del Río: San Diego de los Baños; El Guamá, Pinar del Río (Stejneger 1917); San Vicente (Cochran 1934); Sumidero; Guane; Consolación del Sur; Cabezas; Soroa; Isabel Rubio; Manuel Lazo; San Cristóbal; La Coloma; Viñales (Ruibal and Williams 1961*b*); Valle de San Juan; María La Gorda (Garrido and Schwartz 1968); Pan de Guajaibón; Rangel (Schwartz 1968); Cayo Real (Garrido 1973*b*); Nortey, Sierra del Rosario (Garrido 1973*c*); El Salón, Sierra del Rosario (Silva Rodríguez and Estrada 1982); El Veral (Rodríguez Schettino and Martínez Reyes 1985); El Rubí; El Mulo; Las Peladas; Las Terrazas; El Taburete; La Cañada del Infierno (Martínez Reyes 1995). *La Habana and Ciudad de La Habana:* Madruga (Barbour 1914); Caimito; Mariel (Stejneger 1917); San José de las Lajas; Guanabo; Cabañas; Jibacoa; La Habana; San Antonio de los Baños (Ruibal and Williams 1961*b*); Aguacate; Nueva Paz (Garrido 1973*c*); Sierra de Camarones; Loma del Grillo (Garrido 1973*f*); Tapaste; Managua; Santiago de las Vegas (Llanes

Echevarría 1978); Parque Zoológico Nacional (Arazoza and Novo Rodríguez 1979); Cueva del Vaho, Boca de Jaruco (Silva Taboada 1988); Jardín Botánico Nacional (García Rodríguez 1989); Escaleras de Jaruco (González González 1989); El Narigón, Puerto Escondido (Hechevarría *et al.* 1990); El Chico (L.R.S.). *Matanzas:* Matanzas (Barbour 1914); Pan de Matanzas (Ruibal and Williams 1961*b*); Península de Hicacos (Buide 1966); Alacranes; Cayo Bahía de Cádiz (Schwartz 1968); Santo Tomás, Ciénaga de Zapata (Garrido 1980*b*); Cayo Cinco Leguas (Garrido and Jaume 1984); Cueva de la Pluma, Corral Nuevo (Silva Taboada 1988); Bacunayagua (Soto Ramírez 1994); Soplillar, Península de Zapata; El Cenote, Península de Zapata (L.R.S.). *Villa Clara:* Manicaragua (Schwartz and Ogren 1956); Sagua la Grande (Garrido 1973*c*). *Cienfuegos:* Soledad (Barbour 1914). *Sancti Spíritus:* Topes de Collantes; Sierra de Jatibonico (Ruibal and Williams 1961*b*); Salto del Caburní; Pico de Potrerillo (Rodríguez Schettino *et al.* 1979); Jobo Rosado (Espinosa López, Sosa Espinosa, and Berovides Álvarez 1990). *Ciego de Ávila:* Ciego de Ávila; Majagua (Ruibal and Williams 1961*b*); Loma de Cunagua (Schwartz 1968). *Camagüey:* Camagüey; Sierra de Najasa; Vertientes; Santa Cruz del Sur; Martí; Cuatro Caminos (Ruibal and Williams 1961*b*). *Las Tunas:* Las Tunas (Díaz Castillo *et al.* 1991). *Holguín:* Sagua de Tánamo; Cananova; Moa; Mayarí (Ruibal and Williams 1961*b*); Potosí (Garrido 1967); Farallones de Moa; Cupeyal del Norte (Estrada *et al.* 1987). *Granma:* Cabo Cruz (Barbour 1914); Río Puercos (Cochran 1934); Bueycitos (Schwartz and Ogren 1956); Buey Arriba; Birama; Belic (Ruibal and Williams 1961*b*). *Santiago de Cuba:* Los Negros; Sierra Maestra (Barbour 1914); Santiago de Cuba (Stejneger 1917); Pico Turquino; Sierra del Cobre (Ruibal and Williams 1961*b*); La Cantera; Alto Songo; Sierra de la Gran Piedra; Siboney; Laguna de Baconao; Palma Soriano (Schwartz 1968); Siboney (Fong and del Castillo 1997). *Guantánamo:* Guantánamo (Cochran 1934); Boquerón; Baracoa; Imías (Ruibal and Williams 1961*b*); Bahía de Taco (Schwartz 1968); Cayo Güín; Baitiquirí (Garrido 1973*c*); Base Naval de Guantánamo (Porter *et al.* 1989); Cajobabo; Yumurí del Sur (L.R.S.). *Isla de la Juventud:* Nueva Gerona (Stejneger 1917); Sierra de Casas; Playa Bibijagua; La Fe; Cayo Piedras; Punta Francés; Jacksonville; Punta del Este (Schwartz 1968); Cayo Cantiles (Garrido and Schwartz 1968).

CONTENT

Polytypic endemic species; two subspecies are recognized:

Anolis homolechis homolechis

Anolis homolechis Cope, 1864, *Proc. Acad. Nat. Sci. Philadelphia* 16:169. *Anolis homolechis homolechis:* Barbour, 1937, *Bull. Mus. Comp. Zool.* 82(2):127. Type locality: "West Indies," restricted to La Habana, La Habana province, Cuba

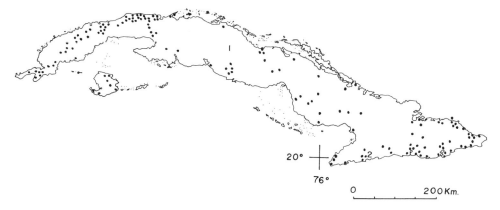

Map 54. Geographic distribution of *Anolis homolechis. 1, A. h. homolechis; 2, A. h. turquinensis.*

(Ruibal and Williams 1961*b*). Geographic range: Isla de Cuba (with the exception of Maisí and Pico Turquino); Isla de la Juventud; Cayo Cantiles; Cayo Real.

Anolis homolechis turquinensis

Anolis homolechis turquinensis Garrido, 1973, *Poeyana,* 120:9; holotype IZ 2900. Type locality: vicinity of Cardero, Pico Turquino, Oriente province, Cuba. Geographic range: Pico Turquino, Sierra Maestra, more than 1,200 m above sea level.

DESCRIPTION

Very small or small size. Mean snout-vent length between 52.0 and 57.4 mm in males and between 33.9 and 43.6 mm in females.

Very dark brown or black. Yellow eyelids. Gray gular region. Pale brown or yellow ventral region. Dewlap in males, white, gray, or white with gray semicircular bars. Green or brown iris.

Triangular, large ear opening. Nuchal, dorsal, and caudal crests in males. Vertically oval section of tail. Small, granular dorsal scales; smooth, circular ventral scales, larger than dorsals. Keeled head scales. Smooth or weakly keeled supracarpal, supratarsal, and supradigital scales. Enlarged post-cloacal scales in males.

Seven pairs of macrochromosomes and seven pairs of microchromosomes (2n = 28).

BIBLIOGRAPHIC SOURCES

Cope (1864): original description. Boulenger (1885): pl. 1, fig. 4. Barbour (1914): dewlap coloration; abundance. Barbour (1916): external morphology; habitat data. Barbour and Ramsden (1916*a*): dewlap coloration; abundance. Stejneger (1917): coloration; drawings of the dorsal scales of the head (fig. 46), of the head in lateral view (fig. 47), of the tail in lateral view in the fifth verticil (fig. 48). Barbour and Ramsden (1919): coloration; drawing of the dorsal scales of the head (pl. 14, fig. 8); habitat data; abundance. Dunn (1926): habitat data. Barbour (1929): coloration. Barbour (1930*a*): habitat data; behavior; abundance. Barbour (1930*b*): habitat data; abundance. Alayo Dalmau (1951): habitat data; abundance; coloration. Alayo Dalmau (1955): coloration; drawing of the dorsal scales of the head (pl. 6, fig. 10, no. 46), of the head in lateral view (pl. 6, fig. 10, no. 47), of the tail in lateral view in the fifth verticil (pl. 6, fig. 10, no. 48); abundance. Schwartz and Ogren (1956): measurements; behavior; habitat data. Etheridge (1960): osteology. Ruibal (1961): habitat data; behavior; territoriality; daily activity; thermic relations; husbandry. Ruibal and Williams (1961*b*): coloration; scalation; habitat data; behavior; drawings of the dorsal scales of the head (fig. 3), of the mental scales (fig. 6a), of the supracarpal and supradigital scales (fig. 7a), of the head in lateral view (fig. 8, left), of the dewlap (fig. 12, center right and lower left), of the dorsal pattern of the female (fig. 15). Ruibal (1964): coloration; scalation; habitat data; altitudinal distribution; drawing of the mental scales (fig. 7a), of the head in lateral view (fig. 8, left). Ruibal and Ernst (1965): structure of the subdigital lamellae. Buide (1966): habitat data; behavior; abundance. Etheridge (1967): tail function. Garrido (1967): habitat data. Ruibal (1967): coloration; drawing of the dewlap (fig. 2, center right); aggressive display between captive males. Gorman and Atkins (1968): karyotype; photograph of the idiogram (fig. 3A). Garrido and Schwartz (1968): coloration; external morphology. Schwartz (1968): coloration; scalation; habitat data; drawing of the head and body in lateral view (pl. 1, upper right). Garrido and Schwartz (1969): coloration; habitat data; abundance. Coy Otero (1970): nematodes. Schoener (1970*a, b*): interspecific relationships in size. Garrido (1973*b*): coloration; habitat data; daily activity. Garrido (1973*c*): coloration; scalation. Garrido (1975*c*): territoriality. Ruiz García (1975): aggressive display between males. Alayón García (1976): feeding. Williams and Rand (1977): coloration. Llanes Echevarría (1978): coloration; habitat data; thermic relationships; daily activity. Ortiz Díaz (1978): coloration; feeding. Arazoza and Novo Rodríguez (1979): habitat data; abundance. Coy Otero and Baruš (1979*a, b*): nematodes. Berovides Álvarez and Sampedro Marín (1980): habitat data; daily activity; feeding. Garrido (1980*a*): reproductive data in captivity. González Suárez (1980): electrophoretic pattern of hemoglobin. Silva Rodríguez (1980): morphometry; electrophoretic pattern of general proteins. Rodríguez Schettino (1981*b*): habitat data; thermic relations. Silva Rodríguez (1981): habitat data;

thermic relations; daily activity; feeding; behavior; reproduction; parasites; electrophoretic pattern of general proteins. Coy Otero and Lorenzo Hernández (1982): helminths. Mugica Valdés *et al.* (1982): electrophoretic pattern of esterases of liver; drawing of the pattern of esterases (fig. 1c). Sampedro Marín *et al.* (1982): coloration; habitat data; feeding. Espinosa López *et al.* (1983): electrophoretic pattern of muscle proteins. Silva Rodríguez and Espinosa López (1983): electrophoretic patterns of general proteins of plasma; genetic variability in relation to environmental variability. Estrada and Silva Rodríguez (1984): morphometry; general index of locomotion. Silva Rodríguez and Estrada (1984): reproductive cycle. Rodríguez Schettino (1985): abundance; habitat data; altitudinal distribution. Rodríguez Schettino and Martínez Reyes (1985): habitat data; thermic relations; abundance. Silva Lee (1985): photograph of a specimen. Menéndez *et al.* (1986): habitat data; thermic relations. Espinosa López *et al.* (1987): electrophoretic patterns of general proteins of plasma; drawing of the pattern of general proteins (fig. 1, left); genetic variability in relation to niche width. Estrada *et al.* (1987): habitat data. Quintana Ferrer (1987): morphometry. García Rodríguez (1989): habitat data; thermic relations; feeding; reproductive cycle; sexual rate; abundance; photograph of a male. Porter *et al.* (1989): karyotype; photograph of the diakinesis (fig. 2c). Alarcón Chávez *et al.* (1990): habitat data; feeding; reproduction; abundance. Burnell and Hedges (1990): phylogenetic relationships. Espinosa López, Posada García *et al.* (1990); electrophoretic pattern of general proteins of muscle. Espinosa López, Sosa Espinosa, and Berovides Álvarez (1990): electrophoretic patterns of eight proteic systems. Hechevarría *et al.* (1990): habitat data; feeding; reproduction; abundance. Quesada Jacob *et al.* (1991): habitat data; feeding; reproduction; abundance. Sampedro *et al.* (1993): habitat data; feeding; electrophoretic patterns of several proteins. Rodríguez Schettino (1994): description of the nemasperms. Rodríguez Schettino (1996): habitat data; Rodríguez Schettino *et al.* (1997): habitat data.

Morphological Variation

Metachromatism from pale brown with yellow lines and spots to dark brown or black. Coloration is quite variable individually and geographically. Dewlap is white, gray, or white with about three gray semicircular bars in the same population. There is great variability and polymorphism in the plasma, muscle, and liver proteins.

Geographic Variation

The two subspecies are described as follows:

Anolis homolechis homolechis (Cuban territory, except Pico Turquino): the same as the species. Brown iris. Mean snout-vent length of 52.0 mm in males and of 33.9 mm in females.

A. h. turquinensis (Pico Turquino): the same as the species. Pale gray dewlap with a basal dark gray suffused spot. Green iris. Mean snout-vent length of 57.4 mm in males and of 43.6 mm in females.

Sexual Variation

Females have brown middorsal rhombs bordered with dark brown, a middorsal longitudinal yellow line, and two lateral longitudinal yellow lines. Terra-cotta head. Yellow ventral region (plate 48). Differences in size are shown in table 7.45. Males are recognized by their dewlaps and their enlarged postcloacal scales.

Age Variation

Juvenile coloration is similar to that of adult females. Males have enlarged postcloacal scales. Snout-vent length of one juvenile was 21.0 mm.

Natural History

This species lives in all kinds of forests and secondary vegetation types, in less dense zones and at the edges; in addition, it lives in some plantations,

Table 7.45. Comparison of snout-vent length *(SVL)*, head length *(HL)*, and tail length *(TL)* between male and female adults of *Anolis homolechis* (symbols as in table 7.1).

Sex	N	X	CV	m	M	t	p
				SVL			
Males	15	53.8	8.89	48.7	65.7	7.20	<0.01
Females	14	40.5	12.64	34.2	55.8		
				HL			
Males	15	16.2	9.97	14.0	19.4	5.96	<0.01
Females	14	12.6	12.79	11.2	17.6		
				TL			
Males	9	91.1	10.13	46.1	107.3	4.77	<0.01
Females	12	64.8	23.35	31.7	110.0		

such as coffee (*Coffea arabica*) and avocado (*Persea americana*). Males perch on trunks of trees and bushes at a mean height above ground of less than 1.5 m, generally with their heads downward. They often use their tails as prehensile organs. On some occasions, they are found on the ground, on fallen trunks, or on rocks. Females and juveniles are found mostly on the ground or on fallen trunks, roots, or stones; they are also found on tree trunks and bushes, at lesser heights than males.

There was no difference between sexes in height above ground at some localities in La Habana province; however, the means varied according to habitat and season of the year (0.51; 0.68; 0.76; 1.29 m). At the edge of the semideciduous forest of Sierra del Rosario, males were found at a mean height of 0.93 m, while at the "mogote" complex of vegetation of Sierra de Casas, both sexes were seen at a mean height of 1.11 m. This species was observed at a mean height of 0.67 m on the limestone floor of the semideciduous forest at Península de Guanahacabibes, independent of sex or season of year. The same was found in the coastal shrubwood of the Jardín Botánico Nacional in 1988, where the mean height was 0.66 m. However, at the same place in 1991, the species was found at greater mean heights: 1.13 m for males and 0.84 m for females. This could be because the vegetation of this artificial habitat has been growing up, thus providing *A. homolechis* with more available perch sites.

Something similar was observed at El Narigón, on the northeastern coast of La Habana province, where the mean height above ground was 0.84 m in the coastal shrubwood, with means of 1.52 m for males and 0.91 m for females, respectively, in the ecotone between coastal shrubwood and dry semideciduous forest. Tall trees such as *"almácigo"* (*Bursera simaruba*) predominate in the latter habitat, allowing lizards to perch at greater heights.

The general index of locomotion calculated for this species is 0.86, which typifies it as a runner, although at the boundary with crawlers; this suggests that it is able to move on the ground as well as on tree and bush trunks.

It is found at altitudes ranging from sea level to the maximum heights of each mountain range, although its abundance diminishes as altitude increases.

These data indicate that *A. homolechis* has a great capacity to adapt to different habitats and environmental conditions. This adaptability is possible because of its high degree of genetic variability, as has been seen in the electrophoretic patterns of plasma, liver, and muscle proteins.

Males are frequently found on the same perches defending their territories. To escape, they run toward the herbaceous stratum or jump to a nearby tree, hiding among stones and dead leaves. Females and juveniles rapidly hide among leaf litter. Aggressive displays of resident males against intruders have been observed under conditions of free life and captivity. The resident male intimidates the intruder with nuchal and dorsal crest risings, slight head bobbings, tail waggings, and forefoot pawings. The intruder responds with some of the same movements. The resident attacks first and usually, after some biting, the intruder escapes.

This species is diurnal and apparently heliothermic, basking in patches of sun filtered by vegetation. Nevertheless, the frequency with which it has been found near those patches of sun has varied according to the places where it has been seen. At Camagüey, Sierra de Casas, El Narigón (semideciduous forest), and Jardín Botánico Nacional (1991), it has been observed with greater frequency in shaded places, while at Sierra del Rosario, Jardín Botánico Nacional (1988), and El Narigón (coastal shrubwood), it has been found with greater frequency in sunny sites.

Body temperature in July and August (X = 31.8°C) was higher than that of air (X = 30.8°C) at Camagüey province; likewise at Sierra del Rosario, where mean body temperature was 32.1°C, higher than that of air: 29.9°C. However, at Sierra de Casas there were no significant differences between them (X_{BT} = 27.9°C; X_{AT} = 27.2°C). At other places, air temperatures where the species were observed have varied between 29.4 and 33.8°C depending on habitat and season of the year. The mean relative humidity reported for localities in La Habana province during a year of observation was 68.3%, varying between 53.4 and 74.2%.

Some individuals can be seen after the first hours of the day and the majority are already active by noon. It seems that daily activity depends mainly on environmental factors and species behavior. At La Habana province, for instance, two peaks of activity were observed—one at 9:00 A.M., the other at 1:00 P.M.—while in Sierra del Rosario this was found only for females, with one peak for males between 12:30 P.M. and 2:30 P.M.

The species feeds on small arthropods, mainly insects, which it detects from perches and rapidly captures. This behavior is more common during the wet season, when there is a great abundance of food; in the dry season, it is also necessary to look for prey under dead leaves and stones on the ground.

Based on the stomach contents of samples from different localities in La Habana province and at Sierra del Rosario, this species feeds on ants with great frequency (they constitute between 57.4 and 75.5% of obtained prey). In decreasing order of frequency were eaten lepidopterans, coleopterans, isopterans, insect larvae, homopterans, spiders, and even juveniles of its own species. The mean number of prey items obtained by males and females is similar, although on some occasions it is higher for females. Prey size is generally higher in males, but they also eat a great amount of small prey. The mean weight of stomach contents was 0.074 g for a sample of males from the coastal shrubwood at El Narigón.

Individuals of this species have been fed in captivity with larvae and adults of *Tenebrio molitor* and larvae of *Galleria* sp., spiders, fruit flies (*Drosophila* spp.), moths, larvae of lepidopterans, cockroaches (*Periplaneta americana*), and domestic flies.

Reproduction is cyclic. Gravid females appear in April and remain in the population until September. However, the months having higher percentages of ovigerous females are April, May, June, and July; these are also the only months in which females have two oviductal eggs, whose mean lengths vary between 6.9 and 8.5 mm. The mean mass of a sample of oviductal eggs was 0.084 g. The calculated reproductive effort varies between 0.13 and 0.32, while the reproductive potential is high (0.92). This indicates that the period between the laying of one egg and another is short; thus the species has high productivity. There is an inverse relationship between size of fat bodies and reproduction: at the beginning and end of the year, fat bodies weigh more (between 0.033 and 0.107 g) than during the months of reproductive activity (between 0.0 and 0.007 g). The mean duration of copulation in captivity was 4.4 min. and an obtained egg was white, measuring 9.0 x 5.2 mm. Two other eggs measured 8.4 x 6.4 and 10.0 x 7.0 mm and had white, flexible shells; both eggs easily dehydrated.

The sexual rate is 0.68 males/female. Adults predominate in populations throughout the year and juveniles are observed from May to October as a result of reproduction. Populations of this species are very numerous throughout its geographic range. Density at Parque Zoológico Nacional in May, calculated by the positive method of marking, liberation, and recapture, was 1,166 individuals/ha. However, using direct count methods, figures for relative abundance and density, calculated at different places and on different dates, vary between 30 and 57.6 individuals/ha and between 49.4 and 713.2 individuals/ha, respectively. This could be because direct count methods underestimate population size. The biomass calculated for a sample obtained in the coastal shrubwood of Jardín Botánico Nacional in June 1991 was 2.068 kg/hour.

To date, ten species of parasitic nematodes are known for *Anolis homolechis: Parapharyngodon cubensis, Parathelandros* sp., *Cyrtosomum sclepori, Porrocaecum* sp. larvae, *Physaloptera squamatae, Skrjabinoptera* (*Didelphysoma*) *phrynosoma, Abbreviata* sp., Physalopteridae gen. sp., *Trichospirura teixeirai,* and *Oswaldocruzia lenteixeirai.*

At the entrance of Cueva de los Majáes, Sierra del Rosario, the Lizard Cuckoo (*Saurothera merlini*) hunts *A. homolechis,* which is attracted by the cockroaches (*Periplaneta americana*) abundant in this cave.

Anolis quadriocellifer (plate 50)

Anolis quadriocellifer Barbour and Ramsden, 1919, *Mem. Mus. Comp. Zool.* 47(2):158; holotype MCZ 11867. Type locality: Cabo de San Antonio, Ensenada de Cajón, Pinar del Río province, Cuba. *Anolis calliurus* Ahl, 1924, *Zool. Archiv. f. Naturgesch.* 90:249; holotype ZMB 9014. Type locality: Cuba. *Anolis homolechis quadriocellifer:* Barbour, 1937, *Bull. Mus. Comp. Zool.* 82(2):127.

GEOGRAPHIC RANGE

Isla de Cuba (Península de Guanahacabibes, from Cabo de San Antonio to the vicinity of Manuel Lazo).

LOCALITIES (MAP 55)

Pinar del Río: Ensenada de Cajón (Barbour and Ramsden 1919); Ensenada de Corrientes (Ruibal and Williams 1961*b*); Cayos de la Leña (Rand 1962); vicinity of El Cayuco (Manuel Lazo); Las Tumbas; Valle de San Juan (Garrido and Schwartz 1968); El Veral; La Bajada; Playa Antonio; Playa Jaimanitas; Caleta del Humo; El Fraile (Rodríguez Schettino and Martínez Reyes 1985); 10.2 km E of Manuel Lazo (L.R.S.).

CONTENT

Monotypic endemic species.

DESCRIPTION

Small size. Mean snout-vent length of 47.2 mm in males and of 41.6 mm in females.

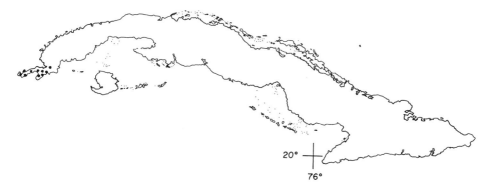

0 _____ 200 Km.

Map 55. Geographic distribution of *Anolis quadriocellifer.*

Pale brown, with a black suprascapular patch bordered by a continuous or discontinuous yellowish white line. Several yellowish white, vertically discontinuous lines on sides. Dark and pale brown ill-defined crossbands on limbs. Dewlap in males, orange yellow, with or without four red semicircular bars and yellow inner scales. Grayish green iris.

Oblique oval ear opening. Small nuchal, dorsal, and caudal crests in males. Vertically oval section of tail. Smooth, circular small dorsal scales; smooth, circular ventral scales, larger than dorsals. Weakly keeled head scales. Keeled supracarpal scales. Enlarged postcloacal scales in males.

Seven pairs of macrochromosomes and seven pairs of microchromosomes (2n = 28).

Bibliographic Sources

Barbour and Ramsden (1919): original description; drawing of the dorsal scales of the head (pl. 10, fig. 1); habitat data. Alayo Dalmau (1955): coloration; drawing of the dorsal scales of the head (pl. 3, fig. 9, no. 1). Etheridge (1960): osteology. Ruibal and Williams (1961b): coloration; scalation; drawing of the dewlap (fig. 12, lower right), of a specimen in lateral view (fig. 16). Garrido and Schwartz (1968): habitat data; behavior; daily activity. Gorman and Atkins (1968): karyotype; photograph of the idiogram (fig. 3B). Schwartz (1968): coloration; scalation; morphological variation; drawing of the head and body in lateral view (pl. 1, upper left). Coy Otero and Baruš (1979a): nematodes. Williams and Rand (1977): coloration; ecomorphology. Rodríguez Schettino and Martínez Reyes (1985): habitat data; thermic relations; abundance. Rodríguez Schettino and Martínez Reyes (1986b): habitat data; thermic relations; feeding; morphological variation; daily activity. Pérez Hernández (1995): habitat data.

Morphological Variation

Metachromatism from pale brown to very pale brown; always with the suprascapular black patches. Coloration is individually variable, as are the distribution and intensity of the lateral lines.

Geographic Variation

Dewlap color and suprascapular patch vary according to locality: in the westernmost places, dewlaps are grayish yellow with three red bars, and there are two patches at each side; at Cabo Corrientes, dewlaps are orange yellow without bars and there are no patches, but white borders; in the easternmost region, dewlaps are orange yellow without bars and there are one or two black patches.

Sexual Variation

Females have pale brown middorsal rhombuses bordered with dark brown, a middorsal longitudinal grayish white line, and two lateral concolor lines. They have black suprascapular patches and a very small dewlap. Mean snout-vent length of adult females is greater than 37.0 mm.

Males are recognized by their dewlaps and their enlarged postcloacal scales. Adult males are larger than females (table 7.46) and their mean snout-vent length is greater than 46.0 mm.

Age Variation

Juvenile coloration is similar to that of adult females. Males have enlarged postcloacal scales; their mean snout-vent length is less than 29.9 mm. In subadult females, snout-vent length ranges from 30.0 to 36.9 mm, subadult male snout-vent length from 30.0 to 45.9 mm. Their gonads are less than 3.0 and 4.0 mm in length, respectively, and are not sexually active.

NATURAL HISTORY

This species lives in semideciduous forests, on limestone floors and cliffs and in the coastal shrubwoods of Península de Guanahacabibes. Males perch on tree and bush trunks at a mean height above ground of less than 1 m, generally with their heads downward. In addition, they can be found on rocks that protrude from the ground a few centimeters. Females and juveniles perch at lesser heights on trunks and are more frequently found on the ground, on stones or roots, or in leaf litter.

When individuals escape, they run toward the herbaceous stratum and hide among dead leaves, stones, and roots. Males, adults as well as juve-

Table 7.46. Comparison of snout-vent length (SVL), head length (HL), and tail length (TL) between male and female adults of Anolis quadriocellifer (symbols as in table 7.1).

Sex	N	X	CV	m	M	t	p
				SVL			
Males	10	47.2	5.34	43.0	51.3	6.26	<0.01
Females	10	41.6	11.95	33.5	48.5		
				HL			
Males	10	14.6	7.32	13.5	17.0	2.38	<0.05
Females	10	13.1	11.38	11.0	16.0		
				TL			
Males	7	76.8	7.90	51.2	84.5	1.77	<0.10
Females	8	65.0	17.22	50.7	84.7		

niles, are aggressive, strongly defending their territories in free life and in captivity. They attack each other and other species with aggressive displays such as head bobbing, dewlap extension, limb extension, and biting on the head and neck. Dominant males kill or displace other males in terraria. They can bite and wrest fragments of tail and limbs from each other in collection bags.

This diurnal heliothermic species usually basks on trunks and stones, in patches of sun filtered by vegetation. Mean body temperature maintained during the wet season was 30.9°C, while mean air and substrate temperatures in the places where the species was found were 28.8 and 28.7°C, respectively. In both cases, body temperature was statistically higher than the environmental temperatures and there was a significant positive correlation between the first temperature and the latter two.

Daily activity appears to depend on environmental conditions, since the first individuals, females and juveniles, begin to appear after 10:00 A.M., with adult males seen much later. This is so in coastal zones with little vegetation, while in forests and shrubwoods lizards are seen after 7:00 A.M., independent of sex or age. The number of active individuals increases during the day in relation to rising air temperature. During the wet season, maximum activity is reached between 2:00 P.M. and 4:00 P.M. at sites where mean air temperature fluctuates between 27.0 and 31.0°C. During the dry season, only isolated individuals are observed in sunny places. When the air temperature is lower than 27.0°C, activity diminishes considerably. No active individuals are found after 6:00 P.M.

This species feeds on different kinds of arthropods, although it exhibits a great preference for scolytid coleopterans and ants. Also included in its diet are other hymenopterans, hemipterans, orthopterans, blatopterans, insect larvae, and spiders. Males eat larger prey than females on average. Most of the 368 prey items found among the stomach contents of thirty-five specimens of both sexes measured less than 3 mm in length. Only males contained prey with lengths greater than 6 mm. Mean prey length was 2.85 mm in males and 2.08 mm in females, with the mean for females statistically less than that for males. Individuals have been fed in captivity with small insects such as fruit flies (*Drosophila* spp.) and moths, among others. When several individuals are placed together, the dominant ones exhibit cannibalism.

Apparently, reproduction is cyclic; although there are not enough data to verify this assumption, the fact that juveniles and subadults have been found only in October indicates that egg production does not occur during the entire year.

Populations are numerous in the semideciduous forests during the wet season. During the wet season, relative abundance varied between 1.0 and 4.4 individuals/hour in cliffs and forests, respectively; it ranged from 0.1 to 0.3 individuals/hour during the dry season.

To date, four species of parasitic nematodes are known for *Anolis quadriocellifer: Parapharyngodon cubensis, Cyrtosomum scelopori, Skrjabinoptera (Didelphysoma) phrynosoma,* and *Oswaldocruzia lenteixeirai.*

Anolis jubar (plate 51)

Anolis homolechis jubar Schwartz, 1968, *Tulane Studies Zool.* 14(4):157; holotype AMNH 96529. Type locality: Paso de la Trinchera, Sierra de Cubitas, Camagüey province, Cuba. *Anolis jubar:* Garrido 1973, *Poeyana* 120:14.

Geographic Range

Isla de Cuba (coastal regions of Villa Clara, Sancti Spíritus, Ciego de Ávila, Camagüey, Las Tunas, Holguín, Granma, Santiago de Cuba, and Guantánamo provinces; also in inner places and moderated uplands); Archipiélago de Sabana-Camagüey.

Localities (map 56)

Villa Clara: Cayo Santa María (Garrido 1973d); Cayo Fragoso (Coy Otero et al. 1987); Cayo Las Brujas (Martínez Reyes 1998). *Sancti Spíritus:* Loma de Platero; Mayajigua; Caguanes (Garrido 1973c); Jobo Rosado (Garrido and Jaume 1984); Arroyo Blanco, Jatibonico (CZACC). *Ciego de Ávila:* Loma de Cunagua (Ruibal and Williams 1961b); Isla de Turiguanó (Garrido 1973c); Cayo Coco (Garrido 1976a); Cayo Guillermo (Estrada and Garrido 1990). *Camagüey:* Sierra de Cubitas (Ruibal 1961); Cayos los Ballenatos; Jaronú; Playa Santa Lucía; Bahía de Nuevitas (Ruibal and Williams 1961b); Cayo Sabinal (Schwartz 1968); Minas; Cayo Romano (Garrido 1973c); Cayo Guajaba (Garrido et al. 1986); Cayo Paredón Grande (Estrada and Garrido 1990). *Las Tunas:* Puerto Manatí (Garrido 1973c); Bahía Malagueta; Puerto Padre (CZACC). *Holguín:* Banes (Ruibal and Williams 1961b); Gibara; Pesquero Nuevo; Guardalavaca (Garrido 1973e); Levisa (Garrido and Jaume 1984); Cayo Saetía; Cayo Carenero (CZACC). *Granma:* Cabo Cruz (Ruibal and Williams 1961b); El Guafe; Pesquero de la Alegría (Estrada and Garrido 1991). *Santiago de Cuba:* southern coast of Pico Turquino; Playa Juraguá (Ruibal and Williams 1961b); Aserradero; Aguadores; Santiago de Cuba (Schwartz 1968); La Mula (Rodríguez Schettino 1985); Siboney (Fong and del Castillo 1997). *Guantánamo:* Maisí; Río Ovando (Ruibal and Williams 1961*b*); Bahía de Guantánamo; Río Yumurí; Baitiquirí; Imías; Boquerón (Schwartz 1968); Tortuguilla; Cueva de la Patana (Garrido 1973*c*); Cajobabo (L.R.S.).

CONTENT

Polytypic endemic species; ten subspecies are recognized:

Anolis jubar jubar

Anolis homolechis jubar Schwartz, 1968, *Tulane Studies Zool.* 14(4):157; holotype AMNH 96529. Type locality: Paso de la Trinchera, Sierra de Cubitas, Camagüey province, Cuba. *Anolis jubar jubar:* Garrido, 1973, *Poeyana* 120:18. Geographic range: Sierra de Cubitas; Isla de Turiguanó; Cayo Romano.

Anolis jubar cuneus

Anolis homolechis cuneus Schwartz, 1968, *Tulane Studies Zool.* 14(4):158; holotype AMNH 96536. Type locality: 1 mi. E of Playa Santa Lucía, Camagüey province, Cuba. *Anolis jubar cuneus:* Garrido, 1973, *Poeyana* 120:22. Geographic range: Playa Santa Lucía; Cayo Sabinal.

Anolis jubar balaenarum

Anolis homolechis balaenarum Schwartz, 1968, *Tulane Studies Zool.* 14(4):161; holotype AMNH 95975. Type locality: smallest cay of Cayos Los Ballenatos, Bahía de Nuevitas, Camagüey province, Cuba. *Anolis jubar balaenarum:* Garrido, 1973, *Poeyana* 120:41. Geographic range: Cayos Los Ballenatos, Bahía de Nuevitas.

Anolis jubar oriens

Anolis homolechis oriens Schwartz, 1968, *Tulane Studies Zool.* 14(4):162; holotype AMNH 95976. Type locality: Cabo Cruz, Oriente province, Cuba.

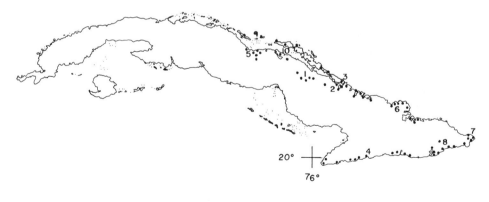

Map 56. Geographic distribution of *Anolis jubar. 1, A. j. jubar; 2, A. j. cuneus; 3, A. j. balaenarum; 4, A. j. oriens; 5, A. j. yaguajayensis; 6, A. j. gibarensis; 7, A. j. maisiensis; 8, A. j. albertschwartzi; 9, A. j. santamariae; 10, A. j. cocoensis.*

Anolis jubar oriens: Garrido, 1973, *Poeyana* 120:139. Geographic range: southern coast of Granma and Santiago de Cuba provinces.

Anolis jubar yaguajayensis

Anolis jubar yaguajayensis Garrido, 1973, *Poeyana* 120:15; holotype IZ 2372. Type locality: El Yaguey, Loma de Platero, about 15 km *E* of Caibarién, Las Villas, Cuba. Geographic range: northern zone of Sancti Spíritus province.

Anolis jubar gibarensis

Anolis jubar gibarensis Garrido, 1973, *Poeyana* 120:23; holotype IZ 2387. Type locality: El Catuco, 2.5 km of Gibara, Oriente province, Cuba. Geographic range: northern coast of Holguín province.

Anolis jubar maisiensis

Anolis jubar maisiensis Garrido, 1973, *Poeyana* 120:28; holotype IZ 1524. Type locality: Punta de Maisí, Baracoa, Oriente province, Cuba. Geographic range: known only from the vicinities of the type locality.

Anolis jubar albertschwartzi

Anolis jubar albertschwartzi Garrido, 1973, *Poeyana* 120:33; holotype IZ 2621. Type locality: Tortuguilla, 13 km *E* of Bahía de Guantánamo, Oriente province, Cuba. Geographic range: southern coast of Guantánamo province.

Anolis jubar santamariae

Anolis jubar santamariae Garrido, 1973, *Poeyana* 120:43; holotype IZ 2643. Type locality: Cayo Santa María, Archipiélago de Sabana-Camagüey, Caibarién, Las Villas province, Cuba. Geographic range: known only from the type locality.

Anolis jubar cocoensis

Anolis jubar cocoensis Estrada and Garrido, 1990, *Rev. Biol.* 4(1):73; holotype MNHN-381. Type locality: Cueva de los Hoyancos, Cayo Coco, Ciego de Ávila, Cuba. Geographic range: known only from Cayo Coco and Loma de Cunagua.

DESCRIPTION

Very small or small size. Mean snout-vent length between 45.6 and 62.0 mm in males and between 38.9 and 47.6 mm in females.

Dark brown or black. Yellow eyelids. Pale brown labials. Grayish yellow gular region. Yellow ventral region. Dewlap in males, pale or dark orange yellow or dark orange, with yellow or white edge and yellow inner scales. Green or brown iris.

Vertically oval ear opening. Nuchal, dorsal, and caudal crests in males. Vertically oval section of tail. Small, granular dorsal scales; smooth, circular ventral scales, larger than dorsals. Weakly keeled head scales. Smooth supracarpal, supratarsal, and supradigital scales. Enlarged postcloacal scales in males.

BIBLIOGRAPHIC SOURCES

Ruibal (1961): dewlap coloration; habitat data; behavior; territoriality; daily activity; thermic relations; egg-laying sites; feeding in captivity. Ruibal and Williams (1961*b*): dewlap coloration; drawing of the dewlap (fig. 12, up and center left); habitat data. Ruibal (1967): drawing of the dewlap (fig. 2, center left); territoriality; aggressive display between captive males. Schwartz (1968): original description; coloration; scalation; drawings of the head and body in lateral view (pl. 1, second row, right and left; third row, left). Lando and Williams (1969): coloration; scalation; feeding in captivity. Garrido (1973*c*): subspecific variations. Garrido (1973*d*, *e*): coloration. Coy Otero (1976): nematodes. Williams and Rand (1977): coloration. Coy Otero (1979): nematode. Coy Otero and Baruš (1979*a*): nematodes. Garrido (1980*a*): reproductive data in captivity. Rodríguez Schettino (1985): habitat data; abundance. Coy Otero *et al.* (1987): nematode. Espinosa López, Posada García *et al.* (1990): electrophoretic patterns of general proteins of muscle. Espinosa López, Sosa Espinosa, and Berovides Álvarez (1990): electrophoretic patterns of eight proteic systems. Estrada and Garrido (1990): habitat data; behavior; abundance; coloration. Socarrás Torres (1995): abundance.

MORPHOLOGICAL VARIATION

Metachromatism from pale brown to dark brown, to black. Coloration is quite variable individually and geographically.

Geographic Variation

The ten subspecies are described as follows:

Anolis jubar yaguajayensis (northern coast of Sancti Spíritus province): reddish brown with two dorsal pale red rhombuses and two reddish lateral longitudinal lines. Ocher dewlap with pale yellow edge and several reddish semicircular bars in some individuals. Mean snout-vent length of 45.6 mm in males and of 38.9 mm in females.

A. j. cocoensis (Cayo Coco and Loma de Cunagua): grayish brown. Ochraceous yellow dewlap with pale yellow washes at the base and grayish white edge. Snout-vent length between 50.0 and 57.0 mm in males and between 36.0 and 41.0 mm in females.

A. j. jubar (Sierra de Cubitas): brown. Yellow or pale orange dewlap, with white or pale yellow edge. Mean snout-vent length of 47.4 mm in males and of 47.6 mm in females.

A. j. balaenarum (Cayos Los Ballenatos): brown. Yellow or orange dewlap. Maximal snout-vent length of 62.0 mm in males; females are not known.

A. j. cuneus (Playa Santa Lucía): brown. Yellow or pale orange dewlap, with three white bars. Mean snout-vent length of 48.2 mm in males and of 45.2 mm in females.

A. j. gibarensis (northern coast of Holguín province): dark brown, with ill-defined dark brown and gray transverse bands on hindlimbs. Dark orange dewlap, with the base and the edge yellow. Mean snout-vent length of 50.5 mm in males and of 45.3 mm in females.

A. j. oriens (southern coast of Granma and Santiago de Cuba provinces): brown. Orange yellow or dark orange dewlap. Mean snout-vent length of 51.0 mm in males and of 44.4 mm in females.

A. j. albertschwartzi (southern coast of Guantánamo province): grayish pale brown with some dark brown spots on dorsum and dark brown crossbands on hindlimbs. Mean snout-vent length of 53.1 mm in males and of 46.2 mm in females.

A. j. maisiensis (Maisí): brown head and anterior part of body. Yellow rest of body, hindlimbs, and tail. Yellow or orange dewlap. Mean snout-vent length of 54.5 mm in males and of 42.2 mm in females.

A. j. santamariae (Cayo Santa María): brown with pale brown dorsal spots. Ochraceous yellow dewlap. Mean snout-vent length of 49.4 mm in males and of 41.8 mm in females.

Sexual Variation

Females have pale brown dorsal rhombuses and a middorsal, very pale brown line. Males are recognized by their dewlaps, their enlarged postcloacal scales, and their larger size (table 7.47).

Table 7.47. Comparison of snout-vent length *(SVL)*, head length *(HL)*, and tail length *(TL)* between male and female adults of *Anolis jubar* (symbols as in table 7.1).

Sex	N	X	CV	m	M	t	p
				SVL			
Males	51	50.4	8.92	40.0	59.8	6.31	<0.01
Females	47	44.0	12.81	33.2	53.2		
				HL			
Males	50	15.6	9.32	13.1	19.3	6.00	<0.01
Females	47	13.7	12.20	10.3	16.9		
				TL			
Males	30	74.3	17.06	40.1	95.0	1.92	<0.10
Females	20	67.6	16.54	37.0	85.5		

Age Variation

Juvenile coloration is similar to that of adult females. Subadult snout-vent length varies between 47.8 and 49.0 mm; males have enlarged postcloacal scales.

NATURAL HISTORY

This species lives in evergreen and microphyllous semideciduous forests, generally near coasts, and in coastal shrubwoods with an abundance of sea grapes (*Coccoloba uvifera*). It can also inhabit inland and upland areas of the central eastern region of the country. It is more frequently found at the edges of forests, at the sides of paths that cross them, and in the ecotones between coastal shrubwoods and microphyllous forests.

Males perch on trunks of trees and bushes at a low height above ground, generally less than 2 m; sometimes they are found on the ground, on fallen trunks or branches, and in leaf litter, together with juveniles. The mean height above ground calculated for twenty-two individuals in the southern coastal zone of Guantánamo province was 0.88 m, while the maximum height (2.6 m) was reached by only one lizard. For the most part, individuals are found with their heads downward. When escaping, they run toward the herbaceous stratum and leaf litter, where they hide; occasionally some males rapidly climb trunks, often jumping onto the ground or another nearby tree. On being captured, they emit acute squeaks.

The territories occupied by males are not very large, and, in general, they stay on the same perches. Captive resident males respond to intruding males with a rapid extension of their dewlaps, which is maintained for 8 sec.; they wag their tails rapidly about ten times and sometimes raise their bodies by flexing and extending their limbs.

This is a diurnal species that is much more frequently observed in shade than basking (72.7 and 27.3%, respectively) on the southeastern coast of Guantánamo province. Nevertheless, the mean body temperature of eleven males captured in that zone was 32.7°C (from 31.0 to 34.0°C) and that of air 31.6°C (from 30.0 to 33.2°C). At Sierra de Cubitas, the mean body temperature was 31.7°C (from 26.8 to 33.4°C). These high figures for body temperature indicate that *A. jubar* is a behavioral thermoregulating species; this is not a very expensive strategy, since the lizard inhabits warm places where air temperature is high and stable during most of the year.

In July the number of active individuals did not vary during the day (between 8:00 A.M. and 4:00 P.M.) at Sierra de Cubitas or on the southern coast of Guantánamo province. However, lizards observed at La Mula in the same month first appeared at 11:00 A.M. and were seen only until 1:00 P.M. This seems to indicate that daily activity is conditioned by environmental and not seasonal factors.

Feeding in free life is unknown; only foraging on the ground has been observed. Individuals have been fed in captivity with larvae of *Tenebrio molitor* and *Galleria* sp., cockroaches, moths, flies, crickets, grasshoppers, and caterpillars.

Eggs of this species have been found deposited among humus and decayed trunks at Sierra de Cubitas. In captivity, the mean duration of the precopulatory phase is from 19.8 to 24.0 sec. and differs according to the subspecies. The mean duration of copulation varies between 3.06 and 5.06 min., also according to the subspecies; the longest period was 12.15 min., for *A. j. balaenarum*.

The population at la Mula was considered very numerous in 1980, but by 1985 it had already diminished; it had declined even more by 1986, relative to the development of a camping base established there in 1982. This seems to indicate that *A. jubar* does not tolerate environmental changes produced by human activity. However, in Cayo Coco the density was calculated to be between 213 and 377 individuals/ha in the evergreen microphyllous forest, where this species is dominant.

To date, five species of parasitic nematodes are known for *Anolis jubar: Parapharyngodon cubensis, Cyrtosomum sclepori, C. longicaudatum, Porrocaecum* sp. larvae, and *Abbreviata* sp. larvae.

Anolis guafe

Anolis guafe Estrada and Garrido, 1991, *Carib. J. Sci.* 27(3–4):148; holotype CZACC 4.60516 (original field number CARE 60516). Type locality: El Guafe (2 km *NE* of Cabo Cruz), Niquero, Granma, Cuba.

GEOGRAPHIC RANGE

Isla de Cuba (vicinity of the type locality).

LOCALITIES (MAP 57)

Granma: El Guafe; Punta Inglés; Pesquero de la Alegría; Monte Gordo; Agua Fina; Farallones de Cabo Cruz (Estrada and Garrido 1991).

CONTENT

Monotypic endemic species.

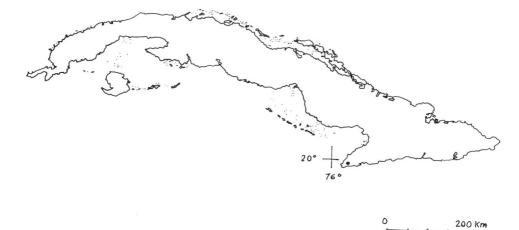

Map 57. Geographic distribution of *Anolis guafe.*

Description

Small size. Mean snout-vent length of 46.0 mm in males and of 36.0 mm in females.

Pale gray with reddish head. Two pale brown lateral lines from the suprascapular region to the groin. Two yellowish brown triangular spots, bordered with dark brown, joined by their tips at the middorsal line. White ventral region. Pale brown limbs. Dewlap in males, small, yellowish white, with white inner scales and edge. Greenish brown iris.

Vertically oval ear opening. Tail longer than body, with caudal crest in males. Keeled, rhomboidal, small dorsal scales; smooth, rhomboidal ventral scales, larger than dorsals. One row of scales between supraorbital semicircles. Smooth supracarpal scales. Enlarged postcloacal scales in males.

Bibliographic Sources

Estrada and Garrido (1991): original description; coloration; habitat data; drawings of the head in dorsal view (fig. 1A), in lateral view (fig. 1B), of one male (fig. 2), of the dewlap (fig. 5A-E).

Morphological Variation

Metachromatism from pale gray or brown, to black and gray, brown and gray, or grayish brown with yellowish brown transverse bands. Coloration varies, as do the intensity of the triangular spots and the presence of yellowish spots. Dewlap can have one, two, or five black suffused lines.

Geographic Variation

Variations previously mentioned are observed within the limited geographic range.

Sexual Variation

Females pale gray with a pale brown middorsal line and the dorsal region of the head reddish brown; they have spots from the scapular region to the sacrum, occasionally over the tail. Males are recognized by their dewlaps, their enlarged postcloacal scales, and their larger size (table 7.48).

Age Variation

Juveniles are not known.

NATURAL HISTORY

This species lives in the xeromorphic coastal shrubwood of the marine terraces at Cabo Cruz and, partially, in the adjacent microphyllous evergreen forest, to an altitude of 275 m above sea level. It is found on karstic rocks and occasionally on bush trunks less than 1 m above ground. According to the general index of locomotion (0.82), it is a runner.

Anolis confusus

Anolis confusus Estrada and Garrido, 1991, *Carib. J. Sci.* 27(3–4):156; holotype CZACC 4.60503 (original field number CARE 60503). Type locality: between Monte Gordo and Vereón, 7 km *NNE* of Cabo Cruz, Niquero, Granma, Cuba.

Table 7.48. Comparison of snout-vent length *(SVL)*, head length *(HL)*, and tail length *(TL)* between male and female adults of *Anolis guafe* (symbols as in table 7.1). Data after Estrada and Garrido, 1991.

Sex	N	X	CV	m	M	t	p
				SVL			
Males	24	45.7	38.8	39.5	48.8	12.04	<0.01
Females	13	36.0	83.0	28.0	40.0		
				HL			
Males	24	11.7	12.2	10.5	13.0	12.48	<0.01
Females	10	9.2	9.5	8.8	9.7		
				TL			
Males	5	75.0	61.0	74.0	80.0		

GEOGRAPHIC RANGE

Isla de Cuba (vicinity of the type locality).

LOCALITIES (MAP 58)

Granma: 7 km *NNE* of Cabo Cruz; Vereón; 4 km *SW* of Alegría de Pío; Bosque Castillo (Estrada and Garrido 1991).

CONTENT

Monotypic endemic species.

DESCRIPTION

Small size. Mean snout-vent length of 47.0 mm in males and of 39.6 mm in females.

Brown, with two gray longitudinal lateral lines between fore- and hind-limbs. Six pairs of black triangular spots, arranged along both sides of the middorsal line. Yellow ventral region. Dewlap in males, pale yellow in the anterior part and white posteriorly, or pale yellow with white edge. Brown iris.

Vertically oval ear opening. Tail longer than body, with caudal crest in males. Keeled, rhomboidal, small dorsal scales; smooth, rhomboidal ventral scales, larger than dorsals. Two rows of scales between supraorbital semicircles. Smooth or weakly keeled supracarpal scales. Enlarged post-cloacal scales in males.

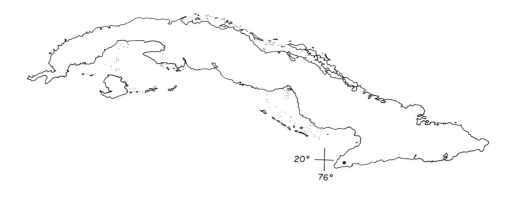

0   200Km.

Map 58. Geographic distribution of *Anolis confusus.*

BIBLIOGRAPHIC SOURCES

Estrada and Garrido (1991): original description; coloration; habitat data; drawings of the head in dorsal view (fig. 4A), in lateral view (fig. 4B), of the dewlap (fig. 5 G–K).

MORPHOLOGICAL VARIATION

Metachromatism from brown to very dark brown. Dewlap is pale yellow in some individuals; in others it is pale yellow and white posteriorly, pale yellow with an orange anterior spot, or pale yellow with one or two semi-circular orange bars.

Geographic Variation

Variations previously mentioned are observed within the limited geographic range.

Sexual Variation

Coloration pattern and scalation are similar in both sexes. Males are larger than females (table 7.49) and are recognized by their dewlaps and their enlarged postcloacal scales.

Age Variation

Juveniles are not known.

NATURAL HISTORY

This species lives in the microphyllous evergreen forest of the Meseta de Cabo Cruz at more than 275 m above sea level. Males perch on trunks of

Table 7.49. Comparison of snout-vent length *(SVL)*, head length *(HL)*, and tail length *(TL)* between male and female adults of *Anolis confusus* (symbols as in table 7.1). Data after Estrada and Garrido (1991).

Sex	N	X	CV	m	M	t	p
			SVL				
Males	12	47.0	91.0	40.0	53.0	3.50	<0.01
Females	3	39.6	86.6	41.0	48.0		
			HL				
Males	12	12.7	26.0	11.1	13.8	4.60	<0.01
Females	3	10.2	23.1	9.8	10.6		
			TL				
Males	4	82.0	95.0	78.0	86.0		

trees and bushes at heights of up to 2 m above ground; females are found on bushes and on the ground. According to the general index of locomotion (0.83), this species is a runner.

Anolis mestrei (plate 52)

Anolis mestrei Barbour and Ramsden, 1916, *Proc. Biol. Soc. Washington* 29:19; holotype MCZ 11285. Type locality: Valle de Luis Lazo, Pinar del Río province, Cuba. *Anolis cubanus* Ahl, 1925, *Zool. Anz.* 62:87; holotype ZMB 27810. Type locality: Cuba. *Anolis allogus mestrei:* Barbour, 1937, *Bull. Mus. Comp. Zool.* 82(2):120.

GEOGRAPHIC RANGE

Isla de Cuba (from the vicinity of Isabel Rubio to Sierra de Anafe).

LOCALITIES (MAP 59)

Pinar del Río: Valle de Luis Lazo; San Diego de los Baños; El Guamá, Pinar del Río (Barbour and Ramsden 1916*a*); Sumidero; San Vicente; Cabezas; Matahambre; Sierra del Rosario (Ruibal and Williams 1961*b*); Soroa; Pan de Azúcar; Rangel (Schwartz 1968); Cueva del Indio, Viñales (Zajicek and Mauri Méndez 1969); San Andrés; Pan de Guajaibón; Sierra de la Güira; Sierra del Milindre (L.R.S.); San Cristóbal (CZACC). *La Habana:* Sierra de Anafe (Buide 1967).

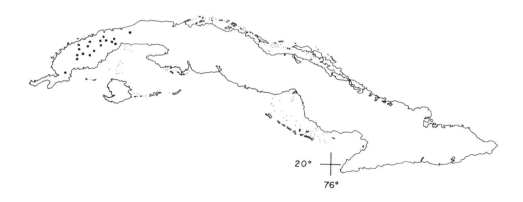

Map 59. Geographic distribution of *Anolis mestrei*.

CONTENT

Monotypic endemic species.

DESCRIPTION

Very small or small size. Mean snout-vent length of 52.8 mm in males and of 44.2 mm in females.

Olivaceous brown with grayish green, ill-defined transverse bands on body, limbs, and tail. Grayish green head. Yellow labials, with green spots. Yellow eyelids. Dewlap in males, red, with two or three green or yellow, small, semicircular bars, white edge and inner scales. Pale gray iris.

Vertically oval, small ear opening. Small nuchal, dorsal, and caudal crests in males. Vertically oval section of tail. Small, granular dorsal scales; smooth, circular ventral scales, larger than dorsals. Keeled head scales. Smooth supracarpal, supratarsal, and supradigital scales. Enlarged postcloacal scales in males.

Seven pairs of macrochromosomes and seven pairs of microchromosomes (2n = 28).

BIBLIOGRAPHIC SOURCES

Barbour and Ramsden (1916a): original description; habitat data. Barbour and Ramsden (1919): coloration in life; scalation; drawing of the dorsal scales of the head (pl. 10, fig. 3). Barbour (1930a, b): abundance; habitat data. Alayo Dalmau (1955): coloration; drawing of the dorsal scales of the head (pl. 3, fig. 9, no. 3). Etheridge (1960): osteology. Ruibal and Williams (1961b): coloration; external morphology; habitat data; behavior; daily activity; drawings of the dorsal scales of the head (fig. 5), of the dewlap (fig. 11, center). Ruibal (1964): coloration; scalation; habitat data; drawing of mental scales (fig. 7a). Buide (1967): habitat data. Gorman and Atkins (1968): karyotype. Schwartz (1968): coloration; scalation; geographic, sexual, and age variation; habitat data; drawing of the lateral region of the head and the forelimb (pl. 1, third row, right). Williams (1969): habitat data. Zajicek and Mauri Méndez (1969): hemoparasites. Schoener (1970a): interspecific relationships in size. Coy Otero (1976): nematodes. Williams and Rand (1977): coloration; ecomorphology. Coy Otero and Baruš (1979a): nematodes. Garrido (1980a): reproductive data in captivity. Espinosa López et al. (1983): electrophoretic patterns of muscle proteins. Estrada and Silva Rodríguez (1984): morphometry; general index of locomotion. Silva Lee (1985): coloration; habitat data; photographs of one male and one female. Espinosa López, Posada García et al. (1990): electrophoretic pattern of general muscle proteins. Espinosa López, Sosa Espinosa, and Berovides Álvarez (1990):

electrophoretic patterns of eight proteic systems. Rodríguez Schettino (1994): description of the nemasperms.

Morphological Variation

Metachromatism from olivaceous brown to dark grayish brown. Shading coloration varies individually, but the pattern is the same.

Geographic Variation

Dewlap can be red, reddish brown, brown, or orange, and the semicircular bars yellowish green, pale orange, or absent in individuals from San Vicente, Viñales.

Sexual Variation

Females have pale grayish green dorsal rhombuses, bordered by olivaceous brown, and a rudimentary red dewlap. Males are recognized by their dewlaps, their enlarged postcloacal scales, and their larger size (table 7.50).

Age Variation

Juvenile coloration is similar to that of adult females. Males have enlarged postcloacal scales.

Natural History

This species lives on the rocky limestone outcrops of the Sierra de los Órganos, Sierra del Rosario, and Sierra de Anafe, in areas of semideciduous, evergreen, and "mogote" forests. It is also found in the forest, on the limestone floors of lowlands, in the vicinity of Isabel Rubio, Guane. Males perch on limestone rocks and walls; females and juveniles are more fre-

Table 7.50. Comparison of snout-vent length *(SVL)*, head length *(HL)*, and tail length *(TL)* between male and female adults of *Anolis mestrei* (symbols as in table 7.1).

Sex	N	X	CV	m	M	t	p
				SVL			
Males	8	52.8	6.37	35.0	56.5	5.40	<0.01
Females	10	44.2	7.46	37.7	48.5		
				HL			
Males	8	16.0	9.34	11.0	18.0	3.24	<0.01
Females	10	13.9	9.38	11.8	15.2		
				TL			
Males	7	78.6	12.11	45.8	92.5	1.08	n.s.
Females	7	73.2	12.13	45.0	83.8		

quently found on the ground, on stones, and sometimes in leaf litter. Males are also found in fallen trunks near the rocks. Males as well as females position themselves horizontally on the rocky substrate. Nevertheless, some individuals have been observed on trunks of trees and bushes. They run rapidly when escaping and hide inside rock fissures. According to the general index of locomotion (0.78), this species is a runner, which is consistent with its free-life behavior.

This diurnal nonheliothermic species is almost always found in shaded sites. On a few occasions some individuals have been observed under the sun filtered by vegetation. All lizards at Sierra de la Güira were found in shade in April, on rocks and at a mean air temperature of 26.0°C. During the first hours of the day few individuals are observed, only some females and juveniles; males begin to be seen at noon, when the activity of the species is greatest. At night, individuals sleep on small twigs near rocks, or in crevices of rocky cliffs and outcrops.

Feeding in free life is unknown. Individuals have been fed in captivity with fruit flies (*Drosophila* spp.), larvae and adults of *Tenebrio molitor*, moths, larvae of lepidopterans, cockroaches, and domestic flies.

Apparently reproductive activity increases during the months of the wet season. A buried cream-colored egg measuring 13.5 x 9.9 mm was found at Pica Pica; two empty shells were found in April at Sierra de la Güira in the rocks of a "mogote," one in a rock hole and the other on a superficial layer of earth at the base of the same rock. Both were white, leathery, and open at one pole. Based on observations of captive adults, the mean duration of the precopulatory phase is 9.04 sec.; copulation has a mean duration of 2.01 min., with the longest one lasting 12 min. When copulation ends, both members of the pair slowly separate. During aggressive displays between males, the two opponents elevate themselves on their four limbs and raise their nuchal and dorsal crests and tails, at the same time extending their dewlaps and raising their heads repeatedly. This behavior continues until one of them pursues the other, which generally escapes by running.

Populations of this species are numerous throughout its geographic range.

To date, protozoa of the genus *Hemogregarina* are known in red cells of *Anolis mestrei*, as well as two species of parasitic nematodes: *Cyrtosomum sclepori* and *Skrjabinoptera* (*Didelphysoma*) *phrynosoma*.

Anolis allogus (plate 53)

Anolis allogus Barbour and Ramsden, 1919, *Mem. Mus. Comp. Zool.* 47(2):159; holotype MCZ 8544. Type locality: Bueycitos, near Bayamo (Sierra

Maestra), Oriente province, Cuba. *Anolis abatus* Ahl, 1924, *Zool. Archiv. f. Naturgesch* 90:248; holotype ZMB 6965. Type locality: Cuba.

GEOGRAPHIC RANGE

Isla de Cuba (with the exception of the Península de Zapata).

LOCALITIES (MAP 60)

Pinar del Río: Sumidero; Cabezas; Rangel; San Vicente; San Diego de los Baños; Pan de Azúcar; Soroa; Pinar del Río (Ruibal and Williams 1961*b*); Pan de Guajaibón (Garrido 1967); El Cayuco (Manuel Lazo) (Garrido and Schwartz 1968); El Taburete (González Bermúdez and Rodríguez Schettino 1982); El Salón, Sierra del Rosario (Silva Rodríguez and Estrada 1982); El Mulo (Martínez Reyes 1995); Sierra del Milindre; Mil Cumbres; Sierra de la Güira (L.R.S.). *Matanzas:* San Miguel de los Baños (Schwartz and Thomas 1975). *Sancti Spíritus:* San Felipe, Jatibonico (Garrido 1973*c*). *Ciego de Ávila:* Loma de Cunagua (Ruibal and Williams 1961*b*); Morón (Schwartz and Thomas 1975). *Camagüey:* 15 km *SW* of Camagüey; Banao; Sierra de Cubitas; Jaronú; Sierra de Najasa; Santa Cruz del Sur (Ruibal and Williams 1961*b*); Martí (Schwartz 1968). *Las Tunas:* Las Tunas (Díaz Castillo *et al.* 1991). *Holguín:* Sagua de Tánamo; Bahía de Nipe; Mayarí (Barbour and Ramsden 1919); Banes (Ruibal and Williams 1961*b*); Cupeyal del Norte; Farallones de Moa; Calentura Arriba; La Melba (Estrada *et al.* 1987); Majayara (J. F. Milera, pers. comm.). *Granma:* Bueycito; Jiguaní (Barbour and Ramsden 1919); Buey Arriba (Ruibal and Williams 1961*b*); Dos Bocas, Pilón (Schwartz 1968). *Santiago de Cuba:* Baire; Los Negros; San Luis (Barbour and Ramsden 1919); Pico Turquino; El Cobre; Jutinicú (Ruibal and Williams 1961*b*); La Cantera; Maffo; Ocujal; La Gran Piedra (Schwartz 1968). *Guantánamo:* Monte Líbano (Barbour and Ramsden 1919); Baracoa; mountains north of Imías (Ruibal and Williams 1961*b*); La Casimba, Maisí; Cayo Güín (Garrido 1967); Sierra del Purial (Garrido and Schwartz 1967); Río Ovando; Río Yumurí; Bahía de Miel; Bahía de Taco (Schwartz 1968); Cayo Probado; springs of Jaguaní River; Ojito de Agua; springs of Yarey River; Cayo Fortuna; Alto del Yarey (Estrada *et al.* 1987); Yumurí del Sur (L.R.S.).

CONTENT

Monotypic endemic species.

DESCRIPTION

Very small or small size. Mean snout-vent length of 54.8 mm in males and of 42.6 mm in females.

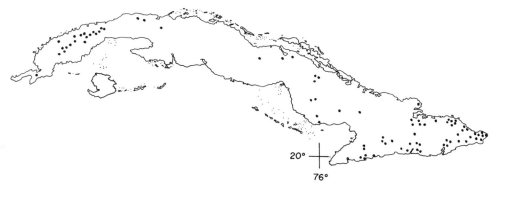

Map 60. Geographic distribution of *Anolis allogus*.

Pale brown with red and yellow spots and dots. Brown and red, ill-defined transverse bands on tail and limbs. Grayish yellow gular region. Pale brown ventral region. Yellow eyelids, with a yellow subocular line. Dewlap in males, orange yellow, with three or four semicircular reddish bars, yellow edge and inner scales. Blue iris.

Vertically oval, large ear opening. Small nuchal and dorsal crests in males. Caudal crest of scales modified as spines. Vertically oval section of tail. Small, granular dorsal scales; smooth, circular ventral scales, larger than dorsals. Keeled head scales. Multicarinate supracarpal, supratarsal, and supradigital scales. Prenasal scales divided into two portions. Postcloacal scales not enlarged.

BIBLIOGRAPHIC SOURCES

Barbour (1914): dewlap coloration. Barbour and Ramsden (1919): original description; habitat data; behavior; drawing of the dorsal scales of the head (pl. 10, fig. 2). Barbour (1930*a, b*): habitat data. Alayo Dalmau (1955): dewlap coloration; drawing of the dorsal scales of the head (pl. 3, fig. 9, no. 2); habitat data; abundance. Ruibal (1961): habitat data; behavior; territoriality; thermic relations; feeding in captivity. Ruibal and Williams (1961*b*): coloration; scalation; drawings of the dorsal scales of the head (fig. 4), of the mental scales (fig. 6b), of the supracarpal and supradigital scales (fig. 7b), of the anterior part of the head in lateral view (fig. 8, right); habitat data; behavior; abundance. Garrido (1967): habitat data. Ruibal (1967): coloration; drawing of the dewlap (fig. 2, upper); behavior; territoriality; aggressive display in captivity. Garrido and Schwartz (1968): external morphology; habitat data; behavior; abundance. Schwartz (1968): coloration; scalation; drawing of the anterior part of the head and body in lateral view (pl. 1, fourth row, right and left, fifth row, left). Coy Otero (1970): helminths. Schoener (1970*a, b*): interspecific relationships in size. Coy Otero (1976): helminths. Williams and Rand (1977): coloration; ecomorphology. Coy Otero (1979): nematode. Coy Otero and Baruš (1979*a*): nematodes. Garrido (1980*a*): reproductive data in captivity. Silva Rodríguez (1980): electrophoretic patterns of general proteins. González Martínez (1981): electrophoretic pattern of hemoglobin. Mugica Valdés (1981): electrophoretic patterns of esterases; photograph of the esterases pattern (fig. 8). Silva Rodríguez (1981): habitat data; thermic relations; daily activity; feeding; behavior; reproduction; electrophoretic patterns of general proteins. Mugica Valdés *et al.* (1982): electrophoretic pattern of esterases of liver; drawing of the esterases pattern (fig. 1f). Silva Rodríguez and Espinosa López (1983): electrophoretic patterns of general proteins. Estrada and Silva Rodríguez (1984): morphometry; general index of locomotion. Silva Rodríguez and Estrada (1984): reproductive cycle. Rodríguez Schettino (1985): habitat data; abundance. Silva Lee (1985): coloration; behavior; habitat data; photographs of two males and one female. Estrada *et al.* (1987): habitat data. Socarrás *et al.* (1988): saurophagy. Espinosa López, Posada García *et al.* (1990): electrophoretic patterns of muscle general proteins. Espinosa López, Sosa Espinosa, and Berovides Álvarez (1990): electrophoretic patterns of eight proteic systems. Rodríguez Schettino (1994): description of the nemasperms.

Morphological Variation

Metachromatism from pale brown with yellow and red spots, to dark reddish brown with yellow spots and dots. Coloration is individually variable for the arrangement of the spots and dots.

Geographic Variation

Little variability throughout its geographic range. Individuals from Pinar del Río province have yellow dewlaps with three reddish bars; those from eastern provinces have yellowish brown dewlaps with three or four red bars or a basal red spot; those from other populations have orange-yellow dewlaps with three or four red bars or a basal red spot.

Sexual Variation

Females have a pale brown dorsal rhombus bordered with dark brown and a middorsal longitudinal yellow line. Males are recognized by their dewlaps and their larger size (table 7.51).

Table 7.51. Comparison of snout-vent length (SVL), head length (HL), and tail length (TL) between male and female adults of Anolis allogus (symbols as in table 7.1).

Sex	N	X	CV	m	M	t	p
				SVL			
Males	10	54.8	9.97	44.5	62.8	6.50	<0.01
Females	10	42.6	5.41	37.9	47.0		
				HL			
Males	10	16.0	10.59	13.1	19.0	4.52	<0.01
Females	10	13.2	5.59	12.0	14.5		
				TL			
Males	3	102.7	6.36	40.1	109.5	2.10	<0.01
Females	3	83.2	17.67	46.0	99.5		

Age Variation

Juvenile coloration is similar to that of adult females.

NATURAL HISTORY

This species lives in semideciduous, evergreen, and gallery forests; submontane rain forests; and coniferous forests, not only in upland but also in lowland areas; it usually lives near streams and glens, where the sunlight scarcely reaches the ground. Males perch on trunks of trees and bushes, at heights above ground of less than 1 m (mean 0.98 m), with their heads downward. Females and juveniles are found on the ground, on stones, and in leaf litter, at lesser heights than males (mean 0.61 m). In addition, males and females are found on rocks in places where they are abundant. When escaping, individuals rapidly run down the trunks to the ground; females are faster than males and hide among dead leaves. According to the calculated general index of locomotion (0.79), this species is a runner, which is consistent with its free-life behavior.

Males defend their territories, not only in free life but also in captivity. The initial response of a resident captive male against a decoy male lasted 12 sec. and included diverse movements: the resident male raised his nuchal and dorsal crests; bobbed his head; extended his dewlap, which was held thus or partially folded, eight or ten times; vertically wagged his tail, which was curled laterally at its end; and kept his mouth open. Similar displays were also observed in free life, where in addition males raised up on their limbs, raised their heads and tails, arched their bodies, and maximally extended their dewlaps.

This is a diurnal nonheliothermic species based on the shaded and moist places where it is found. The mean body temperature was 29.2°C for lizards captured at Camagüey in July and August, while air temperature was 30.0°C; there was a significant positive correlation between both temperatures. This species can also be classified as a thermoconform nonheliothermic species, based on observations at Sierra del Rosario, where mean body temperature was the same as mean air temperature (27.5°C). It was experimentally proved that this species maintained mean body temperatures between 23.0 and 28.0°C under air temperatures between 19.0 and 31.0°C. When air temperature was artificially increased, heat gain rate was 2°C/min.; lizards started to pant at a mean body temperature of 30.3°C and began to move when this mean reached 32.4°C.

The first individuals are observed after 7:00 A.M., with activity increasing from 7:30 A.M. until a maximum level is reached—between 12:30 P.M. and 2:30 P.M. They generally sleep on limbs of bushes or on tree trunks, almost at ground level. In addition, females and juveniles may sleep under stones.

Based on an analysis of the stomach contents of specimens from Sierra del Rosario, this species feeds more frequently on insects than on gastropods, diplopods, wood lice (Isopoda), and spiders. The insect orders most frequently eaten were Coleoptera, Hymenoptera, and Lepidoptera (larvae). Males and females consume the same mean number of prey; males eat larger prey than females, although they also eat small prey. A generalized sit-and-wait feeding strategy is used during the wet season; during the dry season, however, this species also employs the foraging searching mode, whereby prey are sought out under stones, trunks, or leaf litter. Individuals have been fed in captivity with larvae of *Tenebrio* sp. and *Galleria* sp., domestic flies, cockroaches, caterpillars, and crickets. On some occasions, they have been observed to swallow juveniles of *A. sagrei* in terraria, and *A. mimus* in free life.

Reproduction is cyclic in this species. Gravid females appear in the Sierra del Rosario population in April and stay at least until September. Between 12 and 29% of ovigerous females have two oviductal eggs. The reproductive effort is not too large (0.11) and the reproductive potential is 0.84. Fat body weight varies inversely relative to reproduction; in January and February it is higher (from 0.022 to 0.041 g) than in other months, when reproductive activity is increased (from 0.0 to 0.003g).

The precopulatory phase, observed in twenty-six matings in captivity, lasted 28 sec.; the mean duration of copulation was 14.53 min. and the longest copulation lasted 27 min. When copulation ends, both members of the pair slowly separate. One white egg measured 8.5 x 6.1 mm.

Populations are abundant in moist forests.

To date, one species of parasitic cestode is known for *Anolis allogus: Proteocephalus* sp.; in addition, six species of parasitic nematodes are known: *Parapharyngodon cubensis, Cyrtosomum scelopori, Porrocaecum* sp. larvae, *Physaloptera squamatae, Oswaldocruzia lenteixeirai,* and *Abbreviata* sp.

Anolis ahli (plate 54)

Anolis ahli Barbour, 1925, *Occ. Papers Boston Soc. Nat. Hist.* 5:168; holotype MCZ 19905. Type locality: near the hydroelectric plant, 1,500 ft., Sierra de Trinidad, Las Villas province, Cuba. *Anolis allogus ahli:* Barbour, 1937, *Bull. Mus. Comp. Zool.* 82(2):120. *Anolis mestrei ahli:* Hardy, 1958, *Herpetologica* 14(4):205.

Geographic Range

Isla de Cuba (Sierra de Trinidad).

Localities (map 61)

Cienfuegos: San Blas; Sierra de Trinidad (Barbour 1930*a*); Cumanayagua (Ruibal and Williams 1961*b*). *Villa Clara:* Salto del Hanabanilla (Ruibal and Williams 1961*b*); Manicaragua (Schwartz 1968). *Sancti Spíritus:* Topes de Collantes (Hardy 1958*c*); Salto del Caburní; Pico de Potrerillo (Rodríguez Schettino *et al.* 1979).

Content

Monotypic endemic species.

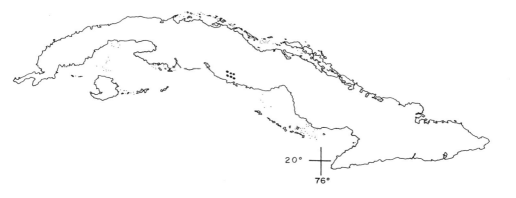

20°

76°

0   200 Km.

Map 61. Geographic distribution of *Anolis ahli.*

DESCRIPTION

Very small or small size. Mean snout-vent length of 53.2 mm in males and of 45.1 mm in females.

Pale brown with red and yellow spots and dots. Brown and reddish transverse bands on tail, ill defined on limbs. Grayish yellow gular region. Pale brown ventral region. Dewlap in males, orange-red at the base and pale ochraceous yellow to the edge, with very pale yellow inner scales. Greenish blue iris.

Small, vertically oval ear opening. Nuchal and dorsal crests in males. Vertically oval section of tail. Smooth, circular, small dorsal scales; smooth, circular ventral scales, larger than dorsals. Smooth head scales. Keeled supracarpal, supratarsal, and supradigital scales. Enlarged postcloacal scales in males.

BIBLIOGRAPHIC SOURCES

Barbour (1925): original description. Barbour (1930a, b 1937): habitat data; abundance. Alayo Dalmau (1955): dewlap coloration. Hardy (1958c): drawings of the resting position (left), of the holding position (center), of the wakefulness position (right); behavior in captivity. Ruibal and Williams (1961b): coloration; external morphology; drawing of the dewlap (fig. 11, center right); habitat data; behavior. Ruibal (1964): coloration; scalation; drawing of the mental scales (fig. 7b); habitat data. Schwartz (1968): coloration; scalation; drawing of the anterior part of the head and body in lateral view (pl. 1, lower right). Williams (1969): habitat data. Schoener (1970a): interspecific relationships in size. Williams and Rand (1977): coloration. Garrido (1980a): reproductive data in captivity. Espinosa López (1981): electrophoretic pattern of muscle proteins.

MORPHOLOGICAL VARIATION

Metachromatism from pale brown with reddish dots, to brown with ill-defined pale brown dots. Coloration is individually variable for the arrangement of the spots and dots.

Geographic Variation

Slightly variable throughout its geographic range.

Sexual Variation

Females have pale brown dorsal rhombuses bordered with dark brown; they do not differ from males in size (table 7.52). Males are recognized by the dewlaps and their enlarged postcloacal scales.

Table 7.52. Comparison of snout-vent length *(SVL)*, head length *(HL)*, and tail length *(TL)* between male and female adults of *Anolis ahli* (symbols as in table 7.1).

Sex	N	X	CV	m	M	t	p
				SVL			
Males	8	53.2	12.46	43.0	61.7	1.76	n.s.
Females	3	45.1	15.96	38.0	52.4		
				HL			
Males	8	16.4	12.65	13.7	19.0	2.04	<0.10
Females	3	13.5	17.71	10.8	15.4		
				TL			
Males	4	82.7	11.07	47.5	92.6		
Females	1	75.0					

Age Variation

Juvenile coloration is similar to that of adult females. Males have enlarged postcloacal scales.

NATURAL HISTORY

This species lives in submontane evergreen forests and rain forests of Sierra de Trinidad, in the most shaded and moist places. In addition, it is found in open areas within the forests and at the edges of paths that cross them. Males perch on trunks of trees, at a low height above ground; females are more frequently found among dead leaves, on fallen trunks, and on rocks on the ground. When escaping, individuals run toward the herbaceous stratum, where they hide among grasses and dead leaves.

Observations of captive individuals revealed that this species assumes three different positions on trunks: a resting position, with the head upward and the tail curled, almost always to the left of the body; a holding position, with the head downward and the tail coiled around the trunk or branch like a prehensile organ; and a wakeful position, with the head upward and the tail extended, but with the tip curled.

This species is diurnal and apparently nonheliothermic based on the shaded places where it has been observed.

Feeding in free life is unknown. Individuals have been fed in captivity with fruit flies (*Drosohila* sp.), cockroaches (*Periplaneta americana*), and wood lice (Isopoda).

In two matings that occurred in captivity, the precopulatory phase lasted 64 sec.; the mean duration of copulation was 28.66 min. and the longest one lasted 31.02 min. When copulation ends, both members of the pair slowly separate.

Populations of this species are numerous throughout its geographic range.

Anolis delafuentei

Anolis delafuentei Garrido, 1982, *Doñana, Acta Vertebrata* 9:132; holotype CZACC 4.7173. Type locality: Topes de Collantes, Sierra de Trinidad, Sancti Spíritus province, Cuba.

GEOGRAPHIC RANGE

Isla de Cuba (Topes de Collantes).

LOCALITIES (MAP 62)

Sancti Spíritus: Topes de Collantes (Garrido 1982*b*).

CONTENT

Monotypic endemic species.

DESCRIPTION

Small size. Snout-vent length of the only known male: 61.0 mm.

BIBLIOGRAPHIC SOURCES

Garrido (1982*b*): original description of the only specimen; habitat data.

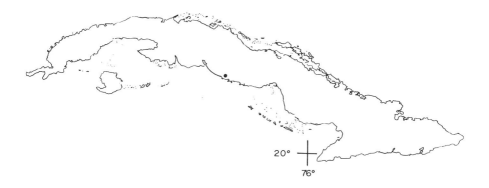

0   200 Km

Map 62. Geographic distribution of *Anolis delafuentei.*

MORPHOLOGICAL VARIATION

It is not possible to evaluate this aspect, since only one male is known.

NATURAL HISTORY

The only known specimen was collected on a glen bed in an evergreen forest, 700 m above sea level; the individual was perched on a tree trunk at 1 m above ground, in a shaded place.

Anolis rubribarbus (plate 55)

Anolis rubribarbus Barbour and Ramsden, 1919, *Mem. Mus. Comp. Zool.* 47(2):156; holotype MCZ 11941. *Anolis homolechis:* Barbour, 1937, *Bull. Mus. Comp. Zool.* 82(1):127. Type locality: Puerto de Cananova, near Sagua de Tánamo, Oriente province, Cuba.

GEOGRAPHIC RANGE

Isla de Cuba (Cananova; Nibujón; Sierra del Toa).

LOCALITIES (MAP 63)

Holguín: Puerto de Cananova (Barbour and Ramsden 1919); Punta Gorda; Moa (Ruibal and Williams 1961b); Potosí (Garrido 1967); Mina Piloto, Sagua de Tánamo (Schwartz 1968); Nibujón; Nuevo Mundo; Monte Iberia (CZACC). *Guantánamo:* Farallones de Cabo Maisí (Barbour and Ramsden 1919); Cupeyal (CZACC).

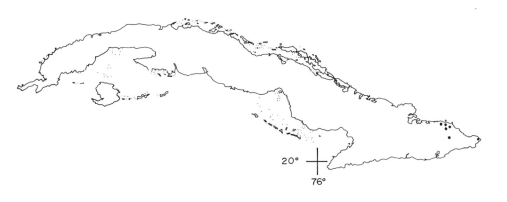

20°
76°

0 _____ 200Km.

Map 63. Geographic distribution of *Anolis rubribarbus*.

CONTENT

Monotypic endemic species.

DESCRIPTION

Very small or small size. Mean snout-vent length of 59.9 mm in males and of 42.2 mm in females.

Pale grayish brown or greenish yellow, with dark brown transverse bands on body, limbs, and tail. Pale gray head; dark brown radial lines, centered in eye, which extends into the gular region. Dewlap in males, yellow with four semicircular red bars, white edge, and black inner scales. Grayish green iris.

Vertically oval, large ear opening. Large nuchal, dorsal, and caudal crests in males. Vertically oval transverse section of tail. Smooth, circular, small dorsal scales; smooth, circular ventral scales, larger than dorsals. Keeled supracarpal, supratarsal, and supradigital scales. Enlarged postcloacal scales in males.

Seven pairs of macrochromosomes and seven pairs of microchromosomes (2n = 28).

BIBLIOGRAPHIC SOURCES

Barbour and Ramsden (1919): original description; drawing of the dorsal scales of the head (pl. 9, fig. 2), of the lateral region of the tail and caudal crest (pl. 9, fig. 3). Alayo Dalmau (1955): coloration; drawing of the dorsal scales of the head (pl. 3, fig. 6, no. 2), of the lateral part of the tail (pl. 3, fig. 6, no. 3). Ruibal and Williams (1961b): coloration; scalation; drawing of the dewlap (fig. 11, lower), of a specimen in lateral view (fig. 14); habitat data; behavior. Ruibal (1964): coloration. Garrido (1967): coloration; habitat data; behavior. Gorman and Atkins (1968): karyotype; photograph of the diakinesis (fig. 4). Schwartz (1968): coloration; scalation. Williams and Rand (1977): coloration; ecomorphology. Coy Otero and Baruš (1979a): nematodes. Estrada et al. (1987): habitat data.

MORPHOLOGICAL VARIATION

Metachromatism from pale brown with dark brown transverse bands to grayish white with black transverse bands, to pale gray without bands.

Geographic Variation

Slightly variable throughout its geographic range.

Sexual Variation

Females have pale brown dorsal rhombuses bordered with dark brown. Males are recognized by their dewlaps, their enlarged postcloacal scales, and their larger size (table 7.53).

Age Variation

Juvenile coloration is similar to that of adult females.

NATURAL HISTORY

This species lives in gallery, coniferous, and evergreen forests, as well as rain forests, of the northeastern zone of Cuba, at more than 300 m above sea level. Males perch on trunks of trees and pines, a few meters above ground, with their heads downward. They aggressively defend their territories. Females and juveniles are found at lesser heights than males and among the herbaceous stratum near the ground.

It is a very shy species that can be approached only to a proximity of 2 m. When escaping, it quickly runs toward the ground. This behavior and its cryptic coloration make detection very difficult.

A diurnal species, it is seemingly nonheliothermic based on the shaded places where it is found.

Feeding in free life is unknown. It has been fed in captivity with fruit flies (*Drosophila* spp.), ants, and wood lice (Isopoda).

Populations are very geographically limited and scarce.

To date, two species of parasitic nematodes are known for *Anolis rubribarbus*: *Cyrtosomum scelopori* and *Porrocaecum* sp. larvae.

Table 7.53. Comparison of snout-vent length *(SVL)*, head length *(HL)*, and tail length *(TL)* between male and female adults of *Anolis rubribarbus* (symbols as in table 7.1).

Sex	N	X	CV	m	M	t	p
				SVL			
Males	10	59.9	8.15	49.0	65.9	6.44	<0.01
Females	4	42.2	9.32	38.5	47.5		
				HL			
Males	10	18.6	5.84	16.0	19.5	8.05	<0.01
Females	4	13.2	9.66	12.0	15.0		
				TL			
Males	7	100.9	6.62	49.3	107.2	5.96	<0.01
Females	4	58.3	29.67	45.7	82.3		

Anolis imias (plate 56)

Anolis imias Ruibal and Williams, 1961, *Bull. Mus. Comp. Zool.* 125(8):237; holotype MCZ 42556. Type locality: Imías, Oriente province, Cuba; emended to the mountains north of Imías (Sierra del Purial), Oriente province, Cuba, by Schwartz 1968, *Tulane Stud. Zool.* 14(4):172; emended to coastal cliffs at Imías by Garrido and Jaume 1984, *Doñana Acta Vertebrata* 11(2):59.

Geographic Range

Isla de Cuba (southern coast of Guantánamo province).

Localities (map 64)

Guantánamo: Imías (Ruibal and Williams 1961*b*); mountains north of Imías (Schwartz 1968); 4.5 km W of Baitiquirí; E Imías (Garrido and Jaume 1984).

Content

Monotypic endemic species.

Description

Very small or small size. Mean snout-vent length of 61.2 mm in males and of 46.3 mm in females.

Grayish green with black or very dark brown transverse bands on head, body, limbs, and tail. Pale grayish green gular and ventral regions, with

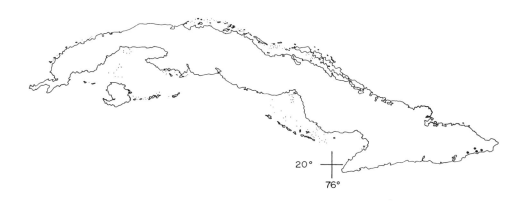

Map 64. Geographic distribution of *Anolis imias*.

brown lines. Dewlap in males, pale orange-yellow, with two semicircular orange yellow bars and yellow edge and inner scales. Brown iris.

Vertically oval, large ear opening. Nuchal, dorsal, and caudal crests in males. Vertically oval section of tail. Smooth, circular, small dorsal scales; smooth, circular ventral scales, larger than dorsals. Smooth head scales. Smooth supracarpal, supratarsal, and supradigital scales. Enlarged postcloacal scales in males.

BIBLIOGRAPHIC SOURCES

Ruibal and Williams (1961b): original description. Ruibal (1964): coloration; scalation; drawing of the mental scales (fig. 7b). Schwartz (1968): coloration; scalation. Williams and Rand (1977): coloration; ecomorphology. Martínez Reyes and Rodríguez Schettino (1987): cannibalism. Socarrás et al. (1988): cannibalism.

MORPHOLOGICAL VARIATION

Metachromatism from grayish green with black or dark brown transverse bands, to pale gray with pale brown ill-defined transverse bands.

Geographic Variation

Slightly variable throughout its limited geographic range.

Sexual Variation

Females have pale brown dorsal rhombus. Males are recognized by their dewlaps, their enlarged postcloacal scales, and their larger size (table 7.54).

Table 7.54. Comparison of snout-vent length (SVL), head length (HL), and tail length (TL) between male and female adults of Anolis imias (symbols as in table 7.1).

Sex	N	X	CV	m	M	t	p
				SVL			
Males	5	61.2	10.75	53.5	67.4	3.02	<0.05
Females	2	46.3	0.61	46.1	46.5		
				HL			
Males	5	17.6	8.47	15.7	19.0	2.29	<0.10
Females	2	15.0	5.20	14.4	15.5		
				TL			
Males	1	104.8					
Females	1	62.3					

Age Variation

Juveniles are not known.

NATURAL HISTORY

This species lives in the rocky outcrops of the semidesertic southeastern coastal zone of Guantánamo province—where vegetation is xerophyllous with a predominance of thorny bushes—inside lapels and fissures of rocks, from ground level to nearly 2 m above ground. Males perch on trunks of bushes and on soil stones. When escaping, individuals very rapidly hide inside crevices and lapels of rocks. They begin to be active when sunbeams no longer directly reach the rocks: no individuals are found outside shelters before 4:00 P.M.

Adult males are very aggressive, defending their territories against lizards of the same or other species by quickly attacking them. The most frequent actions include extension of the huge dewlap and biting. Captive adult males aggressively display against males and females of the same or other species and even against humans, trying to bite and extending their dewlaps.

This is a diurnal nonheliothermic species based on the shaded sites where it has been observed during most of the day.

Feeding in free life is unknown; the only account given was that of F. González Bermúdez (pers. comm.), who observed an adult male attacking an adult female in an attempt at cannibalism. Individuals have been fed in captivity with large insects such as crickets and moths, and with females of *A. allogus, A. homolechis,* and *A. porcatus,* which were easily and avidly eaten by males. This proves that at least males are inclined to saurophagy and even cannibalism. The severe environmental conditions of the semidesertic ecosystem where *A. imias* lives seem to favor this type of trophic behavior.

Apparently populations are not numerous, but the petricolous way of life and the limited activity of the species may allow it to remain unnoticed on some occasions.

Anolis birama

Anolis birama Garrido, 1990, *Rev. Biol.* 4(2):158; holotype MCTH 4. Type locality: shores of Río Cauto, at Ciénaga de Carenas, Granma province.

GEOGRAPHIC RANGE

Isla de Cuba (known only from the type locality).

LOCALITIES (MAP 65)

Granma: shore of Río Cauto, Birama (Garrido 1990).

CONTENT

Monotypic endemic species.

DESCRIPTION

Very small or small size. Mean snout-vent length of 59.5 mm in males and of 41.5 mm in females.

Gray with dark, V-shaped spots on the middorsal longitudinal line and limbs. Pale gray head and tail. White gular and ventral regions, with a yellow wash on the middle of the venter and vent. Dewlap in males yellow, darker at the base and paler at the edge, with white inner scales.

Circular, small ear opening, with a skin fold behind it. Snout anteriorly bulky. Caudal crest in males. Keeled, circular dorsal scales; smooth ventral scales, larger than dorsals. Keeled head scales; bulky scales around the interparietal scale. Smooth supracarpal scales and keeled arm scales. Enlarged postcloacal scales in males.

BIBLIOGRAPHIC SOURCES

Garrido (1990): original description; coloration; scalation; measurements; habitat data.

MORPHOLOGICAL VARIATION

Metachromatism from gray with dark spots to black. Coloration is slightly variable.

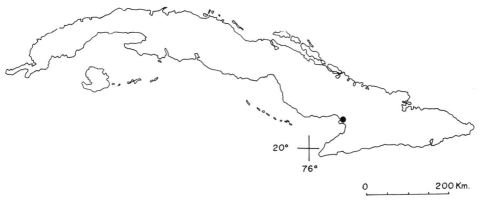

Map 65. Geographic distribution of *Anolis birama*.

Geographic Variation

Little variability throughout its limited geographic range.

Sexual Variation

Females have keeled head scales; their coloration is similar to that of *A. sagrei* females. Males are recognized by their larger size (table 7.55), their dewlaps, and their enlarged postcloacal scales.

Age Variation

Juveniles are not known.

NATURAL HISTORY

This species lives in secondary vegetation on the banks of the Cauto River, on *"marabú"* (*Caillea glomerata*) and *"guásima"* (*Guazuma tomentosa*).

Anolis ophiolepis (plate 57)

Anolis ophiolepis Cope, 1861, *Proc. Acad. Nat. Sci. Philadelphia* 13:211; holotype unlocated. Type locality: Monte Verde, Oriente province, Cuba. *Norops ophiolepis:* Boulenger, 1885, *Cat. Lizards Brit. Mus.* 2:26. *Anolis ophiolepis:* Etheridge, 1960, *University Microfilms:*94.

GEOGRAPHIC RANGE

Isla de Cuba (islandwide, except at Península de Guanahacabibes); Isla de la Juventud.

Table 7.55. Comparison of snout-vent length *(SVL)*, head length *(HL)*, and tail length *(TL)* between male and female adults of *Anolis birama* (symbols as in table 7.1). All measures after Garrido (1990).

Sex	N	X	CV	m	M	t	p
			SVL				
Males	11	59.5	5.90	53.0	65.0	12.92	<0.01
Females	7	41.5	3.00	39.5	43.0		
			HL				
Males	11	15.0	8.70	12.5	16.8	7.60	<0.01
Females	7	10.8	7.40	9.3	12.0		
			TL				
Males	8	91.3	5.00	84.0	97.0	9.03	<0.01
Females	3	61.6	9.20	57.0	68.0		

LOCALITIES (MAP 66)

Pinar del Río: Pinar del Río (Barbour 1914); San Diego de los Baños (Stejneger 1917); Guane (Barbour and Ramsden 1919); La Güira (Cochran 1934); Mil Cumbres (L.R.S.). *La Habana and Ciudad de la Habana:* Habana; Madruga (Barbour 1914); Escaleras de Jaruco (González González 1989); Atabey; Parque Zoológico Nacional (L.R.S.). *Matanzas:* Hanábana (Gundlach 1867); Cárdenas (Buide 1967); San Lorenzo, Ciénaga de Zapata (Garrido 1980b). *Cienfuegos:* Soledad (Barbour 1914). *Granma:* Bayamo (Gundlach 1867). *Guantánamo:* Yateras (Gundlach 1867); Guantánamo (Barbour and Ramsden 1916a); Base Naval de Guantánamo (Johnson 1946). *Isla de la Juventud:* Los Indios; Nueva Gerona (Barbour 1916).

CONTENT

Monotypic endemic species.

DESCRIPTION

Very small size. Mean snout-vent length of 36.6 mm in males and of 32.2 mm in females. Very long, slender tail, twice the body length.

Reddish brown with yellow longitudinal lines. Grayish brown head. A pale brown longitudinal middorsal line from the occipital region to the sacrum. A pale brown longitudinal dorsolateral line from the upper eyelid to the neck, which continues to the first part of the tail, yellow. Very pale brown loreals and sides of head. A yellow longitudinal dorsolateral line from behind the ear opening to the tail. A very narrow, brown labial stripe,

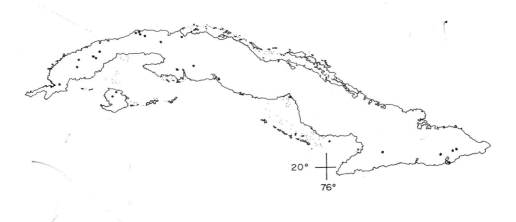

20°

76°

0   200Km.

Map 66. Geographic distribution of *Anolis ophiolepis*.

which continues to the groin. A brown line from the nostrils to the anterior edge of orbit; a dark brown line from behind the eye to the temporal region. White ventral region, with some brown dots. Dewlap in males, small, red with white and yellow inner scales. Yellow iris.

Oblique, oval, small ear opening. Narrow digital pads. Keeled, lanceolate, large dorsal and ventral scales; small, circular, weakly keeled lateral scales. Keeled head scales. Keeled, large inner scales of the dewlap. Enlarged postcloacal scales in males.

BIBLIOGRAPHIC SOURCES

Cope (1861): original description; coloration. Gundlach (1867): habitat data. Gundlach (1880): coloration in life; habitat data; abundance. Barbour (1914): habitat data; abundance; behavior. Barbour (1916): external morphology; habitat data. Barbour and Ramsden (1916a): habitat data; behavior. Stejneger (1917): drawings of the dorsal region of the head (fig. 58), of the lateral region of the head (fig. 59), of the lateral region of the tail in fifth verticil (fig. 60). Barbour and Ramsden (1919): external morphology; coloration; drawing of the lateral region of the head (pl. 14, fig. 10); measurements; habitat data; abundance; behavior; feeding. Barbour (1930b): habitat data; abundance. Alayo Dalmau (1955): drawing of the dorsal region of the head (pl. 7, fig. 6, no. 58), of the lateral region of the head (pl. 7, fig. 6, no. 59), of the lateral region of the tail (pl. 7, fig. 6, no. 60); measurements; habitat data. Neill (1958): habitat data. Williams (1961): habitat data. Ruibal (1964): external morphology; habitat data; behavior; abundance. Schoener (1970b): interspecific relationships in size. Coy Otero and Baruš (1979a): nematodes. Garrido (1980a): reproductive data in captivity. Estrada and Silva Rodríguez (1984): morphometry; general index of locomotion. Silva Lee (1985): behavior; habitat data; photograph of a specimen. Burnell and Hedges (1990): phylogenetic relationships.

MORPHOLOGICAL VARIATION

Metachromatism from reddish brown with yellow lines to dark brown with pale brown lines. It has brighter colors under sun than in shade.

Geographic Variation

Slightly variable throughout its geographic range.

Sexual Variation

Both sexes have similar color patterns. Females have a rudimentary red dewlap. Males are recognized by their enlarged postcloacal scales. Measurements are shown in table 7.56.

Table 7.56. Comparison of snout-vent length *(SVL)*, head length *(HL)*, and tail length *(TL)* between male and female adults of *Anolis ophiolepis* (symbols as in table 7.1).

Sex	N	X	CV	m	M	t	p
			SVL				
Males	10	36.6	8.36	31.8	39.8	1.38	n.s.
Females	20	32.3	29.84	24.2	39.5		
			HL				
Males	10	10.7	7.81	9.5	11.8	2.90	<0.01
Females	20	9.7	9.98	8.1	11.4		
			TL				
Males	3	83.5	12.99	72.7	92.4	3.38	<0.05
Females	5	63.6	9.74	57.6	72.2		

Age Variation

Juvenile coloration is the same as that of adults.

NATURAL HISTORY

This species lives in open savannas, pastures, gardens, uncultivated terrain, grazing fields, marshes, and sugarcane fields, on the stems of grasses and small bushes; it is usually found at a height above ground of less than 0.5 m. According to the calculated general index of locomotion (0.84), this species is a runner, which is consistent with its free-life behavior. Individuals seek shelter on the ground, among dead leaves, or on piles of recently cut grass.

It is a very terrestrial species; when escaping, individuals rapidly hide between the lowest part of the herbaceous stratum and the leaf litter, as well as in holes in the ground. The species' coloration pattern, which includes many longitudinal lines, allows it to remain unnoticed among grass leaves.

The species is diurnal, seemingly heliothermic based on the sunny places where it lives. However, during the wet season, when environmental temperatures are higher, it is found in the lowest parts of grasses, in filtered sun or shade, while during the dry season it is found exposed in sunny places. Individuals hide under the leaf litter of terraria during the night and climb the stems of grasses after approximately 10:00 A.M.

Feeding in free life is unknown; the species has only been observed eating ants and has been fed in captivity with ants, fruit flies (*Drosophila* spp.), domestic flies, and small larvae of *Tenebrio molitor*.

Based on observations of captive adults, the precopulatory phase lasts 38.6 sec.; the mean duration of copulation is 4.31 min., and the longest copulation lasts 9.0 min. After copulation ends, both members of the pair hastily separate. A female laid a cream-colored egg measuring 7.6 x 4.1 mm.

Populations of this species are numerous in gardens, pastures, and grasslands throughout its geographic range.

To date, two species of parasitic nematodes are known for *Anolis ophiolepis: Cyrtosomum scelopori* and *Porrocaecum* sp. larvae.

GLOSSARY

Aggressive display. A threat behavior, generally between two males, which involves limb flexion and extension, lateral compression of the body, and dewlap extension.

Bobbing. Moving the head up and down; a behavior used in courtship and as an aggressive display.

Canthal ridge. Area extending from the nasal opening to the anterior border of the eye and outward to the frontal ridge (figs. 3 and 8).

Caudal autotomy. The ability of the tail backbone, which includes fracture planes, to detach from the vertebrae, muscle fascia, and skin as part of defensive behavior.

Caudal crest. Structure located above the tail in males of some species (fig. 3).

Caudal verticils. Vertebral, muscular, and skin fascia located in some regions of the tail, the fracture planes, which can be detached without hurt (fig. 3).

Cephalic or head casque. Posterior and slightly upward enlargement of the head in *Chamaeleolis* spp. and the *equestris* group of the genus *Anolis* (fig. 7).

Circular scale. A round, smooth scale (fig. 5.2).

Comb. A succession of large scales on the third and fourth toes of *Cyclura*.

Cryptic coloration. Color pattern similar to the substrate most commonly used by a species.

Denticulate scale. A scale having a denticulate border (fig. 5).

Dewlap. A longitudinal skin fold in the gular region, generally featuring contrasting colors. The dewlap is extended during behavioral displays by males, and some females, of species of the genera *Anolis* and *Chamaeleolis* (figs. 3 and 7).

Dewlap base. The part of the dewlap nearest the body (fig. 7.1).

Dewlap edge. The part of the dewlap most distant from the body, extended by erecting the second pair of ceratobranchial arches. The dewlap edge is covered by a cluster of scales sometimes differing in size and color from those on the rest of the dewlap (fig. 7).

Digital pads. Expansions of the second phalanxes in species of the genera *Anolis* and *Chamaeleolis* (fig. 2).

Dorsal crest. Structure extending from the posterior part of the head to the sacrum; males that have the dorsal crest raise it during aggressive or courtship displays (fig. 3).

Dorsal scales. Scales found in the dorsal region (fig. 8).

Ecomorph. A group of species with similar habitat, ecology, morphology, and behavior but not necessarily phyletically related to one another.

Expected or theoretical heterozygosity. The mean of the theoretical heterozygosis in each studied locus.

Frontal ridge. The area extending from the nasal opening to the supraorbital semicircle, inward to the canthal ridge (figs. 3 and 8).

Frontal scales. Scales located behind the supraorbital semicircles (fig. 8).

Genetic distance (Nei 1972). A measurement of the accumulated number of gene substitutions per locus.

Genetic distance (Rogers 1972). The geometric mean of the distances between allelic frequency vectors over all loci.

Genetic identity (Nei 1972). The normalized identity of genes between two populations, with respect to all loci.

Granular scale. A rounded, convex, smooth scale (fig. 5).

Head length. Measured laterally, from the tip of the snout to the posterior end of the jaw (fig. 4).

Head width. Measured dorsally and transversely at the level of the postorbital bones in the broader region of the head (fig. 4).

Imbricate scales. Scales that overlap (fig. 5).

Infralabial scales. Large scales originating behind the mentals and extending over the edge of the lower jaw (fig. 7).

Inner dewlap scales. Scales that cover the dewlap; when it is extended, they give the appearance of a fan (fig. 7).

Interstitial spaces. The skin that can be observed between scales (fig. 7).

Keeled scale. A scale in the center of which is a longitudinal keel (fig. 5).

Labials. The supra- and infralabials as a whole (fig. 7).

Lanceolate scale. A lance- or leaf-shaped scale (fig. 5).

Longitudinal middorsal band of differentiated scales. A band occurring in some species of the genus *Anolis*. It can be composed of several longitudinal rows of scales that are larger or smaller than dorsal scales (fig. 8).

Longitudinal zones. Longitudinal dorsal and lateral color lines, generally of pale or dark brown coloration; found in the majority of species of the genus *Leiocephalus* (fig. 18).

Loreal scales. Scales found in the lateral region of the head, between the canthal ridge and the supralabial scales (fig. 3).

Mental plates. Two plates located at the tip of the lower jaw; species of the genus *Leiocephalus* have only one plate (fig. 3).

Mental scales. See mental plates.

Metachromatism. Total or partial change of body color or shading, in response to external or physiological stimuli.

Mucronate scale. A scale having two or more keels (fig. 5).

Multicarinate scale. See mucronate scale.

Nasal scales. Scales bearing the nasal openings; generally, nasal scales are larger than the surrounding scales (figs. 7 and 20).

Nostrils. Nasal openings (fig. 20).

Nuchal crest. A structure extending over the posterior part of the head and neck; it is occasionally raised in males of some species (fig. 3).

Observed heterozygosity. The frequency of heterozygotic genotypes in the studied sample; the number of heterozygotic specimens per number of analyzed specimens for a given locus.

Paramedian gular scales. Scales of the dewlap edge, differentiated in species of the genus *Chamaeleolis* (fig. 4).

Paramedian lines. A pattern of dark dashes, located in the gular region of some species of the genus *Leiocephalus* (fig. 19).

Parietal scales. Scales located adjacent to the interparietal scale, in the posterior dorsal region of the head (fig. 8).

Perch. A place where individuals of a species usually bask, wait for prey, and engage in social interaction.

Pineal eye. An eye with a lens, retina, and nerve that serves as a light-intensity receptor; it is located under the interparietal scale (fig. 8).

Postcloacal scales. Scales located behind the vent in males of some species; they have different sizes, shapes, and organization patterns (fig. 4).

Postmental scales. Scales located immediately behind the mental plates (fig. 4).

Preauricular scales. Scales located before the ear opening (fig. 3).

Prefrontal scales. Scales located before the frontal scales (fig. 8).

Prefrontal shields. Enlarged prefrontal scales (fig. 20).

Pushup. Raising of the forelimbs and body, a behavioral display performed by males. Occasionally, all limbs are raised simultaneously.

Rostral plate. A plate at the tip of the snout (figs. 7 and 8).

Rostral scale. See rostral plate.

Rows of scales between the supraorbital semicircles. Scales longitudinally aligned between supraorbital semicircles (fig. 8).

Smooth scale. A flat scale, without a keel or protuberance; it can be circular or subcircular (fig. 5).

Snout-vent length. Measured ventrally, from the tip of the snout to the vent or cloacal opening (fig. 4).

Subdigital lamellae. They are located in the lower surface of the fingers and toes of the species of the genera *Anolis* and *Chamaeleolis* (fig. 2).

Suborbital semicircle. A succession of large scales that surrounds the lower eyelid (fig. 21).

Supralabial scales. Scales located behind the rostral plate, at the edge of the upper jaw (fig. 7).

Supraorbital semicircle. A succession of large scales that surrounds the upper eyelid (fig. 8).

Suprascapular stripe. A distinctly colored stripe on the shoulders; found in individuals of species of the *equestris* group in the genus *Anolis* (fig. 7).

Tail length. Measured ventrally, from the cloacal opening to the end of the tail (fig 4).

Temporal scales. Scales located above the ear opening (fig. 3).

Total length. Measured ventrally, from the tip of the snout to the end of the tail (fig. 4).

Ventral scales. Scales located in the ventral region of the body (figs. 3 and 4).

BIBLIOGRAPHY

Abreu, R. M., and de la Cruz, J. 1988. Algunos datos sobre la alimentación del Almiquí. *Garciana* 10:2–3.

Abreu, R. M., de la Cruz, J., Rams, A., and García Romero, M. E. 1989. Vertebrados del complejo montañoso "La Zoilita," Holguín, Cuba. *Poeyana* 370:1–16.

Acevedo, M., Arredondo, O., and N. González Gotera. 1975. *La Cueva del Túnel.* Havana: Editorial Pueblo y Educación.

Acosta Cruz, M., Ibarra Martín, M. E., and Fernández Romero, E. 1985. Aves y reptiles en Cayo Matías, grupo insular de los Canarreos, Cuba. In *Segunda y Tercera Jornadas Científicas del Instituto de Zoología y la Sección de Zoología de la Sociedad Cubana de Ciencias Biológicas,* Havana (Instituto de Zoología, Havana), *Resúmenes,* 20–21.

Adams, S. E., Smith, M. H., and Baccus, R. 1980. Biochemical variation in the American alligator. *Herpetologica* 36(4):289–296.

Aguayo, C. G. 1949. Los orígenes de la fauna cubana. *Anales Acad. Cien. Habana.*

———. 1951. Notas sobre algunos reptiles hallados en residuarios indígenas. *Bol. Hist. Nat. Soc. Felipe Poey.* 2(7):139–140.

Agüero Cobiellas, R., Fernández Velázquez, A., and Vázquez Reyes, E. 1993. Morfometría y algunos aspectos del subnicho estructural de *Chamaeleolis chamaeleonides* (Iguanidae) en El Yayal, Holguín. In *Primer Simposio de Ecología,* Havana (Universidad de La Habana), *Resúmenes,* 112–113.

Ahl, E. 1924. Neue Reptilien und Batrachien aus dem Zoologischen Museum Berlin. *Zool. Archiv f. Naturgesch* 90(5):246–254.

———. 1925. Neue Iguanidem aus dem Zoologischen Museum Berlin. *Zool. Anz.* 62:85–88.

Alarcón Chávez, A., Álvarez Tellechea, M., Ayala Pérez, A., Ayala Percedo, T., Leipzy Cacho, M., and Enjanio González, A. 1990. "Aspectos ecológicos sobre algunas especies de lagartos que habitan en una zona de manigua costera del litoral norte de la Habana" [inedit]. University Students' Report, Facultad de Biología, Universidad de La Habana.

Alayo Dalmau, P. 1951. Especies herpetológicas halladas en Santiago de Cuba. *Bol. Hist. Nat. Soc. Felipe Poey* 2(7):108–109.

———. 1955. *Lista de los reptiles de Cuba.* University of Oriente, Museum Ch. T. Ramsden.

Alayón García, G. 1976. Araneidos depredados por anolinos. *Misc. Zool.* 2:1.

Alberts, A. C. 1994. A reintroduction program for the iguanas of Guantanamo. *J. Intl. Iguanas Soc.* 3(4):10–12.

———. 1995*a*. Use of statistical models based on radiographic measurements to predict oviposition dates and clutch size in rock iguanas (*Cyclura nubila*). *Zoo. Biol.* 14:543–553.

———. 1995*b*. Heading for home. *Zoonoz.* 1995:20–21.

Alberts, A. C., and Grant, T. 1997. Use of a non-contact temperature reader for measuring skin surface temperatures and estimating internal body temperatures in lizards. *Herpetol. Rev.* 28(1):32–33.

Allen, R. E., and Neill, W. T. 1957. The gecko-like habits of *Anolis lucius* a Cuban anole. *Herpetologica* 13(3):246–247.

Andrews, R. M. 1991. Population stability of a tropical lizard. *Ecology* 72(4):1204–1217.

Andrews, R. M., and Rand, A. S. 1974. Reproductive effort in anoline lizards. *Ecology* 55:1317–1327.

Arazoza, F., de, and Novo, J. 1979. Determinación de la densidad de población del *Anolis homolechis*. In *Segundo Evento Científico de la Universidad de La Habana*, La Habana (Universidad de La Habana), *Resúmenes*, 309.

Arcia Rodríguez, M. 1989. El Bosque y el Medio Ambiente. In *Nuevo Atlas Nacional de Cuba* (Academia de Ciencias de Cuba e Instituto Cubano de Geodesia y Cartografía, Havana), Instituto Geográfico Nacional de España, XIII.5.1 (map 5).

Ardines, M., Díaz, L., Díaz., R., Fernández, H., Ferrer, M., Hernández, D., Matos, K., and Riverón, M. 1992. "Estudio comparativo de diferentes especies del genero *Anolis* en la manigua costera y el cuabal del Jardín Botánico Nacional" [inedit], University Students' Report, Facultad de Biología, Universidad de La Habana.

Armas, L. F., de. 1987. Notas sobre la alimentación de *Leiocephalus carinatus cayensis* (Sauria: Iguanidae). *Poeyana* 350:1–7.

Armas, L. F., de, and Alayón García, G. 1986. Depredadores y parasitoides de *Argiope trifasciata* (Araneae: Araneidae) en el sur de La Habana. *Cien. Biol.* 16:114–117.

———. 1987. Observaciones sobre la ecología trófica de una población de *Argiope trifasciata* (Araneae: Araneidae) en el sur de La Habana. *Poeyana* 344:1–18.

Avise, J. C. 1983. Protein variation and phylogenetic reconstruction. In *Protein polymorphism, adaptive, and taxonomic significance* (G. Oxford and D. Rollinson, eds.), Systematic Association, British Mus. Nat. Hist. London:103–130.

———. 1994. *Molecular markers, natural history and evolution.* New York and London: Chapman and Hall.

Avise, J. C., and Aquadro, C. F. 1982. A comparative summary of genetic distances in the vertebrates: Patterns and correlations. *Evol. Biol.* 15:151–185.

Ayala, F. J. 1975. Genetic differentiation during the speciation process. *Evol. Biol.* 8:1–78.

———. 1980. *Evolución molecular.* Editora Omega: Barcelona.

Ayala Castro, N. 1978. Bosques. In *Atlas de Cuba* (Instituto Cubano de Geodesia y Cartografía), Havana, 40.

Baird, I. L. 1970. The anatomy of the reptilian ear. In *Biology of the Reptilia. 2. Morphology B.* (C. Gans and T. S. Parsons, eds.), 193–275. London and New York: Academic Press.

Barbour, T. 1914. A contribution to the zoogeography of the West Indies, with especial reference to amphibians and reptiles. *Mem. Mus. Comp. Zool.* 44(2):209–359.

———. 1916. The reptiles and amphibians of the Isle of Pines. *Ann. Carnegie Mus.* 10(2):297–308.

———. 1925. A new Cuban *Anolis. Occ. Papers Boston Soc. Nat. Hist.* 5:167–168.

———. 1928. Reptiles from the Bay Islands. *Proc. New England Zool. Club* 10:55–61.

———. 1929. Another new Cuban Anolis. *Proc. New England Zool. Club* 11:37–38.

———. 1930a. The anoles. I. The forms known to occur on the neotropical islands. *Bull. Mus. Comp. Zool.* 70(3):106–144.

———. 1930b. A list of Antillean reptiles and amphibians. *Zoologica* 11(4):61–106.

———. 1935. A second list of Antillean reptiles and amphibians. *Zoologica* 19(3):77–141.

———. 1937. Third list of Antillean reptiles and amphibians. *Bull. Mus. Comp. Zool.* 82(2):77–166.

Barbour, T., and Noble, G. K. 1916. A revision of the lizards of the genus *Cyclura. Bull. Mus. Comp. Zool.* 60(4):139–164.

Barbour, T., and Ramsden, C. T. 1916a. A new *Anolis* from Cuba. *Proc. Biol. Soc. Washington* 29:19–20.

———. 1916b. Catálogo de los reptiles y anfibios de la Isla de Cuba. *Mem. Soc. Cubana Hist. Nat. Felipe Poey.* 2(4):124–143.

———. 1919. The herpetology of Cuba. *Mem. Mus. Comp. Zool.* 47(2):71–213.

Barbour, T., and Shreve, B. 1935. Notes on Cuban anoles. *Occ. Papers Boston Soc. Nat. Hist.* 8:249–254.

Baruš, V. 1966. Nemátodos parásitos de aves en Cuba. *Poeyana* 22:1–37.

———. 1969. Nemátodos de codorniz *Colinus virginianus cubanensis* (Gray). *Torreia* 20:1–8.

———. 1973. Nematodes parasitizing hosts of the genus *Bufo* (Amphibia) in Cuba. *Fol. Parasitol.* 20:29–39.

Baruš, V., and Coy Otero, A. 1966. Nota sobre la helmintofauna de ofidios en Cuba. Descripción de tres especies nuevas de nemátodos. *Poeyana* 23:1–16.

———. 1968. *Freitasia teixeirai* gen. n. et sp. n. and other nematodes parasitizing *Anolis equestris* (Squamata: Iguanidae). *Fol. Parasitol.* 15:41–54.

———. 1969a. Nemátodos del género *Parapharyngodon* Chatterfi, 1933 (Oxyuridae) en Cuba. *Torreia* 7:1–10.

———. 1969b. Systematic survey of nematodes parasitizing lizards (Sauria) in Cuba. *Helminthologia* 10:329–346.

———. 1974. Nematodes of the genera *Spauligodon, Skrjabinodon* and *Pharyngodon* (Oxyuridae) parasitizing Cuban lizards. *Vest. Cs. Spol. Zool.* 38:1–22.

———. 1978. Nematodes parasitizing Cuban snakes. *Vest. Cs. Spol. Zool.* 42:85–100.

Baruš, V., Coy Otero, A., and Garrido, O. H. 1969. Helmintofauna de *Cyclura macleayi* Sauria, Iguanidae, en Cuba. *Torreia* 8:1–20.

Baruš, V., and Hubálek, Z. 1995. Phylogenetic systematics of iguanine lizards (Reptilia: Iguaninae). Parasitological arguments. *Folia Zool.* 44(2):123–130.

Baruš, V., Hubálek, Z., and Coy Otero, A. 1996. Testing the context and extent of host-parasite co-evolution: Nematodes parasitizing Cuban lizards (Sauria: Iguanidae). *Folia Zool.* 45(1):57–64.

Bentley, B. J., and Schmidt-Nielsen, K. 1966. Cutaneous water loss in reptiles. *Science* 151:1547–1549.

Berghe G., van de. 1982. Mijo ervaringen met *Anolis equestris,* Merrem, 1820. *Ned. Studies Anol.* 5(3):9–10.

Berovides Álvarez, V. 1980. Notas sobre la ecología de la iguana (*Cyclura nubila*) en Cayo del Rosario. *Cien. Biol.* 5:112–115.

———. 1989. Ecología trófica del Cernícalo Cubano *Falco sparverius sparverioides* (Aves: Falconiformes). *Rev. Biol.* 3(2):167–169.

———. 1995. Situación actual en Cuba de las especies de vertebrados en peligro de extinción. *Rev. Biol.* 9:3–13.

Berovides Álvarez, V., Genaro, J. A., and Sánchez Alonso, C. S. 1988. Nuevas consideraciones acerca del nicho ecológico. *Cien. Biol.* 19–20:3–8.

Berovides Álvarez, V., Rodríguez Schettino, L., and Cubillas Hernández, S. 1996. *Cyclura nubila nubila.* Conservation assessment and management plan for some Cuban species, 93–100. IUCN/SSC/CBSG, Apple Valley, Minnesota.

Berovides Álvarez, V., and Sampedro Marín, A. 1980. Competición en especies de lagartos iguánidos de Cuba. *Cien. Biol.* 5:115–122.

Bezy, R. L., Gorman, G. C., Kim, Y. J., and Wright, J. W. 1977. Chromosomal and genetic divergence in the fossorial lizards of the family Anniellidae. *Syst. Zool.* 26:57–71.

Bocourt, H. 1874. Miss. Scient. Mexique, Zool. Report., livr. 3. [cited by Barbour and Ramsden 1919].

Bogert, C. M. 1949. How reptiles regulate their body temperature. *Sci. Amer.* 200:105–120.

Bogert, C. M., and Cowles, R. B. 1947. Results of the Archbold Expeditions. No. 58. Moisture loss in relation to habitat selection in some Floridian reptiles. *Amer. Mus. Nov.* 1358:1–34.

Boulenger, G. A. 1885. *Catalogue of the lizards in the British Museum (Natural History),* vol. 2, 2nd ed. London: Trustees of the British Museum.

Brach, V. 1976. Structure and function of the ocular conus papillaris of *Anolis equestris* (Sauria: Iguanidae). *Copeia* 3:552–558.

Brattstrom, B. H. (1965): Body temperature of reptiles. *Amer. Midl. Nat.* 73(2):376–422.

Brown, K. M., and Sexton, O. J. 1973. Stimulation of the reproductive activity of female *Anolis sagrei* by moisture. *Physiol. Zool.* 46:168–171.

Bruner, S. C. 1941. Notas sobre dos aves cubanas. *Mem. Soc. Cubana Hist. Nat. Felipe Poey.* 15(2):157–159.

Buide, M. S. 1951. Observation on habits of the Cuban iguana. *Herpetologica* 7(3):124.

———. 1966. Reptiles de la Península de Hicacos. *Poeyana* 21:1–12.

———. 1967. Lista de los anfibios y reptiles de Cuba. *Torreia* 1:1–60.

———. 1985. *Reptiles de Cuba.* Editorial Gente Nueva, Havana.

Buide González, M. S., Fernández Milera, J., García Montaña, F., Garrido Calleja, O. H., de los Santos Izquierdo, H., Silva Taboada, G., and Varona Calvo, L. S. 1974.

Las especies amenazadas de vertebrados cubanos. Havana: Academia de Ciencias de Cuba.

Burgess, G. H., and Franz, R. 1989. Zoogeography of the Antillean freshwater fish fauna. In *Biogeography of the West Indies: Past, Present and Future* (C. A. Woods, ed.), 263–304. Gainesville, Fla.: Sandhill Crane Press.

Burnell, K. L., and Hedges, S. B. 1990. Relationships of West Indian *Anolis* (Sauria: Iguanidae): An approach using slow-evolving loci. *Carib. J. Sci.* 26(1–2):7–30.

Buth, D. G., Gorman, G. C., and Lieb, C. S. 1980. Genetic divergence between *Anolis carolinensis* and its Cuban progenitor *Anolis porcatus. J. Herpetol.* 14:279–284.

Cannatella, D. C., and de Queiroz, K. 1989. Phylogenetic systematics of the anoles: Is a new taxonomy warranted? *Syst. Zool.* 38:57–69.

Capote López, R. P., Ricardo Nápoles, N. E., González Areu, A. V., García Rivera, E. E., Vilamajó Alberdi, D., and Urbino Rodríguez, J. 1989. Vegetación actual. In *Nuevo Atlas Nacional de Cuba* (Academia de Ciencias de Cuba e Instituto Cubano de Geodesia y Cartografía, Havana), X.1.2–3 (map 1). Instituto Geográfico Nacional de España.

Carpenter, C. A. 1967. Aggression and social structure in Iguanid lizards. In *Lizard Ecology, a Symposium* (W. W. Milstead, ed.), 87–105. Columbia: University of Missouri Press.

———. 1978. Ritualistic social behaviors in lizards. In *Behavior and Neurology of Lizards* (N. Greenberg and P. D. MacLean, eds.), 253–267. Rockville, Md.: NIMH.

Case, T. J. 1983. Niche overlap and the assembly of island lizard communities. *Oikos* 41(3):427–433.

Cepero Chaviano, A. L. 1980. "Patrones electroforéticos de esterasas y proteínas totales en *Anolis allisoni* y *Anolis porcatus*" [inedit], University Students' Report, Facultad de Biología, Universidad de La Habana.

Cerny, V. 1966. Nuevas garrapatas (Ixodoidea) en aves y reptiles de Cuba. *Poeyana* 26:1–10.

———. 1967. Some results of tick investigations in Cuba. *Wiad. Parasitol.* 13:533–537.

———. 1969a. Nuevos conocimientos sobre la ixodofauna cubana. *Torreia* 21:1–12.

———. 1969b. The tick fauna of Cuba. *Fol. Parasitol.* 16:279–284.

Chamizo Lara, A., Camacho, A., Rivalta, V., Moreno, L. V., Novo, J., and González, L. 1989. Caracterización electroforética de dos especies del género *Leiocephalus*. In *Quinta Jornada Científica de las Brigadas Técnicas Juveniles*, Havana (Instituto de Ecología y Sistemática, Havana), *Resúmenes*, 16.

Chamizo Lara, A., Camacho, A., Rivalta, V., Novo, J., Moreno, L. V., and González, L. 1991. Comparación de tres especies del género *Leiocephalus* mediante el empleo de marcadores genético-bioquímicos. In *Segundo Simposio de Zoología*, Havana (Instituto de Ecología y Sistemática, Havana), *Resúmenes*, 51.

Chamizo Lara, A., Camacho, A., Torres, A., and González, L. 1989. Caracterización electroforética de dos especies del género *Leiocephalus*. In *Segunda Jornada Nacional de Museos de Ciencias*, Holguín (Museo de Historia Natural Carlos de la Torre y Huerta, Holguín), *Resúmenes*, 17.

Chamizo Lara, A., and Moreno, L. V. 1994. Caracterización motfométrica de *Leiocephalus cubensis*. In *III Simposio de Zoología*, Havana (Sociedad Cubana de Zoología, Havana), *Resúmenes*, 47–48.

Chamizo Lara, A., Rivalta, V., Torres, A., Moreno, L. V., and Daniel, A. 1994. Sistemática bioquímica de 3 especies del género *Leiocephalus* (Iguania: Tropiduridae) de la Ciénaga de Zapata. In *Memorias del II Simposio Internacional Humedales '94,* Ciénaga de Zapata, 208–210.

Chamizo Lara, A., and Rodríguez Schettino, L. 1996. Distribución de los iguánidos en la Isla de la Juventud. Géneros *Leiocephalus* y *Cyclura*. In *25 Años de Ciencias*. Nueva Gerona: Delegación Ministerio de Ciencia, Tecnología y Medio Ambiente, Cuba.

Chamizo Lara, A., and Rodríguez Schettino, L. 1998. *Leiocephalus raviceps klinikowskii*. Conservation assessment and management plan for some Cuban species. IUCN/SSC/CBSG, Apple Valley, Minn.

Christian, K. A. 1987. Aspects of the life history of Cuban iguanas on Isla Magueyes, Puerto Rico. *Carib. J. Sci.* 22(3–4):159–164.

Christian, K. A., Clavijo, I. E., Cordero-López, N., Elías-Maldonado, E. E., Franco, M. A., Lugo-Ramírez, M. V., and Marengo, M. 1986. Thermoregulation and energetics of a population of Cuban iguanas (*Cyclura nubila*) on Isla Magueyes, Puerto Rico. *Copeia* 1:65–69.

Christian, K. A., and Lawrence, W. T. 1991. Microclimatic conditions in nests of the Cuban iguana (*Cyclura nubila*). *Biotropica* 23(3):287–293.

Christian, K. A., Lawrence, W. T., and Snell, H. L. 1991. Effect of soil moisture on yolk and fat distribution in hatchling lizards from natural nests. *Comp. Biochem. Physiol.* 99(1/2):13–19.

Christian, K. A., and Torregrosa, D. 1986. Effect of diet on nitrogenous wastes of the iguana, *Cyclura nubila*. *Comp. Biochem. Physiol.* 85A(4):761–764.

Cihar, J. 1973. *Obojzivelníci a plazi*. Editorial Ludvik Schindler: Prague.

Cochran. D. M. 1928. A second species of *Deiroptyx* from Cuba. *Proc. Biol. Soc. Washington* 41:169–170.

———. 1934. Herpetological collections from the West Indies established by Dr. Paul Bartsch under the Walter Rathbone Bacon Scholarship, 1928–1930. *Smithsonian Misc. Coll.* 92(7):1–48.

———. 1941. The herpetology of Hispaniola. *Bull. U. S. Natl. Mus.* 177:1–398.

Cocteau, J. T. 1836. *Compt. Rend. Acad. Sci. Paris* 3: 226 [cited by Duméril and Bibron 1837].

Cocteau, J. T., and Bibron, G. 1837–1843. *Reptiles*. In *Hist. Fís. Pol. Nat. Isla de Cuba* 4:1–143.

Collette, B. B. 1961. Correlations between ecology and morphology in anoline lizards from Havana, Cuba, and Southern Florida. *Bull. Mus. Comp. Zool.* 125(5):137–162.

Conant, R., and Hudson, R. G. 1949. Longevity records for reptiles and amphibians in the Philadelphia Zoological Garden. *Herpetologica* 5(1):1–8.

Coney, P. J. 1982. Plate tectonic constraints on the biogeography of the Middle America and the Caribbean region. *Ann. Missouri Bot. Garden* 69:432–443.

Cooper, J. E. 1958. Ecological notes on some Cuban lizards. *Herpetologica* 14(19):53–54.

Cope, E. D. 1861. Notes and descriptions of anoles. *Proc. Acad. Nat. Sci. Philadelphia* 13:208–215.

———. 1862. Contributions to Neotropical saurology. *Proc. Acad. Nat. Sci. Philadelphia* 14:176–188.

————. 1864. Contributions on the herpetology of tropical America. *Proc. Acad. Nat. Sci. Philadelphia* 16:166–181.

————. 1900. The crocodilians, lizards and snakes of North America. *Annu. Rep. U. S. Natl. Mus. for 1898*:153–270 [cited by Frost and Etheridge 1989].

Corral, I., del. 1939. La unión de Cuba con el continente americano. *Rev. Soc. Cubana Ing.* 33:582–681.

Corzo, J., Riol-Cimas, J., and Melendes-Hevia, E. 1984. Fish species identification by electrophoresis of muscle proteins in sodium dodecyl sulfate-containing plyacrylamide slab gels. *Electrophoresis* 5:168–170.

Cowles, R. B. 1939. Possible implications of reptilian thermal tolerance. *Science* 90:465–466.

————. 1945. Surface mass ratio, paleoclimate and heat sterility. *Amer. Nat.* 79:561–567.

Cowles, R. B., and Bogert, C. M. 1944. A preliminary study of the thermal requirements of desert reptiles. *Bull. Amer. Mus. Nat. Hist.* 83:262–296.

Coy Otero, A. 1970. Contribución al conocimiento de la helmintofauna de los saurios cubanos. *Ciencias* 4(4):1–50.

————. 1976. "Nemátodos parásitos de los reptiles cubanos, sistemática, especificidad y zoogeografía" [inedit]. Ph.D. thesis, Academy of Sciences of Czechoslovakia, Brno.

————. 1979. Tres nuevos hospederos definitivos de nemátodos en Cuba. *Misc. Zool.* 8:1.

————. 1989. Nuevo género de tremátodo (Plagiorchiidae) parásito de reptiles cubanos. *Poeyana* 376:1–5.

Coy Otero, A., and Baruš, V. 1973a. Notes on nematodes of the genus *Cyrtosomum* (Atractidae) parasitic in Cuban lizards (Sauria). *Fol. Parasitol.* 20:297–305.

————. 1973b. New hosts for *Parapharyngodon cubensis* (Oxyuridae) of the families Geckonidae and Iguanidae. *Fol. Parasitol.* 20:379–380.

————. 1979a. Nematodes parasitizing Cuban reptiles. *Acta Sci. Nat. Brno* 13(2):1–43.

————. 1979b. On the species *Trichospirura teixeirai* (Nematoda: Rabdochonidae) parasitizing Cuban reptiles. *Vest. Cs. Spol. Zool.* 43:90–97.

————. 1980. Nueva especie del género *Physalopteroides* Wu et Liu, 1940 (Nematoda: Physalopteridae) en Cuba. *Poeyana* 200:1–6.

Coy Otero, A., Espinosa, J., and Rams, A. 1987. Contribución al conocimiento de la fauna de Cayo Fragoso, costa norte de Cuba. Parte 1. Parásitos. *Garciana* 4:1–3.

Coy Otero, A., and Lorenzo Hernández, N. 1982. Lista de los helmintos parásitos de los vertebrados silvestres cubanos. *Poeyana* 235:1–57.

Coy Otero, A., and Rodríguez Schettino, L. 1982. Nuevo hospedero definitivo para *Piratuba digiticauda* Lent et Freitas, 1941 (Nematoda: Splendidofilariidae) en Cuba. *Misc. Zool.* 17:3–4.

Crews, D. 1979. The hormonal control of behavior in a lizard. *Sci. Amer.* 241(2):180–187.

Crother, B. I., and Guyer, C. 1996. Caribbean historical biogeography: Was the dispersal-vicariance debate eliminated by an extraterrestrial bolide? *Herpetologica* 52(3):440–465.

Cruz, J., de la. 1978. Composición zoogeográfica de la fauna de garrapatas (Acarina: Ixodoidea) de Cuba. *Poeyana* 185:1–5.

————. 1984a. Nueva especie de garrapata del género *Ornithodoros* (Acarina, Ixodoidea, Argasidae), parásita nasal de la iguana *Cyclura nubila* (Sauria, Iguanidac) de Cuba. *Poeyana* 277:1–6.

————. 1984b. Sistemática de la familia Pterygosomidae (Acarina, Prostigmata), con la descripción de un nuevo género y especie. *Poeyana* 278:1–22.

Cruz, J., de la, and Abreu, R. M. 1986. El abuje o abujo, *Eutrombicula alfreddugesi* (Oudemans, 1910) (Acarina: Trombiculidae), un ácaro de importancia biomédica en las provincias orientales de Cuba. *Rev. Cubana Med. Trop.* 38(1):119–125.

Cruz, J., de la, and Daniel, M. 1991. Incidencia de abujes (Acarina: Trombiculidae) en vertebrados autóctonos de Cuba. In *Segundo Simposio de Zoología*, Havana (Instituto de Ecología y Sistemática, Havana), *Resúmenes*, 108.

Cubillas Hernández, S., and Berovides Álvarez, V. 1991. Características de los refugios de la iguana de Cuba. *Cyclura nubila. Rev. Biol.* 5(1):85–87.

Cubillas Hernández, S., and Polo, J. L. 1997. Introducción de *Leiocephalus carinatus* en el Parque Zoológico Nacional. In *Cuarto Simposio de Zoología*, Havana, *Resúmenes*, 80.

Darlington, P. J., Jr. 1957. *Zoogeography: The geographical distribution of animals.* New York: Wiley and Sons.

de Queiroz, K. 1982. The scleral ossicles of sceloporine iguanids: A reexamination with comments on their phylogenetic significance. *Herpetologica* 38(2):302–11.

————. 1987. Phylogenetic systematics of iguanine lizards. A comparative osteological study. *Univ. California Publ. Zool.* 118:203.

De Smet, W. H. O. 1981. Description of the orcein stained karyotypes of 27 lizard species (Lacertilia Reptilia) belonging to the families Iguanidae, Agamidae, Chamaleontidae and Gekkonidae (Ascalabota). *Acta Zool. Pathol. Antverpiensia* 76:35–72.

Díaz, L. M., Estrada, A. R., and Moreno, L. V. 1996. A new species of *Anolis* (Sauria: Iguanidae) from the Sierra de Trinidad, Sancti Spíritus, Cuba. *Carib. J. Sci.* 32(1):54–58.

Díaz Castillo, R., González Leiva, N., Zayas Pérez, N., and Leiva Camejo, Y. 1991. Sinopsis preliminar de tetrápodos de la provincia Las Tunas. In *Segundo Simposio de Zoología*, Havana (Instituto de Ecología y Sistemática, Havana), *Resúmenes*, 52.

Donnelly, T. W. 1985. Mesozoic and Cenozoic plate evolution of the Caribbean region. In *The Great American Biotic Interchange* (F. G. Stehli and S. D. Webb, eds.), 89–121. New York: Plenum Press.

Dubois, G., and Macko, J. 1972. Contribution a l'etude de Strigeata La Rue 1926 (Trematoda: Strigeida) de Cuba. *Ann. Parasitol. Humana Comp.* 47:51–75.

Duellman, W. E. 1978. The biology of an equatorial herpetofauna in Amazonian Ecuador. *Univ. Kansas Mus. Nat. Hist. Misc. Publ.* 65:1–352.

————. 1987. Lizards in an Amazonian rain forest community: Resource utilization and abundance. *Natl. Geogr. Research* 3(4):489–500.

Duméril, A. M. C., and Bibron, G. 1837. *Erpétologie générale, ou histoire naturelle complète des Reptiles 4.* Paris: Librarie Encyclopédique de Roret.

Dunn, E. R. 1920. A new lizard from Haití. *Proc. New England Zool. Club* 7:33–34.

————. 1926. Notes on Cuban anoles. *Copeia* 157:153–154.

Ehrig, R. W. 1993. The captive husbandry and propagation of the Cuban rock iguana, *Cyclura nubila*. *AAZPA Regional Proc.* Southern Regional Conference, American Association of Zoological Parks and Aquariums, Lake Monroe, Fla.

Espinosa López, G. 1981. Patrones electroforéticos de miógeno en especies de *Anolis* (Sauria: Iguanidae). In *Primer Congreso Nacional de Ciencias Biológicas*, Havana (Sociedad Cubana de Ciencias Biológicas, Havana), *Resúmenes*, 177.

———. 1989. "Polimorfismo bioquímico en especies del género *Anolis* (Sauria: Iguanidae). Aspectos sistemáticos y su relación con el ambiente" [inedit]. Ph.D. thesis, Facultad de Biología, Universidad de La Habana, Havana.

Espinosa López, G., Alonso Biosca, M. E., Berovides Álvarez, V., and Cepero, M. L. 1979. Patrones electroforéticos de esterasas y proteínas totales de cinco especies del género *Anolis*. In *Segundo Evento Científico de la Universidad de La Habana*, Havana (Universidad de La Habana), *Resúmenes*, 313.

Espinosa López, G., Benítez, E., Berovides, V., and Alonso, M. E. 1985. Polimorfismo genético de esterasas de músculo en poblaciones de *Anolis sagrei* (Sauria: Iguanidae). *Cien. Biol.* 14:81–90.

Espinosa López, G., Cárdenas Cepero, Y., and Berovides Álvarez, V. 1983. Desplazamiento de caracteres a través del polimorfismo bioquímico de miógeno en dos especies de lagartos (*Anolis*). *Cien. Biol.* 10:121–123.

Espinosa López, G., González Suárez, C., and Berovides Álvarez, V. 1985. Albúmina de suero y esterasas de músculo en poblaciones de *Anolis porcatus* Gray, 1840, del occidente de Cuba. *Carib. J. Sci.* 21:55–62.

Espinosa López, G., Hernández, I., and García, E. 1991. Variabilidad y distancia genética entre especies de la serie *lucius* del género *Anolis*. In *Segundo Simposio de Zoología*, Havana (Instituto de Ecología y Sistemática, Havana), *Resúmenes*, 111.

Espinosa López, G., Menéndez Alarcón, A., and Berovides Álvarez, V. 1987. Relación entre la amplitud del nicho y los patrones electroforéticos de proteínas plasmáticas en tres especies de *Anolis*. *Cien. Biol.* 17:77–84.

Espinosa López, G., Posada García, A., Berovides Álvarez, V., and Alonso Biosca, M. E. 1990. Relaciones entre algunas especies de lagartos del género *Anolis* sobre la base de la movilidad de las bandas de miógeno. *Rev. Biol.* 4(2):149–156.

Espinosa López, G., Sosa Espinosa, A., and Berovides Álvarez, V. 1990. Relaciones filogenéticas entre especies del género *Anolis* sobre la base de ocho loci de polimorfismo bioquímico. *Rev. Biol.* 4(2):133–148.

Espinosa López, G., Vargas, D., Orúe, G., and Berovides Álvarez, V. 1986. Relaciones genético-bioquímicas entre *Anolis porcatus* y *Anolis allisoni*. In *Quinta Conferencia Científica de Ciencias Naturales*, Havana (Universidad de La Habana), *Resúmenes*, 43.

Estes, R. 1963. Early Miocene salamanders and lizards from Florida. *Quart. J. Florida Acad. Sci.* 26(3):234–256.

———. 1983. The fossil record and early distribution of lizards. In *Advances in Herpetology and Evolutionary Biology* (A. G. Rhodin and K. Miyata, eds.), 365–398. Museum of Comparative Zoology. Cambridge: Harvard University Press.

Estrada, A. R. 1984. Subnicho trófico de *Anolis homolechis* y *A. allogus* (Sauria: Iguanidae) en un bosque de la Sierra del Rosario, Pinar del Río, Cuba. In *Segundo Congreso Nacional de Ciencias Biológicas*, Havana (Universidad de La Habana), *Resúmenes*, 452.

————. 1987*a*. Sintopía de *Anolis argenteolus* y *A. lucius* en la Sierra de Najasa, Camaguey, Cuba. *Misc. Zool.* 31:3–4.

————. 1987*b*. *Anolis argillaceus* (Sauria: Iguanidae): Un nuevo caso de puestas comunales en *Anolis* cubanos. *Poeyana* 353:1–9.

————. 1988. Nido comunal de *Anolis argenteolus* (Sauria: Iguanidae). *Garciana* 9:3–4.

Estrada, A. R., Alayón, G., and Pérez Asso, A. 1987. Lista preliminar de anfibios y reptiles de las Cuchillas de Toa y Moa, Cuba. *Garciana* 8:3–4.

Estrada, A. R., and Garrido, O. H. 1990. Nueva subespecie de *Anolis jubar* (Lacertilia: Iguanidae) para Cayo Coco y la Loma de Cunagua, Ciego de Avila, Cuba. *Rev. Biol.* 4(19):71–79.

————. 1991. Dos nuevas especies de *Anolis* (Lacertilia: Iguanidae) de la región oriental de Cuba. *Carib. J. Sci.* 27(3–4):146–161.

Estrada, A. R., and Hedges, S. B. 1995. A new species of *Anolis* (Sauria: Iguanidae) from eastern Cuba. *Carib. J. Sci.* 31(1–2):65–72.

Estrada, A. R., and Novo Rodríguez, J. 1984*a*. Reptiles y aves de Cayo Inés de Soto, Archipiélago de Los Colorados, Pinar del Río, Cuba. *Misc. Zool.* 23:1.

————. 1984*b*. Ciclo reproductivo y puestas comunales de *Anolis bartschi* (Sauria: Iguanidae). In *Segunda Jornada Científica de la Sociedad Cubana de Ciencias Biológicas, Sección de Zoología*, Havana (Sociedad Cubana de Ciencias Biológicas, Havana).

————. 1986*a*. Subnicho estructural de *Anolis sagrai* en Cayo Inés de Soto, Cuba. Análisis intra-y extrapoblacional. *Poeyana* 320:1–13.

————. 1986*b*. Nuevos datos sobre las puestas comunales de *Anolis bartschi* (Sauria: Iguanidae) en la Sierra de los Organos, Pinar del Río, Cuba. *Cien. Biol.* 15:135–136.

————. 1986*c*. Subnicho estructural de *Anolis bartschi* en la Sierra de los Organos, Pinar del Río, Cuba. *Poeyana* 316:1–10.

————. 1987. Subnicho climático de *Anolis bartschi* (Sauria: Iguanidae). *Poeyana* 341:1–19.

Estrada, A. R., and Rodríguez, R. 1985. Lista de vertebrados terrestres de Cayo Campos, Archipiélago de los Canarreos, Cuba. *Misc. Zool.* 27:2–3.

Estrada. A. R., and Silva Rodríguez, A. 1982. Aspectos ecológicos de una población de *Anolis angusticeps* en el Bosque de La Habana. In *Primera Jornada Científica del Instituto de Zoología*, Havana (Instituto de Zoología, Havana), *Resúmenes*, 7.

————. 1984. Análisis de la ecomorfología de 23 especies de lagartos cubanos del género *Anolis. Cien. Biol.* 12:91–104.

Etheridge, R. 1960. "The relationships of the anoles (Reptilia: Sauria: Iguanidae). An interpretation based on skeletal morphology" [inedit]. Ph.D. diss., University of Michigan, Microfilms 60-2529. Ann Arbor.

————. 1964. The skeletal morphology and systematic relationships of sceloporine lizards. *Copeia* 1964:610–631.

————. 1965. Fossil lizards from the Dominican Republic. *Quart. J. Florida Acad. Sci.* 28(1):83–105.

————. 1966*a*. Pleistocene lizards from New Providence. *Quart. J. Florida Acad. Sci.* 28(4)[1965]:349–358.

————. 1966*b*. An extinct lizard of the genus *Leiocephalus* from Jamaica. *Quart. J. Florida Acad. Sci.* 29(1):47–59.

———. 1966c. The systematic relationships of West Indian and South American lizards referred to the iguanid genus *Leiocephalus. Copeia* (1):79–91.

———. 1967. Lizard caudal vertebrae. *Copeia* 1967:699–721.

Etheridge, R., and de Queiroz, K. 1988. A phylogeny of Iguanidae. In *Phylogenetic relationships of the lizard families. Essays commemorating Charles L. Camp.* (R. Estes and G. Pregill, eds.), 283–367. Stanford, Calif.: Stanford University Press.

Evans, L. T. 1938. Cuban field studies on territoriality of the lizard *Anolis sagrei. J. Comp. Psich.* 25:97–125.

Fernández Méndez, I. 1997a. *Anolis pigmaequestris.* Conservation assessment and management plan for some Cuban species. IUCN/SSC/CBSG, Apple Valley, Minn.

———. 1997b. *Anolis equestris* ssp. Conservation assessment and management plan for some Cuban species. IUCN/SSC/CBSG, Apple Valley, Minn.

Fernández Méndez, I., Alonso, O., Rodríguez Schettino, L., and Chamizo Lara, A. R. 1997. *Anolis equestris potior.* Conservation assessment and management plan for some Cuban species. IUCN/SSC/CBSG, Apple Valley, Minn.

Fitch, H. S. 1940. A field study of the growth and behavior of the fence lizard. *Univ. California Publ. Zool.* 44(2):151–172.

———. 1973. Population structure and survivorship in some Costa Rican lizards. *Occ. Papers Mus. Nat. Hist. Univ. Kansas* 18:1–41.

———. 1975. Sympatry and interrelationships in Costa Rican anoles. *Occ. Papers Mus. Nat. Hist. Univ. Kansas* 40:1–60.

Fleming, T. H., and Hooker, R. S. 1975. *Anolis cupreus:* The response of a lizard to tropical seasonality. *Ecology* 56:1243–1261.

Floyd, H. B., and Jenssen, T. A. 1983. Food habits of the Jamaican lizard *Anolis opalinus:* Resource partitioning and seasonal effects examined. *Copeia* 2:319–331.

Fong, A., and del Castillo, E. B. 1997. Consideraciones preliminares sobre los reptiles de la Reserva Natural de Siboney, Santiago de Cuba. In *Cuarto Simposio de Zoología,* La Habana, *Resúmenes,* 86.

Frost, D. R., and Etheridge, R. 1989. Phylogenetic analysis and taxonomy of iguanian lizards (Reptilia: Squamata). *Univ. Kansas Mus. Nat. Hist. Misc.* Publ. 81:1–65.

———. 1993. A consideration of Iguanian lizards and the objectives of Systematics: A reply to Lazell. *Herpetol. Rev.* 24:50–54.

Furrazola, G., Judoley, C., Mijailovskaia, M. I., Miroliubov, Y. S., Novojatsky, I. P., Núñez Jiménez, A., and Solsona, J. B. 1964. *Geología de Cuba.* Havana: Editorial Nacional de Cuba.

García Avila, I., Gutsevich, A. V., and González Broche, R. 1969. Nuevos datos sobre la familia Phlebotomidae en Cuba. *Torreia* 14:1–7.

García de Francisco, D., González, B., and Milán, G. 1984a. Distribución de diferentes tipos de hepatocitos en el hígado de dos especies del género *Anolis,* según la etapa del desarrollo. In *Segundo Congreso Nacional de Ciencias Biológicas,* Havana (Universidad de La Habana), *Resúmenes,* 464.

García de Francisco, D., López, M., and Milán, G. 1984b. Histología y morfometría de la mucosa gástrica de la especie *Anolis sagrei,* en dos etapas del desarrollo. In

Segundo Congreso Nacional de Ciencias Biológicas, Havana (Universidad de La Habana), *Resúmenes*, 465.

García Rodríguez, N. 1989. "Relaciones ecológicas entre especies de saurios en la manigua costera del Jardín Botánico Nacional" [inedit]. University Students' Report, Facultad de Biología, Universidad de La Habana.

Garman, S. 1887. On West Indian Iguanidae and Scincidae. *Bull. Essex Inst.* 19:1–29.

Garrido, O. H. 1967. Sobre el *Anolis rubribarbus* (Sauria: Iguanidae) en Cuba. *Trab. Divulg. Mus. Felipe Poey* 55:1–6.

———. 1972. *Anolis bremeri* Barbour (Lacertilia: Iguanidae) en el occidente de Cuba e Isla de Pinos. *Carib. J. Sci.* 12(1–2):59–77.

———. 1973*a*. Nueva especie de *Leiocephalus* (Lacertilia: Iguanidae) para Cuba. *Poeyana* 116:1–19.

———. 1973*b*. Anfibios, reptiles y aves de Cayo Real (Cayos de San Felipe), Cuba. *Poeyana* 119:1–50.

———. 1973*c*. Distribución y variación de *Anolis homolechis* Cope (Lacertilia: Iguanidae) en Cuba. *Poeyana* 120:1–68.

———. 1973*d*. Anfibios, reptiles y aves del Archipiélago de Sabana-Camagüey, Cuba. *Torreia* 27:1–72.

———. 1973*e*. Nuevas subespecies de reptiles para Cuba. *Torreia* 30:1–31.

———. 1973*f*. Lista de los anfibios, reptiles, aves y mamíferos colectados en el plan Jibacoa-Cayajabos. Informe del trabajo faunístico realizado en el plan Jibacoa-Cayajabos. *Ser. Biol.* 43:16–25.

———. 1975*a*. Nuevos reptiles del Archipiélago Cubano. *Poeyana* 141:1–58.

———. 1975*b*. Distribución y variación de *Anolis argillaceus* Cope (Lacertilia: Iguanidae) en Cuba. *Poeyana* 142:1–28.

———. 1975*c*. Distribución y variación del complejo *Anolis cyanopleurus* (Lacertilia: Iguanidae) en Cuba. *Poeyana* 143:1–60.

———. 1975*d*. Variación de *Anolis angusticeps* Hallowell (Lacertilia: Iguanidae) en el occidente de Cuba y en la Isla de Pinos. *Poeyana* 144:1–18.

———. 1976*a*. Aves y reptiles de Cayo Coco, Cuba. *Misc. Zool.* 3:4.

———. 1976*b*. Nota sobre *Deiroptyx vermiculatus* Duméril y Bibron (Lacertilia: Iguanidae). *Misc. Zool.* 4:1–2.

———. 1979. Nuevas subespecies de *Leiocephalus macropus* Cope (Lacertilia: Iguanidae) para Cuba. *Poeyana* 188:1–16.

———. 1980*a*. Revisión del complejo *Anolis alutaceus* (Lacertilia: Iguanidae) y descripción de una nueva especie de Cuba. *Poeyana* 201:1–41.

———. 1980*b*. Los vertebrados terrestres de la Península de Zapata. *Poeyana* 203: 1–49.

———. 1981. Nueva subespecie de *Anolis equestris* (Sauria: Iguanidae) para Cuba, con comentarios sobre la distribución y afinidades de otras poblaciones del complejo. *Poeyana* 232:1–15.

———. 1982*a*. Descripción de una especie cubana de *Chamaeleolis* (Lacertilia: Iguanidae), con notas sobre su comportamiento. *Poeyana* 236:1–25.

———. 1982*b*. Nueva especie de *Anolis* (Lacertilia: Iguanidae) para Cuba. *Doñana, Acta Vertebrata* 9:131–137.

———. 1983. Nueva especie de *Anolis* (Lacertilia: Iguanidae) de la Sierra del Turquino, Cuba. *Carib. J. Sci.* 19(3–4):71–76.

———. 1985. Nueva subespecie de *Anolis isolepis* (Lacertilia: Iguanidae) para Cuba. *Doñana, Acta Vertebrata* 12(1):41–49.

———. 1988. Nueva especie para la ciencia de *Anolis* (Lacertilia: Iguanidae) de Cuba perteneciente al complejo *argillaceus*. *Doñana, Acta Vertebrata* 15(1):45–47.

———. 1990. Nueva especie de *Anolis* de la sección Beta (Lacertilia: Iguanidae) para Cuba. *Rev. Biol.* 4(2):157–162.

———. 1994. Dispersión de la bayoya *Leiocephalus cubensis* (Reptilia: Tropiduridae: Leiocephalinae) en la Península de Zapata, inducida por causas antrópicas. In *Tercer Simposio de Zoología*, Havana, *Resúmenes*, 54.

Garrido, O. H., and Estrada, A. R. 1989. Nueva especie del complejo *Anolis alutaceus* (Lacertilia: Iguanidae) para Cuba. *Rev. Biol.* 3(1):57–66.

Garrido, O. H., Estrada, A. R., and Llanes, A. 1986. Anfibios, reptiles y aves de Cayo Guajaba, Archipiélago de Sabana-Camagüey, Cuba. *Poeyana* 328:1–34.

Garrido, O. H., and Hedges, S. B. 1992. Three new grass anoles from Cuba (Squamata: Iguanidae). *Carib. J. Sci.* 28(1–2):21–29.

Garrido, O. H., and Jaume, M. L. 1984. Catálogo descriptivo de los anfibios y reptiles de Cuba. *Doñana, Acta Vertebrata* 11(2):5–128.

Garrido, O. H., and Pareta, L. 1994. Puestas comunales de *Anolis allisoni* (Lacertilia: Iguanidae) en Cuba. *Rev. Biol.* 8:145–147.

Garrido, O. H., Pérez-Beato, O., and Moreno, L. V. 1991. Nueva especie de *Chamaeleolis* (Lacertilia: Iguanidae) para Cuba. *Carib. J. Sci.* 27(3–4):162–168.

Garrido, O. H., and Schwartz, A. 1967. Cuban lizards of the genus *Chamaeleolis*. *Quart. J. Florida Acad. Sci.* 30(3):197–220.

———. 1968. Anfibios, reptiles y aves de la Península de Guanahacabibes, Cuba. *Poeyana* 53:1–68.

———. 1969. Anfibios, reptiles y aves de Cayo Cantiles. *Poeyana* 67:1–44.

———. 1972. The Cuban *Anolis spectrum* complex (Sauria: Iguanidae). *Proc. Biol. Soc. Washington* 85(45):509–522.

Gilles-Baillien, M. 1981. Osmoregulation in reptiles. *Acta Zool. Pathol. Antverpiensia* 76:29–33.

González, A., Berovides Álvarez, V., and Castañeira, M. 1995. Variación espacio-temporal de la densidad de *Cyclura nubila nubila* (Sauria, Iguanidae) en Cayo del Rosario, Archipiélago de los Canarreos, Cuba. In *Bioeco, Segundo Taller de Biodiversidad*, Santiago de Cuba (Centro Oriental de Ecosistemas y Biodiversidad), *Resúmenes*, 13.

González Bermúdez, F., and Rodríguez Schettino, L. 1982. Datos etoecológicos sobre *Anolis vermiculatus* (Sauria: Iguanidae). *Poeyana* 245:1–18.

González González, O. 1989. *Las biocenosis de las Escaleras de Jaruco y áreas cercanas, Cuba.* Havana: Editorial Academia.

González Grau, A., Manójina, N., Valdés Lafont, O., and Hernández Marrero, A. 1989. Reporte de una cueva de calor en la Reserva de la Biosfera Sierra del Rosario. *Misc. Zool.* 43:3–4.

González Martínez, B. 1981. Estudio de la histoestructura del hígado en dos especies del género *Anolis*. In *Décima Jornada Científica Estudiantil*. Havana: Universidad de La Habana.

———. 1982. "Particularidades histofisiológicas del hígado de dos especies del género *Anolis* en dependencia de la edad" [inedit]. University Students' Report, Facultad de Biología, Universidad de La Habana.

González Suárez, C. 1980. Patrones electroforéticos de proteínas en cuatro especies del género *Anolis* (Sauria: Iguanidae). In *Novena Jornada Científica Estudiantil*, Havana (Universidad de La Habana), *Resúmenes*, 102.

Gorman, G. C. 1965. Interespecific karyotypic variation as a systematic character in the genus *Anolis* (Sauria: Iguanidae). *Nature* 208:95–97.

Gorman, G. C., and Atkins, L. 1966. Chromosomal heteromorphism in some male lizards of the genus *Anolis*. *Amer. Nat.* 100:579–583.

———. 1967. The relationships of *Anolis* of the *roquet* species group (Sauria: Iguanidae). II. Comparative chromosome cytology. *Syst. Zool.* 16:137–143.

———. 1968. New karyotypic data on 16 species of *Anolis* from Cuba, Jamaica and the Cayman Islands. *Herpetologica* 24(1):13–21.

Gorman, G. C., Atkins, L., and Holzinger, T. 1967. New karyotypic data on 15 genera of lizards in the family Iguanidae, with a discussion of taxonomic and cytological implications. *Cytogenetics* 6:286–299.

Gorman, G. C., Buth, D. G., Soulé, M., and Yang, S. Y. 1980. Relationships of the *Anolis cristatellus* species group: Electrophoresis analysis. *J. Herpetol.* 14(3):269–278.

———. 1983. The relationships of the Puerto Rican *Anolis*. Electrophoretic and karyotypic studies. In *Advances in Herpetology and Evolutionary Biology: Essays in Honor to Ernest E. Williams* (A. G. J. Rhodin and K. Miyata, eds.), 626–642. Museum of Comparative Zoology. Cambridge: Harvard University Press.

Gorman, G. C., Buth, D. G., and Wyles, J. S. 1980. *Anolis* lizards of the Eastern Caribbean: A case study in evolution. III. A cladistic analysis of albumin inmunological data, and the definition of species groups. *Syst. Zool.* 29(2):143–158.

Gorman, G. C., Huey, R. B., and Williams, E. E. 1969. Cytotaxonomic studies on some unusual iguanid lizards assigned to the genera *Chamaeleolis*, *Polychrus*, *Polichroides*, and *Phenacosaurus* with behavioral notes. *Breviora* 316:1–17.

Gorman, G. C., and Kim, Y. J. 1976. *Anolis* lizards of the Eastern Caribbean: A case study in evolution. II. Genetic relationships and genetic variation of the *bimaculatus* group. *Syst. Zool.* 25:62–77.

Gorman, G. C., and Licht, P. 1974. Seasonality in ovarian cycles among tropical lizards. *Ecology* 55:360–369.

———. 1975. Differences between the reproductive cycles of sympatric *Anolis* lizards on Trinidad. *Copeia*:332–337.

Gorman, G. C., Lieb, C. S., and Hardwood, R. H. 1984. The relationships of *Anolis gadovi*: Albumin immunological evidence. *Carib. J. Sci.* 20:145–152.

Granda Martínez, M. A. 1987. "Determinación de un coeficiente de arboricidad en especies cubanas del género *Anolis* (Sauria: Iguanidae)" [inedit]. University Students' Report, Facultad de Biología, Universidad de La Habana.

Gray, J. E. 1827. A description of a new genus and some new species of saurian reptiles, with a revision of the species of chamaeleons. *Phil. Mag.* 2(2):207–209.

———. 1831. *A synopsis of the species of the Class Reptilia. Appendix to E. Griffith, Cuvier's Animal Kingdom.* London: Whittaker, Treacher.

———. 1840. Catalogue of the species of reptiles collected in Cuba by W. S. Mac Leay, Esq.; with some notes on their habits extracted from his MS. *Ann. Mag. Nat. Hist.* 1(5):108–115.

Groschaft J., and del Valle, M. T. 1968. Tremátodos de los murciélagos de Cuba. *Torreia* 18:1–20.

Guerra, F. 1986. El monstruo más tímido del mundo. *Bohemia* 78(3):3–7.

Gundlach, J. C. 1867. Revista y catálogo de los reptiles cubanos. In *Repertorio Físico Natural de la Isla de Cuba* 2:102–119.

———. 1880. *Contribución a la Erpetología Cubana.* Impresora G. Montiel, La Habana.

Guyer, C., and Savage, J. M. 1986. Cladistic relationships among anoles (Sauria: Iguanidae). *Syst. Zool.* 35(4):509–531.

Hadley, C. E. 1929. Color changes in two Cuban lizards. *Bull. Mus. Comp. Zool.* 69(5):107–114.

———. 1931. Color changes in excised and intact reptilian skin. *J. Exp. Zool.* 58:321–331.

Hallowell, E. 1856. Notes on the reptiles in the collection of the Museum of the Academy of Natural Sciences. *Proc. Acad. Nat. Sci. Philadelphia* 8:146–153.

Hardy, J. D., Jr. 1956. Notes on the Cuban iguana. *Herpetologica* 12(4):323–324.

———. 1957. Observations on the life history of the Cuban lizard *Anolis lucius. Herpetologica* 13(3):241–245.

———. 1958a. A geographic variant gradient in the Cuban lizard *Leiocephalus macropus* Cope. *Herpetologica* 13(4):275–276.

———. 1958b. A new lizard of the genus *Leiocephalus* from Cuba (Squamata: Iguanidae). *J. Washington Acad. Sci.* 48(9):294–300.

———. 1958c. Tail prehension and related behavior in nine New World lizards. *Herpetologica* 14(4):205–206.

———. 1966. Geographic variation in the West Indian lizard, *Anolis angusticeps,* with the description of a new form, *Anolis angusticeps paternus,* from the Isle of Pines, Cuba (Reptilia: Iguanidae). *Carib. J. Sci.* 6(1–2):23–31.

Harris, H. 1966. Enzyme polymorphisms in man. *Proc. Royal Soc.* Ser. 164:298–310.

Hass, C. A. 1991. Evolution and biogeography of West Indian *Sphaerodactylus* (Sauria: Gekkonidae): a molecular approach. *J. Zool. London* 225:525–561.

Hass, C. A., Hedges, S. B., and Maxson, L. R. 1993. Molecular insights into the relationships and biogeography of West Indian anoline lizards. *Biochem. Syst. Ecol.* 21(1):97–114.

Hechevarría, G., Molinea, M., Otero, L., Padrón, M., Paneque, Y. Pérez, A., and Reyes, T. 1990. "Estructura de la comunidad de saurios y algunos aspectos ecológicos sobre *Anolis homolechis* en El Narigón" [inedit]. University Students' Report, Facultad de Biología, Universidad de La Habana.

Hedges, S. B. 1982. Caribbean biogeography: Implications of recent plate tectonics studies. *Syst. Zool.* 31:518–522.

———. 1989. Evolution and biogeography of West Indian frogs of the genus

Eleutherodactylus: slow-evolving loci and the major groups. In *Biogeography of the West Indies: Past, Present and Future* (C. A. Woods, ed.), 305–369. Gainesville, Fla.: Sandhill Crane Press.

———. 1996*a*. Vicariance and dispersal in Caribbean biogeography. *Herpetologica* 52(3):466–473.

———. 1996*b*. The origin of West Indian amphibians and reptiles. In *Contribution to West Indian Herpetology: A Tribute to Albert Schwartz* (R. Powell and R. W. Henderson, eds.). Ithaca, N.Y.: Society for the Study of Amphibians and Reptiles. *Contributions to Herpetology* 12:95–128.

———. 1996*c*. Historical biogeography of West Indian vertebrates. *Annu. Rev. Ecol. Syst.* 27:163–196.

Hedges, S. B., Hass, C. A., and Maxson, L. R. 1992. Caribbean biogeography: Molecular evidence for dispersal in West Indian terrestrial vertebrates. *Proc. Natl. Acad. Sci. USA* 89:1009–1013.

Hedrick, P. W. 1971. A new approach to measuring genetic similarity. *Evolution* 25:276–280.

Henderson, R. W., and Crother, B. I. 1989. Biogeographic patterns of predation in West Indian colubrid snakes. In *Biogeography of the West Indies. Past, Present and Future* (C. A. Woods, ed.), 479–517. Gainesville, Fla.: Sandhill Crane Press.

Hertz, P. E. 1980. Response to dehydration in *Anolis* lizards sampled along altitudinal transects. *Copeia* 3:440–446.

Hillman, S. S., and Gorman, G. C. 1977. Water loss, desiccation and survival under desiccating conditions in 11 species of Caribbean *Anolis*. Evolutionary and ecological implications. *Oecologia* 29:105–116.

Huey, R. B. 1974. Behavioral thermoregulation in lizards: Importance of associated costs. *Science* 184:1001–1003.

Huey, R. B., and Pianka, E. R. 1981. Ecological consequences of foraging mode. *Ecology* 62:991–999.

Huey, R. B., and Slatkin, M. 1976. Costs and benefits of lizard thermoregulation. *Quart. Rev. Biol.* 51(3):363–384.

Hunter, R. C., and Markert, C. 1957. Histochemical demonstration of enzymes separated by starch gel. *Science* 125:294.

I.C.G.C. 1978. Mapa geográfico general. In *Atlas de Cuba* 106–143. Havana: Instituto Cubano de Geodesia y Cartografía.

Iturralde-Vinent, M. A. 1975. Problemas en la aplicación de dos hipótesis tectónicas modernas a Cuba y la región Caribe. *Rev. Tecnol.* 13(1):46–63.

———. 1981. Nuevo modelo interpretativo de la evolución geológica de Cuba. *Cien. Tierra Espacio* 3:51–90.

———. 1982. Aspectos geológicos de la biogeografía de Cuba. *Cien. Tierra Espacio* 5:85–100.

———. 1988. *Naturaleza geológica de Cuba*. Havana: Editorial Científico-Técnica.

Iturralde-Vinent, M. A., Hubbell, G., and Rojas, R. 1996. Catalogue of Cuban fossil Elasmobranchii (Paleocene to Pliocene) and paleogeographic implications of their Lower to Middle Miocene occurrence. *J. Geol. Soc. Jamaica* 31:7–21.

Iverson, J. B. 1978. The impact of feral cats and dogs on population of the West Indian rock iguana *Cyclura carinata*. *Biol. Cons.* 14:63–73.

———. 1979. Behavior and ecology of the rock iguana, *Cyclura carinata*. *Bull. Florida State Mus. Biol. Sci.* 24(3):175–358.

Jenssen, T. A. 1970. The ethoecology of *Anolis nebulosus* (Sauria: Iguanidae). *J. Herpetol.* 4(1–2):1–38.

———. 1978. Display diversity in anoline lizards and problems of interpretation. In *Behavior and Neurology of Lizards* (N. Greenberg and P. D. MacLean, eds.), 269–285. Rockville, Md.: NIMH.

Jenssen, T. A., and Hover, E. L. 1976. Display analysis of the signature display of *Anolis limifrons* (Sauria: Iguanidae). *Behaviour* 57(3–4):227–240.

Jiménez, H. 1981. Reporte de *Leishmania* sp. (Zoomastigophorea: Kinetoplastida) en *Anolis lucius* en Cuba. In *Primer Congreso Nacional de Ciencias Biológicas,* La Habana (Sociedad Cubana de Ciencias Biológicas, La Habana), *Resúmenes,* 431.

Joglar, R. L. 1989. Phylogenetic relationships of the West Indian frogs of the genus *Eleutherodactylus:* A morphological analysis. In *Biogeography of the West Indies: Past, Present and Future* (C. A. Woods, ed.), 371–408. Gainesville, Fla.: Sandhill Crane Press.

Johnson, G. 1974. Enzyme polymorphism and metabolism. *Science* 184:28–37.

Johnson, M. L. 1946. Herpetological notes from the West Indies. *Copeia* 1:50–51.

Judoley, C., and Meyerhoff, A. A. 1971. Paleogeography and geological history of Greater Antilles. *G. S. A. Mem.* 129:1–199.

Keiraus, J. E. 1985. *Amblyomma antillarum* Kohls, 1969 (Acari: Ixodoidea): Description of the inmature stages from the rock iguana, *Iguana pinguis* (Sauria: Iguanidae) in the British Virgin Islands. *Proc. Entomol. Soc. Washington* 87(4):821–825.

Kimura, M. 1968. Evolutionary rate at the molecular level. *Nature* 217:624–626.

Kirckconnell, A., and Posada, R. M. 1987. Algunas observaciones sobre conducta alimentaria de *Tyrannus caudifasciatus* (Aves: Passeriformes: Tyrannidae). *Misc. Zool.* 31:4.

Koopman, K. F., and Ruibal, R. 1955. Cave fossil vertebrates from Camagüey, Cuba. *Breviora* 46:1–8.

Kreitman, M. 1991. Detecting selection at the level of DNA. In *Evolution at the Molecular Level* (R. Selander, A. G. Clark, and T. S. Whittam, eds.) Sinauer Associates, Sunderland, Mass.

Ladd, J. W. 1976. Relative motion of South America and Caribbean tectonics. *Bull. Geol. Soc. America* 87:969–976 [cited by Perfit and Williams 1989].

Lambert, J. B., Frye, J. S., and Poinar, G. O., Jr. 1985. Amber from Dominican Republic: Analysis by nuclear magnetic resonance spectroscopy. *Archaeometry* 27:43–51. [cited by Williams 1989*a*].

Lando, R. V., and Williams, E. E. 1969. Notes on the herpetology of the U.S. Naval Base at Guantánamo Bay, Cuba. *Studies Fauna Curacao Carib. Islands* 31(116):159–201.

Lazell, J. D., Jr. 1992. The family Iguanidae: Disagreement with Frost and Etheridge (1989). *Herpetol. Rev.* 23:109–112.

Leal Díaz, H., and Morales Palmero, M. 1991. Cariotipo de *Leiocephalus cubensis* (Sauria: Iguanidae). Identificación de banda C y de la región del organizador nucleolar (NoRS). In *Segundo Simposio de Zoología,* Havana (Instituto de Ecología y Sistemática, Havana), *Resúmenes,*113.

Leal Díaz, H., Morales Palmero, M., and Cabrera, M. E. 1991. Análisis del patrón de banda C y de la región del organizador nucleolar (NORs) en dos especies de la familia Iguanidae (*A. porcatus, A. allisoni*). In *Segundo Simposio de Zoología,* Havana (Instituto de Ecología y Sistemática, Havana), *Resúmenes,*113.

Leclair, R., Jr. 1978. Water loss and microhabitats in three sympatric species of lizards (Reptilia, Lacertilia) from Martinique, West Indies. *J. Herpetol.* 12(2):177–182.

Lee, J. C. 1980. An ecogeographic analysis of the herpetofauna of the Yucatán Península. *Univ. Kansas Mus. Nat. Hist. Misc. Publ.* 67:1–75.

Lewontin, R. 1974. *The genetic basis of evolutionary change.* New York: Columbia University Press.

Lewontin, R., and Hubby, J. 1966. A molecular approach to the study of genic heterozygosity in natural populations. II. Amount of variation and degree of heterozygosity in natural populations of *Drosophila pseudobscura. Genetics* 54:595–609.

Licht, P. 1971. Regulation of the annual testis cycle by photoperiod and temperature in the lizard *Anolis carolinensis. Ecology* 52(2):240–252.

Licht, P., and Gorman, G. C. 1970. Reproductive and fat cycles in Caribbean *Anolis* lizards. *Univ. California Publ. Zool.* 95:1–52.

Lieb, C. S., Buth, D. G., and Gorman, G. C. 1983. Genetic differentiation in *Anolis sagrai:* A comparison of Cuban and introduced Florida populations. *J. Herpetol.* 17(1):90–94.

Lister, B. C. 1976. The nature of niche expansion in West Indian *Anolis* lizards. I: Ecological consequences of reduced competition. *Evolution* 30:659–676.

Llanes Echevarría, J. R. 1978. "Estudio de algunos aspectos ecológicos en cinco especies del género *Anolis* en la provincia de La Habana (Sauria: Iguanidae)" [inedit]. University Students' Report, Facultad de Biología, Universidad de La Habana.

Losos, J. B., and de Queiroz, K. 1997. Darwin's lizards. *Nat. Hist.* 12/97–1/98:33–39.

Losos, J. B., Jackman, T., Larson, A., de Queiroz, K., and Rodríguez Schettino, L. 1998. Contingency and determinism in replicated adaptive radiations of island lizards. *Science* 279:2115–2118.

Lynch, J. D., and Smith, H. M. 1964. Tooth replacement in a senile lizard, *A. equestris* Merrem. *Herpetologica* 20(1):70–71.

Mace, G. M., Collar, N., Cooke, J., Gaston, K., Ginsberg, J., Leader Williams, N., Maunder, M., and Milner-Gulland, E. J. 1992. The development of new criteria for listing species on the IUCN Red List. *Species* 19:16–22.

Mace, G. M., and Stuart, S. M. 1994. Draft IUCN Red List Categories, Version 2.2. *Species* 21–22:13–24.

Macey, J. R., Larson, A., Ananjeva, N. B., and Papenfuss, T. J. 1997. Evolutionary shifts in three major structural features of the mitochondrial genome among Iguanian lizards. *J. Mol. Evol.* 44:660–674.

MacFadden, B. 1980. Rafting mammals or drifting islands?: Biogeography of the Greater Antillean insectivores *Nesophontes* and *Solenodon. J. Biogeog.* 7:11–22.

———. 1981. Comments on Pregill's appraisal of historical biogeography of Caribbean vertebrates: Vicariance, dispersal or both? *Syst. Zool.* 30(3):270–272.

MacPhee, R. D. E., and Iturralde-Vinent, M. A. 1994. First Tertiary land mammal from Greater Antilles: An Early Miocene sloth (Xenarthra, Megalonychidae) from Cuba. *Amer. Mus. Nov.* 3094:1–13.

Malfait, B., and Dinkelman, M. 1972. Circum-Caribbean tectonic and igneous activity and the evolution of the Caribbean plate. *G. S. A. Bull.* 83(2):251–272.

Manójina, N., González, A., and Abreu, R. M. 1989. Datos sobre alimentación de la jutía conga (*Capromys pilorides*) en Guanahacabibes. *Poeyana* 369:1–13.

Marcellini, D. L., and Rodríguez Schettino, L. 1987. Notes on the natural history of the unusual Cuban lizard, *Anolis lucius. Herpetol. Rev.* 18(3):52–53.

Martínez Reyes, M. 1989. Patrones de coloración y diseño dorsal de *Anolis sagrai* (Sauria: Iguanidae) en una zona suburbana. *Cien. Biol.* 21–22:65–69.

———. 1994*a*. Hábitats y alimentación de tres especies del género *Leiocephalus* en Cuba. In *III Simposio de Zoología,* Havana, *Resúmenes,* 54.

———. 1994*b*. Aspectos reproductivos de *Leiocephalus cubensis cubensis* (Sauria: Iguanidae) en una localidad de Ciudad de La Habana. *Cien. Biol.* 27:83–89.

———. 1994*c*. Diversidad de lagartos en uverales de la Península de Zapata, Cuba. En *Memorias del Segundo Simposio Internacional Humedales '94,* Ciénaga de Zapata, 191–194.

———. 1995. Saurios de la Reserva de la Biosfera "Sierra del Rosario" Pinar del Río, Cuba. Evaluación ecológica de tres comunidades. *Inv. Geogr. Bol.* 30(2):50–77.

———. 1998. "Riqueza de reptiles terrestres del Archiiélago de Sabana-Camagüey, Cuba" [inedit]. Master's thesis, Instituto de Ecología y Sistemática, Havana.

Martínez Reyes, M., Castillo Chávez, F., González Medina, A., and Rodríguez Schettino, L. 1989. Queratoconjuntivitis ulcerativa en *Cyclura nubila nubila* (Sauria: Iguanidae). *Misc. Zool.* 45:1–2.

Martínez Reyes, M., Estrada, A. R., and Novo Rodríguez, J. 1990. Aspectos ecológicos y reproductivos de *Leiocephalus s. stictigaster* (Sauria: Iguanidae) en la Península de Guanahacabibes, Cuba. *Poeyana* 403:1–20.

Martínez Reyes, M., and Fernández García, I. 1988. Subnicho trófico y reproducción de *Leiocephalus cubensis* en Ciudad de La Habana. In *Primer Simposio de Zoología y Segundo de Botánica,* Havana (Instituto de Ecología y Sistemática, Havana), *Resúmenes,* 140.

———. 1994. Hábitat y alimentación de *Leiocephalus cubensis cubensis* (Iguania: Tropiduridae) en una localidad de Ciudad de La Habana, Cuba. *Cien. Biol.* 26:21–30.

Martínez Reyes, M., and Rodríguez Schettino, L. 1987. Canibalismo en *Leiocephalus carinatus* (Gray) (Sauria: Iguanidae). *Misc. Zool.* 29:1–2.

Menéndez, A., Espinosa, G., and Berovides, V. 1986. Amplitud del nicho y termorregulación en poblaciones de dos especies de *Anolis* cubanos. In *Quinta Conferencia Científica de Ciencias Naturales,* Havana (Universidad de La Habana), *Resúmenes,* 42.

Meshaka, W. E., Jr., Clause, R. M., Butterfield, B. P., and Hauge, J. B. 1997. The Cuban green anole, *Anolis porcatus:* A new anole established in Florida. *Herpetol. Rev.* 28(2):101–102.

Milera, J. F. 1984. Agresividad de *Leiocephalus cubensis* Gray 1840 (Reptilia: Sauria: Iguanidae). *Misc. Zool.* 22:2.

Milton, T. H., and Jenssen, T. A. 1979. Description and significance of vocalizations by *Anolis grahami grahami* (Sauria: Iguanidae). *Copeia* 3:481–489.

Moermond, T. C. 1979. Habitat constraints on the behavior, morphology and community structure of *Anolis* lizards. *Ecology* 60:152–164.

Montañez Huguez, L., Berovides Álvarez, V., Sampedro Marín, A., and Mugica Valdés, L. 1985. Vertebrados del Embalse "Leonero," provincia Granma. *Misc. Zool.* 25:1–2.

Moreno, L. V., and Valdés, E. 1991. Distribución altitudinal de la familia Iguanidae en Cuba. In *Segundo Simposio de Zoología,* Havana (Instituto de Ecología y Sistemática, Havana), *Resúmenes,* 44.

Morgan, G. S. 1977. Late Pleistocene vertebrates from the Cayman Islands, British West Indies. Master's thesis, University of Florida, Gainesville [cited by Pregill 1981*b*].

Morgan, G. S., Franz, R., and Crombie, R. I. 1993. The Cuban crocodile, *Crocodylus rhombifer,* from Late Quaternary fossil deposits on Grand Cayman. *Carib. J. Sci.* 29(3–4):153–164.

Mugica Valdés, L. 1981. "Estudios electroforéticos de esterasas en varias especies de *Anolis* cubanos (Sauria: Iguanidae)" [inedit]. University Students' Report, Facultad de Biología, Universidad de La Habana.

Mugica Valdés, L., Espinosa López, G., and Berovides Álvarez, V. 1982. Patrones electroforéticos de esterasas hepáticas en siete especies de lagartos del género *Anolis. Cien. Biol.* 8:37–48.

Navarro, N., and Peña, C. 1995. Nuevas contribuciones a la distribución de *Anolis anfiloquioi* (Garrido 1980). In *Bioeco, Segundo Taller de Biodiversidad,* Santiago de Cuba (Centro Oriental de Ecosistemas y Biodiversidad), *Resúmenes,* 13.

Nei, M. 1972. Genetic distance between populations. *Amer. Nat.* 106:283–292.

Neill, W. T. 1958. The occurrence of amphibians and reptiles in saltwater areas, and a bibliography. *Bull. Marine Sci. Gulf Carib.* 8(1):1–97 [cited by Schwartz and Henderson 1991].

Neill, W. T., and Allen, R. E. 1957. *Deiroptyx*—Cuba's reptilian oddity. *Nat. Mag.* 50:39–41.

Neumann, L. G. 1899. Révision de la famille des Ixodidae. III. Ixodes. *Mem. Soc. Zool. France* 12(2):107–294.

Nevo, E. 1983. Adaptative significance of protein variation. In *Systematic Association 24. Protein Polymorphism: Adaptative and Taxonomic Significance* (G. S. Oxford and D. Rollison, eds.), 239–282. London and New York: Academic Press.

Nevo, E., Beiles, A., and Benshlomo, R. 1984. The evolutionary significance of genetic diversity: Ecological, demographic and life history correlates. *Evol. Dynam. Gen. Divers.* 53:13–213.

Noble, G. K., and Hassler, W. G. 1935. A new giant *Anolis* from Cuba. *Copeia* 3:113–115.

Norell, M. A., and de Queiroz, K. 1991. The earliest iguanine lizard (Reptilia: Squamata) and its bearing on iguanine phylogeny. *Amer. Mus. Novitates* 2997:1–16.

Novo Rodríguez, J. 1985. Nido comunal de *Anolis angusticeps* (Sauria: Iguanidae) en Cayo Francés, Cuba. *Misc. Zool.* 26:3–4.

Novo Rodríguez, J., and Estrada, A. R. 1986. Ciclo reproductivo de *Anolis bartschi.* *Poeyana* 318:1–5.

Novo Rodríguez, J., and Rodríguez Schettino, L. 1982. *Anolis bartschi* (Sauria: Iguanidae): Datos sobre su ecología. In *Primera Jornada Científica del Instituto de Zoología,* Havana (Instituto de Zoologia, Havana), *Resúmenes,* 4.

Oliver, J. A. 1948. The anoline lizards of Bimini, Bahamas. *Amer. Mus. Nov.* 1383:1–36.

Ortiz Díaz, A. R. 1978. "Estudio del nicho trófico en cinco especies del género *Anolis* en la provincia y Ciudad de La Habana (Sauria: Iguanidae)" [inedit]. University Students' Report, Facultad de Biología, Universidad de La Habana.

Otero, A. R. 1950. Sobre la alimentación de *Anolis porcatus. Bol. Hist. Nat.* 1(4):186–187.

Pacala, S., and Roughgarden, J. 1982. Resource partitioning and interspecific competition in two two-species insular *Anolis* lizard communities. *Science* 217:444–446.

Palau Rodríguez, C. M. 1997. Morfometría y alimentación de *Anolis allisoni* (Sauria: Polychridae) en áreas urbanas de la provincia de Sancti Spíritus. In *Cuarto Simposio de Zoología,* Havana, *Resúmenes,* 33.

Palau Rodríguez, C. M., and Pérez Silva, B. 1995. Variación morfométrica en dos especies de anolinos. In *Biosfera '95,* Havana, *Resúmenes,* 41.

Paull, D., Williams, E. E., and Hall, W. P., III. 1976. Lizard karyotypes from the Galapagos Islands: Chromosome in phylogeny and evolution. *Breviora* 441:1–31.

Perera, A. 1984. Aspectos de la ecomorfología de *Cyclura n. nubila* (Sauria: Iguanidae). *Cien. Biol.* 11:117–128.

———. 1985a. Datos sobre la abundancia y actividad de *Cyclura nubila nubila* (Sauria: Iguanidae) en la región de Cayo Largo del Sur. *Poeyana* 288:1–17.

———. 1985b. Datos sobre la alimentación de *Cyclura nubila nubila* (Sauria: Iguanidae) en la región de Cayo Largo del Sur. *Poeyana* 291:1–12.

Perera, A., Berovides Álvarez, V., Garrido, O. H., Estrada, A. R., González, A., and Álvarez, M. 1994. Criterios para la selección de especies amenazadas de vertebrados cubanos. In *Tercer Simposio de Zoología,* Havana, *Resúmenes,* 96.

Pérez Hernández, A. 1995. Actividad y subnicho estructural de *Anolis quadriocellifer.* In *Biosfera '95,* Havana (Instituto de Ecología y Sistemática, Havana), *Resúmenes,* 41.

Pérez Vigueras, I. 1934. On the ticks of Cuba, with description of a new species, *Amblyomma torrei,* from *Cyclura macleayi* Gray. *Psyche* 41(1):13–18.

———. 1935. Sobre la validez de la especie *Atractis cruciata* Leistom 1902. *Rev. Parsitol. Clin. Lab. Habana* 1:188–190.

———. 1936. Notas sobre la fauna parasitológica de Cuba. Parte 1. Vermes. *Mem. Soc. Cubana Hist. Nat.* 10:53–86.

———. 1940. Nota sobre algunas especies nuevas de tremátodos y sobre otras poco conocidas. *Rev. Univ. La Habana.* 28–29:1–28.

———. 1956. *Los ixódidos y culícidos de Cuba, su historia natural y médica.* Havana: Universidad de La Habana.

Pérez-Beato, O. 1982a. Algunos indicadores en el desarrollo de *Anolis porcatus* (Sauria: Iguanidae). *Cien. Biol.* 7:127–129.

―――. 1982*b*. Medición de la habilidad de locomoción en lagartos mediante el uso de un índice femoral. *Cien. Biol.* 8:131–134.

Pérez-Beato, O., and Arencibia, R. 1991. Sistemática y variación fenotípica de *Anolis porcatus* y *Anolis allisoni* (Sauria: Iguanidae) en el occidente de Cuba. In *Segundo Simposio de Zoología*, Havana (Instituto de Ecología y Sistemática, Havana), *Resúmenes,* 45.

Pérez-Beato, O., and Berovides Álvarez, V. 1979. Diferenciación de *Anolis porcatus* (Sauria: Iguanidae) en el occidente de Cuba. In *Segundo Evento Científico de la Universidad de La Habana,* Havana (Universidad de La Habana), *Resúmenes,* 315.

―――. 1981. Estudio de la diferenciación en dos especies del complejo *carolinensis* (Sauria: Iguanidae) en Cuba. In *Primer Congreso Nacional de Ciencias Biológicas,* Havana (Sociedad Cubana de Ciencias Biológicas, Havana), *Resúmenes,* 183.

―――. 1982. Estudio electroforético de las proteínas del plasma en dos especies de *Anolis* (Sauria: Iguanidae). *Poeyana* 242:1–7.

―――. 1984. Indice cefálico y patrón de placa mental: Dos elementos fenotípicos en la diferenciación de *Anolis porcatus* (Sauria: Iguanidae) en el occidente de Cuba. *Poeyana* 262:1–8.

―――. 1986*a*. Clasificación poblacional para *Anolis porcatus* y *Anolis allisoni* (Sauria: Iguanidae) mediante análisis multivariado. In *Quinta Conferencia Científica de Ciencias Naturales,* Havana (Universidad de La Habana), *Resúmenes,* 42.

―――. 1986*b*. Límites de variación de la coloración en *Anolis allisoni:* Hibridación o polimorfismo? *Poeyana* 327:1–5.

Perfit, M. R., and Heezen, B. C. 1978. The geology and evolution of the Cayman Trench. *Bull. Geol. Soc. Amer.* 89:1155–1174.

Perfit, M. R., and Williams, E. E. 1989. Geological constraints and biological retrodictions in the evolution of the Caribbean Sea and its islands. In *Biogeography of the West Indies: Past, Present and Future* (C. A. Woods, ed.), 47–102. Gainesville, Fla.: Sandhill Crane Press.

Peters, G. 1970. Zür Taxonomie und Zoogeographie der Kubanischen Anolinen Eideschen (Reptilia, Iguanidae). *Mitt. Zool. Mus.* Berlin 46(1):197–234.

Peters, W. 1863. Uber einige neue arten der Saurier-Gatung *Anolis. Monatsb. Akad. Wiss. Berlin:*135–149.

Peterson, J. A., and Williams, E. E. 1981. A case history in retrograde evolution: The *onca* lineage in anoline lizards. II. Subdigital fine structure. *Bull. Mus. Comp. Zool.* 149(4):215–268.

Petzold, H. G. 1962. Successful breeding of *Leiocephalus carinatus* Gray. *Internatl. Zool. Yearbook* 4:97–98.

Phillips, J. A. 1994. On the trail of large lizards: Rover rolls over. *Zoonooz* 1994:6–11.

Pindell, J. L., and Dewey, J. F. 1982. Permo-Triassic reconstruction of Western Pangea and the evolution of the Gulf of Mexico, Caribbean region. *Tectonics* 1:179–212 [cited by Perfit and Williams 1989].

Polo, J. L., and Moreno, L. 1986. Observaciones realizadas en *Leiocephalus cubensis* (Sauria: Iguanidae) (Gray 1840) en cautiverio. In *Primera Jornada Científica de Animales de Zoológicos y Fauna Silvestre,* Havana (Academia de Ciencias de Cuba, Havana), *Resúmenes,* 8.

Porter, C. A., Crombie, R. I., and Baker, R. J. 1989. Karyotypes of five species of Cuban lizards. *Occ. Papers Mus. Texas Tech Univ.* 130:1–6.

Porter, K. R. 1972. *Herpetology*. Philadelphia, London, and Toronto: W. B. Saunders.

Powell, R. 1992. *Anolis porcatus*. *Cat. Amer. Amph. Rept.* 541.1–541.5.

Powell, R., Henderson, R. W., Adler, K., and Dundee, H. A. (1996): An annotated checklist of West Indian amphibians and reptiles. In *Contributions to West Indian Herpetology: A Tribute to Albert Schwartz* (R. Powell and R. W. Henderson, eds.). Ithaca, N.Y.: Society for the Study of Amphibians and Reptiles. *Contributions to Herpetology* vol. 12, 51–93.

Powell, R., Smith, D. D., Parmerlee, J. S., Jr., Taylor, C. V., and Jolley, M. L. 1990. Range expansion by an introduced anole: *Anolis porcatus* in the Dominican Republic. *Amphibia-Reptilia* 11:421–425.

Pregill, G. K. 1981*a*. An appraisal of the vicariance hypothesis of the Caribbean biogeography and its applications to West Indian terrestrial vertebrates. *Syst. Zool.* 30(2):147–155.

———. 1981*b*. Late Pleistocene herpetofaunas from Puerto Rico. *Univ. Kansas Mus. Nat. Hist. Misc. Publ.* 71:1–72.

———. 1982. Fossil amphibians and reptiles from New Providence Island, Bahamas. In *Fossil vertebrates from the Bahamas* (S. L. Olson, ed.), *Smithsonian Contrib. Paleobiol.* 48:8–21.

———. 1992. Systematics of the West Indian lizards of the genus *Leiocephalus* (Squamata: Iguania: Tropiduridae). *Univ. Kansas Mus. Nat. Hist. Misc. Publ.* 84: 1–69.

Pregill, G. K., and Olson, S. L. 1981. Zoogeography of West Indian vertebrates in relation to Pleistocene climatic cycles. *Ann. Rev. Ecol. Syst.* 12:75–98.

Quesada Jacob, S. M., Quintana Vázquez, D., Rodríguez Muñoz, A., Sánchez de Céspedes, I-S., and Santana Méridas, O. 1991. "Utilización de algunos recursos ambientales por cuatro especies del género *Anolis* en la manigua costera del Jardín Botánico Nacional" [inedit]. University Students' Report, Facultad de Biología, Universidad de La Habana.

Quintana Ferrer, V. B. 1987. "Relaciones filogenéticas entre especies de *Anolis* cubanos (Sauria: Iguanidae)" [inedit]. University Students' Report, Facultad de Biología, Universidad de La Habana.

Ramos Guadalupe, L. E. 1997. La naturaleza del folklore cubano. Historias olvidadas. In *Cuarto Simposio de Zoología*, Havana, *Resúmenes*, 64.

Ramos Vecín, M. 1970. Estudio métrico-morfológico del nemaspermo de *Crocodylus rhombifer* Cuvier. Reporte preliminar. In *Tercera Jornada Técnica Nacional*, Havana (Combinado Avícola Nacional, Havana), *Memorias*, 1–6.

Rand, A. S. 1962. Notes on Hispaniolan herpetology. 5. The natural history of three sympatric species of *Anolis*. *Breviora* 154:1–15.

———. 1964. Ecological distribution in anoline lizards of Puerto Rico. *Ecology* 45(4):745–752.

———. 1967*a*. Communal egg laying in anoline lizards. *Herpetologica* 23(3):227–230.

———. 1967*b*. The ecological distribution of anoline lizards around Kingston, Jamaica. *Breviora* 272:1–18.

————. 1967c. Ecology and social organization in the iguanid lizard *Anolis lineatopus. Proc. U.S. Natl. Mus.* 122(3595):1–79.

————. 1967d. The adaptive significance of territoriality in iguanid lizards. In *Lizard Ecology: A Symposium* (W. W. Milstead, ed.), 106–115. Columbia: University of Missouri Press.

Rand, A. S., and Humphrey, S. S. 1968. Interspecific competition in the tropical rainforest: Ecological distribution among lizards at Belém, Pará. *Proc. U.S. Natl. Mus.* 125(3658):1–17.

Rand, A. S., and Williams, E. E. 1969. The anoles of La Palma: Aspects of their ecological relationships. *Breviora* 327:1–17.

Read, C. P., and Amrein, U. 1953. North American nematodes of the genus *Pharyngodon* Diesing (Oxyuridae). *Parasitology* 39:365–370.

Regalado Morera, R. 1985. Inmovilidad tónica en el lagarto *Anolis sagrei* (Sauria: Iguanidae). *Cien. Biol.* 14:33–48.

Regalado, R., and Garrido, O. H. 1993. Diferencias en el comportamiento social de dos especies gemelas de anolinos cubanos (Lacertilia: Iguanidae). *Carib. J. Sci.* 29(1/2):18–23.

Rieppel, O. 1980. Green anole in Dominican amber. *Nature* 286:486–487.

Rodríguez González, M. E. 1981. Datos sobre el nicho estructural y trófico de *Anolis porcatus* y *A. allisoni* (Sauria: Iguanidae) en el occidente de Cuba. In *Décima Jornada Científica Estudiantil.* Havana: Universidad de La Habana.

————. 1982. "Ciclo reproductivo en *Anolis porcatus* (Gray), 1840 y *Anolis allisoni* Barbour, 1928" [inedit]. University Students' Report, Facultad de Biología, Universidad de La Habana.

Rodríguez Schettino, L. 1981a. Consideraciones sobre el nicho trófico de *Anolis vermiculatus* (Sauria: Iguanidae). In *Primer Congreso Nacional de Ciencias Biológicas,* Havana (Sociedad Cubana de Ciencias Biológicas, Havana), *Resúmenes,* 178.

————. 1981b. Nicho estructural y climático de *Anolis homolechis* (Sauria: Iguanidae) en la Sierra de Casas, Isla de la Juventud. In *Primer Congreso Nacional de Ciencias Biológicas,* Havana (Sociedad Cubana de Ciencias Biológicas, Havana), *Carteles libres.*

————. 1982. Variación de indicadores taxonómicos en dos especies del género *Anolis* (Sauria: Iguanidae) causada por el método de preservación. *Misc. Zool.* 13:1–2.

————. 1985. Distribución altitudinal de los iguánidos en la Sierra del Turquino, Cuba. *Cien. Biol.* 14:59–66.

————. 1986a. Algunos patrones distribucionales y ecológicos de los reptiles cubanos. *Poeyana* 305:1–15.

————. 1986b. Algunos aspectos ecológicos sobre la iguana (*Cyclura nubila nubila*) en la Península de Guanahacabibes. In *Quinta Conferencia Científica de Ciencias Naturales,* Havana (Universidad de La Habana), *Resúmenes,* 137–138.

————. 1989. Reptiles Terrestres. In *Nuevo Atlas Nacional de Cuba,* XI.1.3 (map 8) (Academia de Ciencias de Cuba e Instituto Cubano de Geodesia y Cartografía, Havana), Instituto Geográfico Nacional de España.

————. 1990. La iguana cubana (*Cyclura nubila nubila*) en condiciones de cautiverio.

In *Cuarto Encuentro Técnico del Parque Zoológico Nacional,* Havana (Parque Zoológico Nacional, Havana), *Resúmenes,* 19.

————. 1993. Areas faunísticas de Cuba según la distribución ecogeográfica actual y el endemismo de los reptiles. *Poeyana* 436:1–17.

————. 1994. Descripción de los nemaspermos de seis especies cubanas del género *Anolis* (Iguania: Polychridae). *Cien. Biol.* 26:114–116.

————. 1996. Distribución de los iguánidos en la Isla de la Juventud. Géneros *Anolis* y *Chamaeleolis.* In *25 Años de Ciencias.* Nueva Gerona: Delegación Ministerio de Ciencia, Tecnología y Medio Ambiente de Cuba.

Rodríguez Schettino, L., and Chamizo Lara, A. 1995. Reptiles. In *Estudio Nacional de Biodiversidad* (M. Vales, A. Álvarez, L. Montes, and H. Ferrás, eds.). Havana: Proyecto GEF/PNUMA, Ministerio de Ciencia, Tecnología y Medio Ambiente de Cuba.

————. 1997. *Anolis vermiculatus.* Conservation assessment and management plan for some Cuban species. IUCN/SSC/CBSG, Apple Valley, Minn.

————. 1998. *Anolis bartschi.* Conservation assessment and management plan for some Cuban species. IUCN/SSC/CBSG, Apple Valley, Minn.

————. In press. Reptiles cubanos con algún grado de amenaza de extinción. *Poeyana.*

Rodríguez Schettino, L., Chamizo Lara, A., Echenique, L., and González, A. 1998. *Anolis juangundlachi.* Conservation assessment and management plan for some Cuban species. IUCN/SSC/CBSG, Apple Valley, Minn.

Rodríguez Schettino, L., de Queiroz, K., Losos, J. B., Chamizo Lara, A., Hertz, P. E., Fleishman, L., Leal, M., Jackman, T., Rivalta González, V., Torres Barbosa, A., and Daniel Álvarez, A. 1997. In *Cuarto Simposio de Zoología,* Havana, *Resúmenes,* 34.

Rodríguez Schettino, L., Espinosa Sáez, J., and Valdés de la Osa, A. 1979. "Consideraciones generales sobre el aprovechamiento de la fauna silvestre para el turismo en Topes de Collantes" [inedit]. Work report, Instituto de Ecología y Sistemática, Havana.

Rodríguez Schettino, L., and González Alonso, H. 1984. Algunos aspectos sobre la zoogeografía de los vertebrados cubanos. In *Taller Latinoamericano de Zoología,* México, D. F. (Facultad de Ciencias, Universidad Autónoma de México, México, D. F.), *Memorias,* 92–131.

Rodríguez Schettino, L., and Lizana, M. 1997. Historia natural del Lagarto caimán cubano, *Anolis vermiculatus* (Iguania: Polychridae). *Bol. Asoc. Herpetol. Española* 8:23–26.

Rodríguez Schettino, L., Marcellini, D. L., and Novo Rodríguez, J. 1987. Algunos aspectos ecológicos sobre *Anolis vermiculatus* (Sauria: Iguanidae) en Soroa, Pinar del Río, Cuba. *Poeyana* 343:1–9.

Rodríguez Schettino, L., and Martínez Reyes, M. 1985. Composición por especies de la familia Iguanidae en la Península de Guanahacabibes. In *Segunda y Tercera Jornadas Científicas del Instituto de Zoología y de la Sección de Zoología de la Sociedad Cubana de Ciencias Biológicas,* Havana (Instituto de Zoología, Havana), *Resúmenes,* 24–25.

————. 1986a. Algunos aspectos ecológicos sobre la iguana (*Cyclura nubila nubila*) en la Península de Guanahacabibes. In *Tercer Foro Científico Técnico Provincial del Medio Ambiente,* Pinar del Río (Academia de Ciencias de Cuba, Pinar del Río).

————. 1986b. Caracterización ecológica de *Anolis quadriocellifer* (Sauria: Iguanidae). In *Cuarta Jornada Científica BTJ-ANIR,* Havana (Instituto de Ecología y Sistemática, Havana), *Resúmenes,* 1–2.

————. 1988. Estructura, densidad y biomasa de una población de *Anolis argenteolus.* In *Primer Simposio de Zoología y Segundo de Botánica,* Havana (Instituto de Ecología y Sistemática, Havana), *Resúmenes,* 141.

————. 1989a. "Algunos aspectos ecológicos sobre cuatro especies endémicas del género *Anolis* (Sauria: Iguanidae)" [inedit]. Work report, Instituto de Ecología y Sistemática, Havana.

————. 1989b. Estructura poblacional de *Anolis argenteolus* en una localidad del sur de la Sierra Maestra. In *Cuarta Jornada Científica de la Sociedad Cubana de Zoología.* Havana (Sociedad Cubana de Zoología, Havana).

————. 1991. Hábitos alimentarios de *Anolis argenteolus* (Sauria: Iguanidae) en la desembocadura del río La Mula, Santiago de Cuba. In *Segundo Simposio de Zoología,* Havana (Instituto de Ecología y Sistemática, Havana), *Resúmenes,* 65.

————. 1992. Hábitos alimentarios de *Anolis bartschi* (Sauria: Iguanidae) en San Vicente, Pinar del Río, Cuba. *Cien. Biol.* 25:30–40.

————. 1994. Características tróficas de una población de *Anolis lucius* (Iguania: Polychridae) en la costa septentrional de Cuba. *Avicennia* 1:67–77.

————. 1996. Algunos aspectos de la ecología trófica de *Anolis argenteolus* (Sauria: Polychridae) en una población de la costa suroriental de Cuba. *Biotropica* 28(2):252–257.

————. In press. Ciclo reproductivo y algunos datos sobre la estructura de una población de *Anolis argenteolus* (Sauria: Iguanidae). *Rep. Inv.*

Rodríguez Schettino, L., and Novo Rodríguez, J. 1982. *Anolis vermiculatus* (Sauria: Iguanidae) en el río San Marcos, Mil Cumbres. In *Primera Jornada Científica del Instituto de Zoología,* Havana (Instituto de Zoología, Havana), *Resúmenes,* 3.

————. 1985. Nuevos datos etoecológicos sobre *Anolis vermiculatus* (Sauria: Iguanidae). *Poeyana* 296:1–11.

Rodríguez Schettino, L., and Valderrama Puente, M. J. 1986. Algunos aspectos del nicho estructural y climático de *Anolis lucius* (Sauria: Iguanidae). *Poeyana* 319:1–12.

Rogers, J. S. 1972. Measures of genetic similarity and genetic distance. *Univ. Texas Publ.* 7213:145–153.

Romer, A. S. 1956. *Osteology of the reptiles.* Chicago: University of Chicago Press.

Rosen, D. E. 1976. A vicariance model of Caribbean biogeography. *Syst. Zool.* 24:431–464.

Roughgarden, J. 1995. *Anolis lizards of the Caribbean: Ecology, Evolution, and Plate Tectonics.* Oxford: Oxford University Press.

Roughgarden, J., Heckel, D., and Fuentes, E. R. 1983. Coevolutionary theory and the biogeography and community structure of *Anolis.* In *Lizard Ecology. Studies of a Model Organism* (R. B. Huey, E. R. Piamka, and T. W. Schoener, eds.), 371–410. Cambridge and London: Harvard University Press.

Ruibal, R. 1961. Thermal relations of five species of tropical lizards. *Evolution* 15(1):98–111.

———. 1964. An annotated checklist and key to the anoline lizards of Cuba. *Bull. Mus. Comp. Zool.* 130(2):475–520.

———. 1967. Evolution and behavior in West Indian anoles. In *Lizard Ecology A Symposium* (W. W. Milstead, ed.), 116–140. Columbia: University of Missouri Press.

Ruibal, R., and Ernst, V. 1965. The structure of the digital setae of lizards. *J. Morphol.* 117(3):271–293.

Ruibal, R., and Philibosian, R. 1970. Eurythermy and niche expansion in lizards. *Copeia* 4:645–653.

Ruibal, R., and Williams, E. E. 1961a. Two sympatric Cuban anoles of the *carolinensis* group. *Bull. Mus. Comp. Zool.* 125(7):183–208.

———. 1961b. The taxonomy of the *Anolis homolechis* complex of Cuba. *Bull. Mus. Comp. Zool.* 125(8):211–246.

Ruiz García, F. N. 1975. Observaciones etológicas sobre *Anolis homolechis* (Lacertilia: Iguanidae) en Cuba. *Misc. Zool.* 1:4.

———. 1977. Observaciones sobre *Anolis sagrei* Duméril y Bibron (Lacertilia: Iguanidae) en cautiverio. *Misc. Zool.* 6:2.

Ruiz Urquiola, A., Garrido, O. H., González Blanco, T., Espinosa López, G., and Ibarra Martín, M. E. 1997a. *Chamaeleolis barbatus*. Conservation assessment and management plan for some Cuban species. IUCN/SSC/CBSG, Apple Valley, Minn.

———. 1997b. *Chamaeleolis guamuhaya*. Conservation assessment and management plan for some Cuban species. IUCN/SSC/CBSG, Apple Valley, Minn.

Ruiz Urquiola, A., and Gutiérrez, A. 1997. Cortejo, apareamiento y puesta de los huevos en especies de *Chamaeleolis* (Sauria: Polychrotidae). In *Cuarto Simposio de Zoología,* Havana, *Resúmenes,* 66.

Rundquist, E. M. 1981. Longevity records at the Oklahoma City Zoo. *Herpetol. Rev.* 12(3):87.

Sampedro Marín, A., Berovides Álvarez, V., and Torres Fundora, O. 1979. Hábitat, alimentación y actividad de dos especies de *Leiocephalus* (Sauria: Iguanidae) en dos localidades de la región suroriental de Cuba. *Cien. Biol.* 3:129–139.

Sampedro Marín, A., Berovides Álvarez, V., and Rodríguez Schettino, L. 1982. Algunos aspectos ecológicos sobre dos especies cubanas del género *Anolis* (Sauria: Iguanidae). *Cien. Biol.* 7:87–103.

Sampedro Marín, A., Espinosa López, G., and Calvo, P. D. 1993. Relación entre el nicho ecológico y la variabilidad genética en dos especies simpátricas de *Anolis* (Sauria: Iguanidae). In *Primer Simposio de Ecología,* Havana, *Resúmenes,* 108.

Sánchez Oria, B. 1980. "Variación y selección natural en *Anolis porcatus* Gray, 1840" [inedit]. University Students' Report, Facultad de Biología, Universidad de La Habana.

Sánchez Oria, B., and Berovides Álvarez, V. 1985. Termorregulación en *Anolis porcatus* (Sauria: Iguanidae). *Cien. Biol.* 14:29–32.

———. 1987. Variación y selección natural en *Anolis porcatus* Gray, 1840 (Sauria: Iguanidae). *Cien. Biol.* 18:111–118.

Sanz, A., Uribe Aranzábal, M. C., Almaguer Cuenca, N., and Díaz, L. 1997. Histología del ciclo reproductivo del testis de *A. sagrei* y *A. porcatus* (Sauria: Iguanidae). Poster in *Cuarto Simposio de Zoología,* La Habana.

Savage, J. M. 1958. The iguanid lizard genera *Urosaurus* and *Uta,* with remarks on related groups. *Zoologica* (New York) 43:41–54.

———. 1982. The enigma of the Central American herpetofauna: Dispersals or vicariance? *Ann. Missouri Bot. Garden* 69:464–547.

Schoener, T. W. 1967. The ecological significance of sexual dimorphism in size in the lizard *Anolis conspersus.* Science 155:474–476.

———. 1968. The *Anolis* lizards of Bimini: Resource partitioning in a complex fauna. *Ecology* 49(4):704–726.

———. 1970a. Size patterns in West Indian *Anolis* lizards. II. Correlations with the size of particular sympatric species-displacement and convergence. *Amer. Nat.* 104(936):155–174.

———. 1970b. Nonsynchronous spatial overlap of lizards in a patchy habitat. *Ecology* 51:408–418.

———. 1971. Theory of feeding strategies. *Ann. Rev. Ecol. Syst.* 2:369–404.

———. 1977. Competititon and the niche. In *Biology of the Reptilia 7* (C. Gans and D. W. Tinkle, eds.), 35–136. New York: Academic Press.

———. 1983. Field experiments on interespecific competition. *Amer. Nat.* 122:240–285.

Schoener, T. W., and Gorman, G. C. 1968. Some niche differences among three species of Lesser Antillean anoles. *Ecology* 49(4):819–830.

Schwartz, A. 1958. A new subspecies of *Anolis equestris* from eastern Cuba. *Herpetologica* 14(1):1–7.

———. 1959a. Variation in lizards of the *Leiocephalus cubensis* complex in Cuba and the Isla de Pinos. *Bull. Florida State Mus.* 4(4):97–143.

———. 1959b. The Cuban lizards of the species *Leiocephalus carinatus* (Gray). *Reading Public Mus. Art Gallery Sci. Publ.* 10:1–47.

———. 1960a. Variation in the Cuban lizard *Leiocephalus raviceps* Cope. *Proc. Biol. Soc. Washington* 73:67–82.

———. 1960b. A new subspecies of *Leiocephalus stictigaster* Schwartz from Central Cuba. *Proc. Biol. Soc. Washington* 73:103–106.

———. 1964a. *Anolis equestris* in Oriente province, Cuba. *Bull. Mus. Comp. Zool.* 131(12):403–428.

———. 1964b. New subspecies of *Leiocephalus* from Cuba. *Quart. J. Florida Acad. Sci.* 27(3):211–222.

———. 1967. The *Leiocephalus* (Lacertilia: Iguanidae) of the southern Bahama Islands. *Ann. Carnegie Mus.* 39(12):153–185.

———. 1968. The Cuban lizards of the *Anolis homolechis* complex. *Tulane Studies Zool.* 14(4):140–184.

———. 1969. The status of the name *Holotropis microlophus* Cocteau. *Copeia* 3:620–621.

———. 1978. A new species of aquatic *Anolis* (Sauria: Iguanidae) from Hispaniola. *Ann. Carnegie Mus.* 47(11):261–279.

Schwartz, A., and Carey, M. 1977. Systematics and evolution in the West Indian iguanid genus *Cyclura*. *Studies Fauna Curaçao Carib. Islands* 53(173):16–97.

Schwartz, A., and Garrido, O. H. 1967. A review of the Cuban iguanid lizard *Leiocephalus macropus* Cope. *Reading Public Mus. Art Gallery Sci. Publ.* 14:1–41.

———. 1968a. An undescribed subspecies of *Leiocephalus raviceps* Cope (Sauria: Iguanidae) from western Cuba. *Proc. Biol. Soc. Washington* 81:23–30.

———. 1968b. Four new subspecies of *Leiocephalus stictigaster* from Cuba. *Natl. Mus. Canada Nat. Hist. Papers* 37:1–23.

———. 1971. The status of *Anolis alutaceus clivicolus* Barbour and Shreve. *Carib. J. Sci.* 11(1–2):11–15.

———. 1972. The lizards of the *Anolis equestris* complex in Cuba. *Studies Fauna Curaçao Carib. Islands* 39(134):1–86.

Schwartz, A., and Henderson, R. W. 1985. *A guide to the identification of the amphibians and reptiles of the West Indies eclusive of Hispaniola*. Milwaukee Public Museum. Milwaukee: Inland Press.

———. 1988. West Indian amphibians and reptiles: A checklist. *Milwaukee Public Mus. Contr. Biol. Geol.* 74:1–264.

———. 1991. *Amphibians and reptiles of the West Indies. Descriptions, distributions, and natural history*. Gainesville: University of Florida Press.

Schwartz, A., and Ogren, L. H. 1956. A collection of reptiles and amphibians from Cuba with the description of two new forms. *Herpetologica* 12(2):91–110.

Schwartz, A., and Thomas, R. 1968. A review of *Anolis angusticeps* in the West Indies. *Quart. J. Florida Acad. Sci.* 31(1):52–59.

———. 1975. A check-list of West Indian amphibians and reptiles. *Carnegie Mus. Nat. Hist. Special Publ.* 1:1–216.

Schwartz, A., Thomas, R., and Ober, L. D. 1978. First supplement to a check-list of West Indian amphibians and reptiles. *Carnegie Mus. Nat. Hist. Special Publ.* 5:1–35.

Schwenk, K. 1988. Comparative morphology of the Lepidosaur tongue and its relevance to Squamate phylogeny. In *Phylogenetic relationships of the lizard families. Essays commemorating Charles L. Camp* (R. Estes and G. Pregill, eds.), 569–598. Stanford, Calif.: Stanford University Press.

———. 1994. Systematics and subjectivity: The phylogeny and classification of iguanian lizards revisited. *Herpetol. Rev.* 25:53–57.

Selander, R. K., and Johnson, W. E. 1973. Genetic variation among vertebrate species. *Ann. Rev. Ecol. Syst.* 4:75–91.

Sexto, L. 1989. Los lagartos sí comen huevo. *Bohemia* 81(42):36–38.

Sexton, O. J. 1967. Population changes in a tropical lizard, *Anolis limifrons* on Barro Colorado Island, Panama, Canal Zone. *Copeia* 1:219–222.

———. 1980. Comments on the reproductive cycle of *Anolis limifrons* (Sauria: Iguanidae) at Turrialba, Costa Rica. *Carib. J. Sci.* 16(1–4):13–17.

Sexton, O. J., Bauman, J., and Ortleb, E. 1972. Seasonal food habits of *Anolis limifrons*. *Ecology* 53(1):182–186.

Sexton, O. J., Ortleb, E., Hathaway, L. M., Ballinger, R. E., and Licht, P. 1971. Reproductive cycles of three species of anoline lizards from the Isthmus of Panama. *Ecology* 52(2):201–215.

Shaw, C. E. 1954. Captive-bred Cuban iguana *Cyclura macleayi macleayi*. *Herpetologica* 10(2):73–78.

Shochat, D., and Dessauer, H. C. 1981. Comparative immunological study of albumins of *Anolis* lizards of the Caribbean islands. *Comp. Biochem. Physiol.* 68A:67–73.

Silva Lee, A. 1984. Lagartos cubanos. *Juv. Téc.* 196:10–13.

———. 1985. *Chipojos, bayoyas y camaleones*. Havana: Editorial Científico Técnica.

Silva Rodríguez, A. 1980. Determinación de los patrones de proteínas totales en cinco especies del género *Anolis*. In *Novena Jornada Científica Estudiantil*, Havana (Universidad de La Habana), *Resúmenes*, 100.

———. 1981. "Utilización de recursos ambientales por dos especies del género *Anolis* (Sauria: Iguanidae) en la estación ecológica 'Sierra del Rosario,' Pinar del Río (Cuba)" [inedit]. University Students' Report, Facultad de Biología, Universidad de La Habana.

Silva Rodríguez, A., Berovides Álvarez, V., and Estrada, A. R. 1982. Sitios de puesta comunales de *Anolis bartschi* (Sauria: Iguanidae). *Misc. Zool.* 15:1.

Silva Rodríguez, A., and Espinosa López, G. 1983. Variabilidad genética y estabilidad ambiental en dos especies de lagartos del género *Anolis* (Sauria: Iguanidae). *Cien. Biol.* 10:63–68.

Silva Rodríguez, A., and Estrada, A. R. 1982. Vertebrados de la estación ecológica Sierra del Rosario. *Misc. Zool.* 15:1–2.

———. 1984. Ciclo reproductivo de dos lagartos del género *Anolis* (*A. homolechis* y *A. allogus*) en la Estación Ecológica Sierra del Rosario, Cuba. *Cien. Biol.* 12:81–89.

Silva Rodríguez, A., Estrada, A. R., and Garrido, O. H. 1982. Nueva localidad para *Anolis spectrum* (Peters). *Misc. Zool.* 16:2–3.

Silva Taboada, G. 1974. Sinopsis de la espeleofauna cubana. *Serie Espeleol. Carsol.* 43:1–65.

———. 1988. *Sinopsis de la espeleofauna cubana*. Havana: Editorial Científico Técnica.

Simon, C. A. 1983. A review of lizard chemoreception. In *Lizard Ecology. Studies of a Model Organism* (R. B. Huey, E. R. Pianka, and T. W. Schoener, eds.), 443–447. Cambridge: Harvard University Press.

Simpson, G. G. 1940. Mammals and land bridges. *J. Washington Acad. Sci.* 30:137–163.

———. 1956. Zoogeography of the West Indian land mammals. *Amer. Mus. Novitates* 1759:1–28.

Smith, D. C. 1929. The direct effect of temperature changes upon the melanophores of the lizard *Anolis equestris*. *Proc. Nat. Acad. Sci. Washington* 15:48–56.

Smith, H. M., and Grant, C. 1958. The proper names of some Cuban snakes: An analysis of dates of publication of Ramon de la Sagra's Historia Natural de Cuba, and of Fitzinger's Systema Reptilium. *Herpetologica* 14(4):215–222.

Smith. H. M., and Willis, E. 1955. Interspecific variation in compression of tail in a Cuban lizard. *Herpetologica* 11(2):86–88.

Socarrás, A. A., Cruz, J., de la, Garcés, G., and Ruiz, A. 1988. Saurofagia en *Anolis* (Sauria: Iguanidae). *Misc. Zool.* 38:4.

Socarrás Torres, E. 1995. Aspectos ecológicos de *Anolis jubar cocoensis* (Lepidosauria: Polychridae) en un bosque siempreverde micrófilo de Cayo Coco, Cuba. In *Bioeco*,

Segundo Taller de Biodiversidad, Santiago de Cuba (Centro Oriental de Ecosistemas y Biodiversidad), *Resúmenes,* 13.

Sonin, M. D., and Baruš, V. 1968. Filariid nematodes in birds and reptiles of Cuba. *Fol. Parasitol.* 15:55–65.

Soto Ramírez, E. 1994. Los vertebrados terrestres de Bacunayagua. In *Memorias del Segundo Simposio Internacional Humedales '94,* Ciénaga de Zapata, 204–208.

Stebbins, R. C., and Eakin, E. M. 1958. The role of the "third eye" in reptilian behavior. *Amer. Mus. Novitates* 1870:1–40.

Stejneger, L. 1917. Cuban amphibians and reptiles collected for the United States National Museum from 1899 to 1902. *Proc. U.S. Natl. Mus.* 53:259–291.

Sutcliffe, R. 1952. Results of the Gatherwood Chaplin West Indies Expedition 1948. Part 6. Amphibia and Reptilia. *Notulae Naturae* 243:1–8.

Thomas, R. 1989. The relationships of the Antillean *Typhlops* (Serpentes: Typhlopidae) and the description of three new Hispaniolan species. In *Biogeography of the West Indies: Past, Present and Future* (C. A. Woods, ed.), 409–432. Gainesville, Fla.: Sandhill Crane Press.

Tokarz, R. R. 1985. Body size as a factor determining dominance in staged agonistic encounters between male brown anoles (*Anolis sagrei*). *Anim. Behav.* 33(3):746–753.

Turner, F. B. 1977. The dynamics of populations of squamates, crocodilians and rynchocephalians. In *Biology of the Reptilia 7* (C. Gans and D. W. Tinkle, eds.), 157–264. New York: Academic Press.

Underwood, G. 1970. The eye. In *Biology of the Reptilia. 2. Morphology B* (C. Gans and T. S. Parsons, eds.), 1–97. London and New York: Academic Press.

Underwood, G., and Williams, E. E. 1959. The anoline lizards of Jamaica. *Bull. Inst. Jamaica Sci. Ser.* 9:1–48.

Valderrama Puente, M. J. 1977. Algunos datos sobre el ciclo reproductivo en *Anolis sagrei.* In *Cuarta Jornada Científica Estudiantil.* Havana (Facultad de Biología, Universidad de La Habana).

———. 1979. "Algunos aspectos morfométricos, reproductivos y del nicho estructural y climático de *Anolis lucius* (Sauria: Iguanidae) [inedit]. University Students' Report, Facultad de Biología, Universidad de La Habana.

Valderrama Puente, M. J., González, D., and Moncada, F. 1976. Relaciones térmicas de un *Anolis* cubano (Sauria: Iguanidae). Informe preliminar. In *Quinta Jornada Científica Estudiantil.* Havana (Facultad de Biología, Universidad de La Habana).

Valderrama Puente, M. J., and Rodríguez Schettino, L. 1988. Algunas características reproductivas de *Anolis lucius* (Sauria: Iguanidae). *Poeyana* 358:1–15.

Valdés Zamora, G., Berovides Álvarez, V., and Fernández Milera, J. 1986. Ecología de *Polymita picta roseolimbata* Torre, 1950, en la región de Maisí, Cuba. *Cien. Biol.* 15:77–93.

Vanzolini, P. E., and Heyer, W. R. 1985. The American herpetofauna and the interchange. In *The Great American Biotic Interchange* (F. G. Stehli and S. D. Webb, eds.), vol. 4:475–487. New York: Plenum Press.

Varona, L. S. 1985. Sistemática de iguanidae, sensu lato, y de anolinae en Cuba (Reptilia; sauria). *Doñana, Acta Vertebrata* 12(1):21–39.

Varona, L. S., and Arredondo, O. 1979. Nuevos táxones fósiles de Capromyidae (Rodentia: Caviomorpha). *Poeyana* 195:1–51.

Varona, L. S., and Garrido, O. H. 1970. Vertebrados de los Cayos de San Felipe, Cuba, incluyendo una nueva especie de jutía. *Poeyana* 75:1–26.

Vitt, L. J. 1990. The influence of foraging mode and phylogeny on seasonality of tropical lizard reproduction. *Papeis Avulsos Zool.* 37(6):107–123.

Wadge, G., and Burke, K. 1983. Neogene Caribbean plate rotation and associated Central American tectonic evolution. *Tectonics* 2:633–643 [cited by Perfit and Williams 1989].

Weber, W. 1983. Photosensitivity of cromatophores. *Amer. Zool.* 23:495–506.

Webster, T. P., Hall, W. P., and Williams, E. E. 1972. Fission in the evolution of a lizard karyotype. *Science* 177(4049):611–613.

Webster, T. P., Selander, R., and Yang, S. 1972. Genetic variability and similarity in the *Anolis* lizards of Bimini. *Evolution* 26:523–535.

Williams, E. E. 1961. Notes on Hispaniolan herpetology. 3. The evolution and relationships of the *Anolis semilineatus* group. *Breviora* 136:1–8.

———. 1969. The ecology of colonization as seen in the zoogeography of anoline lizards on small islands. *Quart. Rev. Biol.* 44(4):345–389.

———. 1972. The origin of faunas: Evolution of lizard congeners in a complex island fauna—a trial analysis. *Evol. Biol.* 6:47–89.

———. 1976. West Indian anoles: A taxonomic and evolutionary summary. 1. Introduction and species list. *Breviora* 440:1–21.

———. 1983. Ecomorphs, faunas, island size, and diverse end points in island radiation of *Anolis*. In *Lizard Ecology: Study of a Model Organism* (R. B. Huey, E. R. Pianka, and T. W. Schoener, eds.), 326–370, 481–483. Cambridge: Harvard University Press.

———. 1989a. Old problems and new opportunities in West Indian biogeography. In *Biogeography of the West Indies: Past, Present and Future* (C. A. Woods, ed.), 1–46. Gainesville, Fla.: Sandhill Crane Press.

———. 1989b. A critique of Guyer and Savage (1986): Cladistic relationships among anoles (Sauria: Iguanidae): Are the data available to reclassify the anoles? In *Biogeography of the West Indies: Past, Present and Future* (C. A. Woods, ed.), 433–478. Gainesville, Fla.: Sandhill Crane Press.

Williams, E. E., and Hecht, M. K. 1955. "Sunglasses" in two anoline lizards from Cuba. *Science* 122:1–3.

Williams, E. E., and Rand, A. S. 1977. Species recognition, dewlap function and faunal size. *Amer. Zool.* 17:261–270.

Wilson, E. G. 1957. Behavior of the Cuban lizard *Chamaeleolis chamaeleontides* (Duméril et Bibron) in captivity. *Copeia* 2:145.

Wotzkow, C. 1989. Datos sobre ecología de la reproducción de *Buteo jamaicensis solitudinis* (Aves: Accipitridae) en sistemas orográficos de mogotes. In *Cuarta Jornada Científica de la Sociedad Cubana de Zoología.* Havana (Universidad de La Habana).

Wyles, J. S., and Gorman, G. C. 1980. The classification of *Anolis*. Conflict between genetic and osteological interpretation as exemplified by *Anolis cybotes*. *J. Herpetol.* 14:149–153.

Yang, S., Soulé, M., and Gorman, G. C. 1974. *Anolis* lizards of the Eastern Caribbean: A case study in evolution. I. Genetic relationships, phylogeny, and colonization sequence of the *roquet* group. *Syst. Zool.* 23(3):387–399.

Zajicek, D., and Mauri Méndez, M. 1969. Hemoparásitos de algunos animales de Cuba. *Poeyana* 66:1–10.

Zouros, E., and Hertz, P. E. 1984. Enzyme function and polymorphism: A test in two *Anolis* lizards species. *Biochem. Gen.* 22 (1–2):89–97.

Zug, G. R. 1959. Three new subspecies of the lizard *Leiocephalus macropus* Cope from Cuba. *Proc. Biol. Soc. Washington* 72:139–150.

———. 1971. The distribution and patterns of the major arteries of the iguanids and comments on the intergeneric relationships of iguanids (Reptilia: Lacertilia). *Smithsonian Contr. Zool.* 83:1–23.

CONTRIBUTORS

Lourdes Rodríguez Schettino is associate researcher of reptile biology at the Institute of Ecology and Systematics, Havana.

Alberto Coy Otero is principal researcher at the Institute of Ecology and Systematics Institute, Havana.

Georgina Espinosa López is principal professor of biochemical systematics and molecular biology at the Faculty of Biology, University of Havana.

Ada R. Chamizo Lara is assistant researcher of biochemical systematics at Institute of the Ecology and Systematics, Havana.

Mercedes Martínez Reyes is associate researcher of reptile ecology at the Institute of Ecology and Systematics, Havana.

Luis V. Moreno García is curator of the herpetological collection at the Institute of Ecology and Systematics, Havana.

INDEX

References to all Latin names (genera, species, and subspecies) are printed in italic type.